Lecture Notes in Artificial Intelligence   4180

Edited by J. G. Carbonell and J. Siekmann

Subseries of Lecture Notes in Computer Science

Michael Kohlhase

OMDoc –
An Open Markup Format
for Mathematical Documents
[version 1.2]

Foreword by Alan Bundy

 Springer

Series Editors

Jaime G. Carbonell, Carnegie Mellon University, Pittsburgh, PA, USA
Jörg Siekmann, University of Saarland, Saarbrücken, Germany

Author

Michael Kohlhase
International University Bremen
Computer Science
Campus Ring 1, 28759 Bremen, Germany
E-mail: m.kohlhase@iu-bremen.de

Library of Congress Control Number: 2006931135

CR Subject Classification (1998): I.2, F.4, F.3.1, G.4, H.3, I.1, I.7

LNCS Sublibrary: SL 7 – Artificial Intelligence

ISSN   0302-9743
ISBN-10   3-540-37897-9 Springer Berlin Heidelberg New York
ISBN-13   978-3-540-37897-6 Springer Berlin Heidelberg New York

Springer is a part of Springer Science+Business Media

springer.com

© by author 2006

Typesetting: Camera-ready by author, data conversion by Markus Richter, Heidelberg
Printed on acid-free paper   SPIN: 11826095   06/3142   5 4 3 2 1 0

To Andrea — my wife, collaborator, and best friend — for all her support

Foreword

Computers are changing the way we think. Of course, nearly all desk-workers have access to computers and use them to email their colleagues, search the Web for information and prepare documents. But I'm not referring to that. I mean that people have begun to think about what they do in computational terms and to exploit the power of computers to do things that would previously have been unimaginable.

This observation is especially true of mathematicians. Arithmetic computation is one of the roots of mathematics. Since Euclid's algorithm for finding greatest common divisors, many seminal mathematical contributions have consisted of new procedures. But powerful computer graphics have now enabled mathematicians to envisage the behaviour of these procedures and, thereby, gain new insights, make new conjectures and explore new avenues of research. Think of the explosive interest in fractals, for instance. This has been driven primarily by our new-found ability rapidly to visualise fractal shapes, such as the Mandelbrot set. Taking advantage of these new opportunities has required the learning of new skills, such as using computer algebra and graphics packages.

The argument is even stronger. It is not just that computational skills are a useful adjunct to a mathematician's arsenal, but that they are becoming essential. Mathematical knowledge is growing exponentially: following its own version of Moore's Law. Without computer-based information retrieval techniques it will be impossible to locate relevant theories and theorems, leading to a fragmentation and slowing down of the field as each research area rediscovers knowledge that is already well-known in other areas. Moreover, without the use of computers, there are potentially interesting theorems that will remain unproved. It is an immediate corollary of Gödel's Incompleteness Theorem that, however huge a proof you think of, there is a short theorem whose smallest proof *is* that huge. Without a computer to automate the discovery of the bulk of these huge proofs, we have no hope of proving these simple-stated theorems. We have already seen early examples of this phenomenon in the Four-Colour Theorem and Kepler's Conjecture on sphere packing. Perhaps computers can also help us to navigate, abstract and, hence, understand these huge proofs.

Realising this dream of computer access to a world repository of mathematical knowledge, visualising and understanding this knowledge, and reusing and combining it to discover new knowledge, presents a major challenge to mathematicians and informaticians. The first part of this challenge arises because mathematical knowledge will be distributed across multiple sources and represented in diverse ways. We need a lingua franca that will enable this babel of mathematical languages to communicate with each other. This is why this book — proposing just such a lingua franca — is so important. It lays the foundations for realising the rest of the dream.

OMDOC is an open markup language for mathematical documents. The 'markup' aspect of OMDOC means that we can take existing knowledge and annotate it with the information required to retrieve and combine it automatically. The 'open' aspect of OMDOC means that it is extensible, so future-proofed against new developments in mathematics, which is essential in such a rapidly growing and complex field of knowledge. These are both essential features. Mathematical knowledge is growing too fast and is too distributed for any centrally controlled solution to its management. Control must be distributed to the mathematical communities that produce it. We must provide lightweight mechanisms under local control that will enable those communities to put the produce of their labours into the commonwealth with minimal effort. Standards are required to enable interaction between these diverse knowledge sources, but they must be flexible and simple to use. These requirements have informed OMDOC's development. This book will explain to the international mathematics community what they need to do to contribute to and to exploit this growing body of distributed mathematical knowledge. It will become essential reading for all working mathematicians and mathematics students aspiring to take part in this new world of shared mathematical knowledge.

OMDOC is one of the first fruits of the Mathematical Knowledge Management (MKM) Network (http://www.mkm-ig.org/). This network combines researchers in mathematics, informatics and library science. It is attempting to realise the dream of creating a universal digital mathematics library of all mathematical knowledge accessible to all via the World-Wide-Web. Of course, this is one of those dreams that is never fully realised, but remains as a source of inspiration. Nevertheless, even its partial realisation would transform the way that mathematics is practised and learned. It would be a dynamic library, providing not just text, but allowing users to run computer software that would provide visualisations, calculate solutions, reveal counter-examples and prove theorems. It would not just be a passive source of knowledge but a partner in mathematical discovery. One major application of this library will be to teaching. Many of the participants in the MKM Network are building teaching aids that exploit the initial versions of the library.

There will be a seamless transition between teaching aids and research assistants — as the library adjusts its contribution to match the mathematical user's current needs. The library will be freely available to all: all nations, all age groups and all ability levels.

I'm delighted to write this foreword to one of the first steps in realising this vision.

May 2006   Alan Bundy

Preface

Mathematics is one of the oldest areas of human knowledge[1]. It forms the basis of most modern sciences, technology and engineering disciplines. Mathematics provides them with modeling tools such as statistical analysis or differential equations. Inventions like public-key cryptography show that no part of mathematics is fundamentally inapplicable. Last, but not least, we teach mathematics to our students to develop abstract thinking and hone their reasoning skills.

However, mathematical knowledge is far too vast to be understood by one person, moreover, it has been estimated that the total amount of published mathematics doubles every ten to fifteen years [Odl95]. Thus the question of supporting the management and dissemination of mathematical knowledge is becoming ever more pressing but remains difficult. Even though mathematical knowledge can vary greatly in its presentation, level of formality and rigor, there is a level of deep semantic structure that is common to all forms of mathematics and that must be represented to capture the essence of the knowledge.

At the same time it is plausible to expect that the way we do (i.e., conceive, develop, communicate about, and publish) mathematics will change considerably in the years to come. The Internet plays an ever-increasing role in our everyday life, and most of the mathematical activities will be supported by mathematical software systems connected by a commonly accepted distribution architecture, which makes the combined systems appear to the user as one homogeneous application. They will communicate with human users and amongst themselves by exchanging structured mathematical documents, whose document format makes the context of the communication and the meaning of the mathematical objects unambiguous.

Thus the inter-operation of mathematical services can be seen as a knowledge management task between software systems. On the other hand, math-

---

[1] We find mathematical knowledge written down on Sumerian clay tablets, and even Euclid's *Elements*, an early rigorous development of a larger body of mathematics, is over 2000 years old.

ematical knowledge management will almost certainly be web-based, distributed, modular, and integrated into the emerging math services architecture.
So the two fields constrain and cross-fertilize each other at the same time.
A shared fundamental task that has to be solved for the vision of a "web of
mathematical knowledge" (MATHWEB) to become reality is to define an open
markup language for the mathematical objects and knowledge exchanged
between mathematical services. The OMDOC format (Open Mathematical
Documents) presented here is an answer to this challenge, it attempts to provide an infrastructure for the communication and storage of mathematical
knowledge.

Mathematics – with its long tradition in the pursuit of conceptual clarity
and representational rigor – is an interesting test case for general knowledge
management, since it abstracts from vagueness of other knowledge without
limiting its inherent complexity. The concentration on mathematics in OM-
DOC and this book does not preclude applications in other areas. On the
contrary, all the material directly extends to the STEM (science, technology,
education, and mathematics) fields, once a certain level of conceptualization
has been reached.

This book tries to be a one-stop information source about the OMDOC
format, its applications, and best practices. It is intended for authors of mathematical documents and for application developers. The book is divided into
four parts: an introduction to markup for mathematics (Part I), an OMDOC
primer with paradigmatic examples for many kinds of mathematical documents (Part II), the rigorous specification of the OMDOC document format
(Part III), and an XML document type definition and schema (Part IV).

The book can be read in multiple ways:

– for users that only need a casual exposure to the format, or authors that
  have a specific text category in mind, it may be best to look at the examples
  in the OMDOC primer (Part II of this book),
– for an in-depth account of the format and all the possibilities of modeling
  mathematical documents, the rigorous specification in Part III is indispensable. This is particularly true for application developers, who will also
  want to study the external resources, existing OMDOC applications and
  projects, in Part IV.
– Application developers will also need to familiarize themselves with the
  OMDOC Schema in the Appendix.

Acknowledgments

Of course the OMDOC format has not been developed by one person alone. The original proposal was taken up by several research groups, most notably the ΩMEGA group at Saarland University, the MAYA and ACTIVEMATH projects at the German Research Center of Artificial Intelligence (DFKI), the MOWGLI EU Project, the RIACA group at the Technical University of Eindhoven, and the COURSECAPSULES project at Carnegie Mellon University. They discussed the initial proposals, represented their materials in OMDOC and in the process refined the format with numerous suggestions and discussions.

The author specifically would like to thank Serge Autexier, Bernd Krieg-Brückner, Olga Caprotti, David Carlisle, Claudio Sacerdoti Coen, Arjeh Cohen, Armin Fiedler, Andreas Franke, George Goguadze, Alberto González Palomo, Dieter Hutter, Andrea Kohlhase, Christoph Lange, Paul Libbrecht, Erica Melis, Till Mossakowski, Normen Müller, Immanuel Normann, Martijn Oostdijk, Martin Pollet, Julian Richardson, Manfred Ricm, and Michel Vollebregt for their input, discussions, and feedback from implementations and applications.

Special thanks are due to Alan Bundy and Jörg Siekmann. The first triggered the work on OMDOC, has lent valuable insight over the years, and has graciously consented to write the foreword to this book. Jörg continually supported the OMDOC idea with his abundant and unwavering enthusiasm. In fact the very aim of the OMDOC format: openness, cooperation, and philosophic adequateness came from the spirit in his ΩMEGA group, which the author has had the privilege to belong to for more than 10 years.

The work presented in this book was supported by the "Deutsche Forschungsgemeinschaft" in the special research action "Resource-adaptive cognitive processes" (SFB 378), and a three-year Heisenberg Stipend to the author. Carnegie Mellon University, SRI International, and the International University Bremen have supported the author while working on revisions for versions 1.1 and 1.2.

Table of Contents

## Part I

Setting the Stage for Open Mathematical Documents

In this part of the book we will look at the problem of marking up mathematical knowledge and mathematical documents in general, situate the OMDOC format, and compare it to other formats like OPENMATH and MATHML.

The OMDOC format is an open markup language for mathematical documents and the knowledge encapsulated in them. The representation in OMDOC makes the document content unambiguous and their context transparent.

OMDOC approaches this goal by embedding control codes into mathematical documents that identify the document structure, the meaning of text fragments, and their relation to other mathematical knowledge in a process called *document markup*. Document markup is a communication form that has existed for many years. Until the computerization of the printing industry, markup was primarily done by a copy editor writing instructions on a manuscript for a typesetter to follow. Over a period of time, a standard set of symbols was developed and used by copy editors to communicate with typesetters on the intended appearance of documents. As computers became widely available, authors began using word processing software to write and edit their documents. Each word processing program had its own method of markup to store and recall documents.

Ultimately, the goal of all markup is to help the recipient of the document better cope with the content by providing additional information e.g. by visual cues or explicit structuring elements. Mathematical texts are usually very carefully designed to give them a structure that supports understanding of the complex nature of the objects discussed and the argumentations about them. Such documents are usually structured according to the argument made and enhanced by specialized notation (mathematical formulae) for the particular objects.[2] In contrast, the structure of texts like novels or poems normally obey different (e.g. aesthetic) constraints.

In mathematical discourses, conventions about document form, numbering, typography, formula structure, choice of glyphs for concepts, etc. and the corresponding markup codes have evolved over a long scientific history and by now carry a lot of the information needed to understand a particular text. But since they pre-date the computer age, they were developed for the consumption by humans (mathematicians) and mainly with "ink-on-paper" representations (books, journals, letters) in mind, which turns out to be too limited in many ways.

In the age of Internet publication and mathematical software systems, the universal accessibility of the documents breaks an assumption implicit in the design of traditional mathematical documents: namely that the reader will come from the same (scientific) background as the author and will directly understand the notations and structural conventions used by the author. We can also rely less and less on the premise that mathematical documents are primarily for human consumption as mathematical software systems are more and more embedded into the process of doing mathematics. This, together with the fact that mathematical documents are primarily produced and stored on computers, places a much heavier burden on the markup format, since it has to make all of this implicit information explicit in the communication.

In the next two chapters we will set the stage for the OMDOC approach. We will first discuss general issues in markup formats (see Section 1.1), existing solutions (see Section 1.2), and the current XML-based framework for markup languages on the web (see Section 1.3). Then we will elaborate the special requirements for marking up the content of mathematics (see Chapter 2).

---

[2] Of course this holds not only for texts in pure mathematics, but for any argumentative text, including texts from the sciences and engineering disciplines. We will use the adjective "mathematical" in an inclusive way to make this distinction on text form, not strictly on the scientific labeling.

1

Document Markup for the Web

Document markup is the process of adding codes to a document to identify the structure of a document and to specify the format in which its fragments are to appear. We will discuss two conflicting aspects — structure and appearance — in document markup. As the Internet imposes special constraints imposed on markup formats, we will reflect its influence.

In the past few years the XML format has established itself as a general basis for markup languages. As OMDoc and all mathematical markup schemes discussed here are XML applications (instances of the XML framework), we will go more into the technical details to supply the technical prerequisites for understanding the specification. We will briefly mention XML validation and transformation tools, if the material reviewed in this section is not enough, we refer the reader to [Har01].

1.1 Structure vs. Appearance in Markup

Text processors and desktop publishing systems (think for example of Microsoft Word) are software systems aiming to produce *"ink-on-paper"* or *"pixel-on-screen"* representations of documents. They are very well-suited to execute typographic conventions for the appearance of documents. Their internal markup scheme mainly defines presentation traits like character position, font choice and characteristics, or page breaks. We will speak of **presentation markup** for such markup schemes. They are perfectly sufficient for producing high-quality presentations on paper or on screen, but for instance it does not support document reuse (in other contexts or across the development cycle of a text). The problem is that these approaches concentrate on the *form* and not the *function* of text elements. Think e.g. of the notorious section renumbering problems in early (WYSIWYG[1]) text processors. Here, the text form of a numbered section heading was used to express

---

[1] "What you see is what you get"; in the context of markup languages this means that the document markup codes are hidden from the user, who is presented with a presentation form of the text even during authoring.

the function of identifying the position of the respective section in a sequence of sections (and maybe in a larger structure like a chapter).

This perceived weakness has lead to markup schemes that concentrate more on function than on form. We will call them **content markup** to distinguish them from presentation markup schemes, and discuss TEX/LATEX [Knu84, Lam94] as an example.

TEX is a typesetting markup language that uses explicit markup codes (strings beginning with a backslash) in a document, for instance, the markup `$\sqrt{\sin x}$` stands for the mathematical expression $\sqrt{\sin x}$ in TEX. To determine from this functional specification the visual form (e.g. the character placement and font information), we need a document formatting engine. This program will transform the document that contains the content markup (the "source" document) into a presentation markup scheme that specifies the appearance (the "target" document) like DVI [Knu84], POSTSCRIPT [Rei87], or PDF [PDF06] that can directly be presented on paper or on screen. This two-stage approach allows the author to mark up the function of a text fragment and leave the conversion of this markup into presentation information to the formatter. The specific form of translation is either hard-wired into the formatter, or given externally in *style files* or *style sheets*.

LATEX [Lam94] is a comprehensive set of style files for the TEX formatter, the heading for a section with the title "The Joy of TEX" would be marked up as

`\section[{\TeX}]{The Joy of {\TeX}\index{tex@\TeX}}\label{sec:TeX}`

This piece of markup specifies the function of the text element: The title of the section should be "The Joy of TEX", which (if needed e.g. in the table of contents) can be abbreviated as "TEX", the glyph "TEX" is inserted into the index, where the word `tex` would have been, and the section number can be referred to using the label `sec:TeX`. Note that renumbering is not a problem in this approach, since the actual numbers are only inferred by the formatter at run-time. This, together with the ability to simply change style file for a different context, yields much more manageable and reusable documents, and has led to a wide adoption of the function-based approach. So that even word-processors like MS Word now include functional elements. Pure presentation markup schemes like DVI or POSTSCRIPT are normally only used for document delivery. On the other hand, many form-oriented markup schemes allow to "fine-tune" documents by directly controlling presentation. For instance, LATEX allows to specify traits such as font size information, or using

`{\bf proof}:...\hfill\Box`

to indicate the extent of a proof (the formatter only needs to "copy" them to the target format). The general experience in such mixed markup schemes is that presentation markup is more easily specified, but that content markup will enhance maintainability and reusability. This has led to a culture of style file development (specifying typographical and structural conventions), which now gives us a wealth of style options to choose from in LATEX.

1.2   Markup for the World Wide Web

The Internet, where screen presentation, hyperlinking, computational limitations, and bandwidth considerations are much more important than in the "ink-on-paper" world of publishing, has brought about a whole new set of markup schemes. The problems that need to be addressed are that

- the size, resolution, and color depth of a given screen are not known at the time the document is marked up,
- the structure of a text is no longer limited to a linear text with (e.g. numbered) cross-references as in a traditional book or article: Internet documents are usually hypertexts,
- the computational resources of the computer driving the screen are not known beforehand. Therefore the distribution of work (e.g. formatting steps) between the client and the server has to be determined at run-time. Finally, the related problem that
- the bandwidth of the Internet is ever-growing but always limited.

These issues impose somewhat conflicting demands on markup languages for the Web. The first two seem to favor content markup languages, since low-level presentational traits like glyph placement and font availability cannot be pre-meditated on the server. However, the amount of formatting that can be delegated to the client, and the availability of style files is limited by the latter two concerns.

In response the "Hypertext Markup Language" (HTML [RHJ98]) evolved as the original markup format for the World Wide Web. This is a markup scheme that addresses the problem of variable screen size and hyperlinking by exporting the decision of character placement and page order to a browser running on the client. It ensures a high degree of reusability of documents on the Internet while conserving bandwidth, so that HTML carries most of the text markup on the Internet today.

The major innovation in HTML was the use of **uniform resource locators** (**URL**) to reference documents provided by web servers. URLs are strings in a special format that can be interpreted by browsers or other web agents to request documents from web servers, e.g. to be displayed to the user in the browser as a new node in the current hypertext document. Since URLs are global references, they are the means that make the Internet into a "*world-wide*" web (of references). Since uniform resource *locators* are closely tied to the physical location of a document on the Internet, which can change over time, they have since been generalized to **uniform resource identifier** (**URI**; see [BLFM98]). These are strings of similar structure, that only identify resources on the Internet, see [Har01], i.e. their structure need not be directly translatable to an Internet location (we call this act **de-referencing**). Indeed, URIs need not even correspond to a physical manifestation of a resource at all, they can identify a virtual resource, that is produced by a web service on demand.

The concrete syntax and architecture of HTML is derived from the "Simple Generalized Markup Language" SGML [Gol90], which is similar to TEX/LaTeX in spirit, but tries to give the markup scheme a more declarative semantics (as opposed to the purely procedural – and rather baroque – semantics of TEX) to make it simpler to reason about (and thus reuse) documents. In particular unlike TEX, SGML separates content markup codes from directives to the formatting engine. SGML has a separate style sheet language DSSSL [DuC97], which was not adopted by HTML, because of resource limitations in the client. Instead, HTML has been augmented with its own (limited) style sheet language CSS [Bos98] that is executed by the browser.

1.3 XML, the eXtensible Markup Language

The need for content markup schemes for maintaining documents on the server, as well as for specialized presentation of certain text parts (e.g. for mathematical or chemical formulae), has led to a profusion of markup schemes for the Internet, most of which share the basic SGML syntax with HTML. To organize this zoo of markup languages, the World Wide Web Consortium (W3C [W3C], an international interest group of universities and web industry) has developed a language framework for Internet markup languages called XML (eXtensible Markup Language) [BPSM97]. XML is a set of grammar rules that allows to interpret certain sequences of Unicode [Inc03] characters as document trees. These grammar rules are shared by all XML-based markup languages (called XML applications) and are very well-supported by a great variety of XML processors. The XML format is accompanied by a set of specialized vocabularies (most of them XML applications) that standardize various aspects of document management and web services. These are canonicalized by the W3C as "recommendations". We will briefly review the ones that are relevant for understanding the OMDOC format and make the book self-contained. For details see one of the many XML books, e.g. [Har01].

1.3.1 XML Document Trees

Conceptually speaking, XML views a document as a tree whose nodes consist of elements, attributes, text nodes, namespace declarations, XML comments, etc. (see Figure 1.1 for an example[2]). For communication this tree is serialized into a balanced bracketing structure (see the listing at the top of Figure 1.1), where an element el is represented by the brackets <el> (called the **opening tag**) and </el> (called the **closing tag**). The leaves of the tree

---

[2] This tree representation glosses over namespace nodes in the tree, but the conceptual tree is sufficient for the application in this book.

```
<omtext xml:id="foo" xmlns="http://www.mathweb.org/omdoc"
        xmlns:om="http://www.openmath.org/OpenMath">
  <CMP xml:lang='en'>
  The number
  <om:OMOBJ><om:OMS cd="nums1" name="pi"/><om:OMOBJ>
  is irrational .
  </CMP>
</omtext>
```

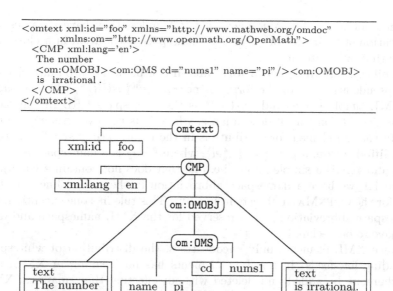

Fig. 1.1. An XML Document as a Tree

are represented by **empty elements** (serialized as `<el></el>`, which can be abbreviated as `<el/>`), and text nodes (serialized as a sequence of UNICODE characters). An element node can be annotated by further information using **attribute nodes** — serialized as an **attribute** in its opening tag: for instance `<el visible="no">` might add the information for a formatting engine to hide this element. As a document is a tree, the XML specification mandates that there must be a unique **document root**.

Let us now come to a feature that we have glossed over so far: XML **namespace**s [Bra99]. In many XML applications, we need to mix several XML vocabularies or languages. In our example in Figure 1.1 we have three: the OMDoc vocabulary with the elements `omtext` and `CMP`, the OPENMATH vocabulary with the elements `om:OMOBJ` and `om:OMS`, and the general XML vocabulary for the attributes `xml:id` and `xml:lang`.

To allow a safe mixing of independent XML vocabularies, XML can associate elements and attributes[3] with a **namespace**, which is simply a URI that uniquely identifies the intended vocabulary[4]. In XML syntax, namespace membership is represented by namespace declarations and qualified names.

A **namespace declaration** is a pseudo-attribute with name `xmlns` whose value is a namespace URI ⟨*nsURI*⟩ (see e.g. the first line in Figure 1.1). In a nutshell, a namespace declaration specifies that this element and all its

---

[3] Traditionally most XML applications use attributes that are not namespaced.
[4] Note that it need not be a valid URL (uniform resource locator; i.e. a pointer to a document provided by a web server).

descendants are in the namespace ⟨⟨nsURI⟩⟩, unless they have a namespace declaration of their own or there is a namespace declaration in a closer ancestor that overwrites it.

Similarly, a **namespace abbreviation** can be declared on any element by a pseudo-attribute of the form xmlns:⟨⟨nsa⟩⟩="⟨⟨nsUR⟩⟩", where ⟨⟨nsa⟩⟩ is an XML simple name, and ⟨⟨nsURI⟩⟩ is the namespace URI. In the scope of this declaration (in all descendants, where it is not overwritten) we can specify that an element or attribute is in the namespace ⟨⟨nsURI⟩⟩ by using a **qualified name**: a pair ⟨⟨nsa⟩⟩:⟨⟨el⟩⟩, where ⟨⟨nsa⟩⟩ is a namespace abbreviation and ⟨⟨el⟩⟩ is a simple name (i.e. one that does not contain a colon). In Figure 1.1, we have a namespace abbreviation in the second line, which is used for the OPENMATH objects in line five. This rule has one exception: the namespace abbreviation xml is reserved for the XML namespace and does not have to be declared.

Since XML elements only encode trees, the distribution of whitespace (including line-feeds) in non-text elements has no meaning in XML, and can therefore be added and deleted without effecting the semantics. XML considers anything between <!-- and --> in a document as a comment. They should be used with care, since they are not necessarily passed on by the XML parser, and therefore might not survive processing by XML applications.

Material that is relevant to the document, but not valid XML, e.g. binary data or data that contains angle brackets or elements that are unbalanced or not part of the XML application can be encoded by embedding it into CDATA **sections**. A CDATA section begins with the string <[CDATA[and suspends the XML parser until the string]]> is found. The result of parsing a CDATA section is equivalent to escaping the five XML-specific characters <, > ", ', and & to the XML entities <, >, ", ', and &. For instance, we have the following correspondence between a CDATA section and XML-escaped content:

<[CDATA[a<b³]]>	≙	a<b<sup>3</sup>

As a consequence, an XML application is free to choose the form of its output and the particular form should not be relied upon.

1.3.2 Validating XML Documents

XML offers various mechanisms for specifying a subset of trees (or well-bracketed XML documents) as admissible in a given XML application: the most commonly used ones are **document type definitions** (**DTD** [BPSM97]), XML **schemata** [XML], and RELAXNG schemata [Vli03]. All of these are context-free grammars for trees, that can be used by a **validating parser** to reject XML documents that do not conform. Note that DTDs and schemata cannot enforce all constraints that a particular XML application may want to impose on documents. Therefore validation is only a necessary condition for **validity** with respect to that application. Since the XML schema languages can express slightly stronger sets of constraints and are namespace-aware, they allow stronger document validation, and usually take normative precedence over the DTD if present.

Listing 1.2 shows part of an OMDOC document. The first line identifies the document as an XML document (version 1.0 of the XML specification). The second and third lines constitute the **document type declaration** which specifies the DTD and the document root element. In this case the omdoc element starting in line 4 is the root element and will be validated against the DTD identified by the **public Identifier**[5] in line two and which can be found at the URI in line three. See Chapter 24 for an in-depth discussion of the OMDOC DTD and validation.

Listing 1.2. The Structure of an XML Document with DTD

```
   <?xml version="1.0"?>
   <!DOCTYPE omdoc PUBLIC "-//OMDoc//DTD OMDoc V1.2//EN"
                  "http://www.mathweb.org/omdoc/omdoc.dtd">
   <omdoc xml:id="example-omdoc" xmlns="http://www.mathweb.org/omdoc">
 5 ...
   </omdoc>
```

Note that it is not mandatory to have a document type declaration in an XML document, or that an XML parser even read it (we call an XML parser **validating** if it does). If no document type declaration is present, then a parser will just check for XML-well-formedness, and possibly rely on some schema for further validation[6]. Note that if a validating parser reads an XML document with a document type declaration, then it must process it and validate the document.

But a DTD not only contains information for validation, it also

declares XML entities XML entities are strings of the form &⟨⟨*abbr*⟩⟩;, which abbreviate sequences of UNICODE characters and are expanded by the parser as it reads the document.

supplies default values for attributes which are added to the representation of the parsed document by the parser as it reads the document.

declares types of attributes This is is relevant for attribute types ID and IDREF. The former are required to be document-unique (as well as being XML simple names [BPSM97, section 2.3]) and the latter must point to an existing ID-type attribute in the same document.

ID-type attributes are commonly used to identify elements in XML documents (see the discussion in Subsection 1.3.3), which raises a subtle point with

---

[5] A string that allows to identify an XML resource, it can be mapped to a concrete URI via the XML catalog; see Section 23.4 for details.

[6] Note that RELAXNG schemata do not have a specified in-document means for associating a schema with elements. For the way to associate an XML schema with a document we refer to XML schema recommendation [XML] or the XML literature.

respect to DTDs. If an XML document is processed without a document type declaration or by a non-validating parser, the information which attributes are ID-type ones is lost, and referencing does not work as as expected. Fortunately, there is a recent W3C-solution to this problem: Following the XML ID recommendation [MVW05] XML parsers must recognize attributes of the form xml:id as ID-type attributes, even if no DTD is present.

However DTDs may still serve an important role, even if they are superseded by schema-based approaches for pure validation. For instance a format like Presentation-MATHML (see Subsection 2.1.1) seems dependent on a DTD, since it needs to define a rich set of mnemonic entities for mathematical symbols in UNICODE and uses ID-type attributes for cross-referencing. Formats like Content-MATHML (Subsection 2.1.1), OPENMATH (Subsection 2.1.2) or OMDOC proper can live without DTDs, since they do not.

1.3.3   XML Fragments and URI References

As documents are construed as trees in XML, the notion of a document fragment becomes definable simply as a sets of well-formed sub-trees. Building on this, URLs and URIs can be extended to references of document fragments. These **URI references** are traditionally considered to consist of two parts: A proper URI and a specific **fragment identifier** separated by the hash character #. The URI identifies an XML document on the web, whereas the fragment identifier identifies a specific fragment of that document.

XML provides the XPOINTER framework [GMMW03b] for fragment identifiers. It specifies multiple schemes for fragment identifiers. Fragment identifiers of the form xpointer($\langle\!\langle path\rangle\!\rangle$) use an XPATH [Cla99b] expression $\langle\!\langle path\rangle\!\rangle$ to specify a path through the document tree leading to the desired element (see [DMD02]). Fragment identifiers in the element() scheme [GMMW03a] use expressions of the form element($\langle\!\langle cpath\rangle\!\rangle$), where $\langle\!\langle cpath\rangle\!\rangle$ is an ID-type identifier together with a simple child-path; e.g. element(foo/3/7) identifies the 7th child of the 3rd child of the (unique) element that has ID-type attribute with value foo.

URI references of the form $\langle\!\langle uri\rangle\!\rangle$#$\langle\!\langle id\rangle\!\rangle$ as they are used in HTML to refer to named anchors (<a name="$\langle\!\langle id\rangle\!\rangle$"/>) are regained as a special case (the shorthand xpointer): If $\langle\!\langle uri\rangle\!\rangle$ is a URI of an XML document D then $\langle\!\langle uri\rangle\!\rangle$#$\langle\!\langle id\rangle\!\rangle$ refers to the unique element in D, that has an attribute of type ID with value $\langle\!\langle id\rangle\!\rangle$.

1.3.4   Summary

In summary, XML provides a widely standardized infrastructure for defining Internet markup languages based on tree structures rather than on sequences of characters. XML processors like parsers, serializers, XML databases, and XSLT transformation engines are widely deployed and incorporated into many programming languages. Building XML applications on top of this infrastructure frees the implementers from dealing with low-level details of parsing, validation, and mass storage. It is no surprise that XML has become one of the most successful interoperability formats in information technology.

Note that the use of XML does not give any support for mathematics in itself, since the tree models are completely general. It is the role of specific XML applications like the ones we will present in the next two chapters to specialize the XML tree structures to representations that can be interpreted as mathematical objects and documents.

2

Markup for Mathematical Knowledge

Mathematicians make use of various kinds of documents (e.g. e-mails, letters, pre-prints, journal articles, and textbooks) for communicating mathematical knowledge. Such documents employ specialized notational conventions and visual representations to convey the mathematical knowledge reliably and efficiently. The respective representations are supported by pertinent markup systems like TeX/LaTeX.

Even though mathematical documents can vary greatly in their level of presentation, formality and rigor, there is a level of deep semantic structure that is common to all forms of mathematics and that must be represented to capture the essence of the knowledge. As John R. Pierce has written in his book on communication theory [Pie80], mathematics and its notations should not be viewed as one and the same thing. Mathematical ideas exist independently of the notations that represent them. However, the relation between meaning and notation is subtle, and part of the power of mathematics to describe and analyze derives from its ability to represent and manipulate ideas in symbolic form. The challenge in putting mathematics on the World Wide Web is to capture both notation and content (that is, meaning) in such a way that documents can utilize the highly-evolved notational forms of written and printed mathematics, and the potential for interconnectivity in electronic media.

In this chapter, we present the state of the art for representing mathematical documents on the web and analyze what is missing to mark up mathematical knowledge. We posit that there are three levels of information in mathematical knowledge: formulae, mathematical statements, and the large-scale theory structure (constructing the context of mathematical knowledge). The first two are immediately visible in marked up mathematics, e.g. textbooks, the third is largely left to an implicit meta-level of mathematical communication, or the organization of mathematical libraries. We will discuss these three levels in the next sections.

2.1 Mathematical Objects and Formulae

A distinguishing feature of mathematical documents is the use of a complex and highly evolved system of two-dimensional symbolic notations, commonly called (mathematical) **formulae**. Formulae serve as representations of mathematical objects, such as functions, groups, or differential equations, and also of statements about them, like the "Fundamental Theorem of Algebra".

The two best-known open markup formats for representing mathematical formulae for the Web are MATHML [ABC⁺03] and OPENMATH [BCC⁺04]. There are various other formats that are proprietary or based on specific mathematical software packages like Wolfram Research's MATHEMATICA® [Wol02]. We will not concern ourselves with them, since we are only interested in open formats. Furthermore, we will only give a general overview for the open formats here to survey the state of the art, since content MATHML and OPENMATH are used for formula representation in the OMDOC format and thus the technical details of the two markup schemes are covered in more detail in the OMDOC specification in Chapter 13. Figure 2.1 gives an overview over the current state of the standardization activities.

language	MATHML	OPENMATH
by	W3C Math WG	OPENMATH society
origin	math for HTML	integration of CAS
coverage	content + presentation; K-14	content; extensible
status	Version 2.2e (VI 2003)	Version 2 (VI 2004)
activity	maintenance	maintenance
Info	`http://w3c.org/Math/`	`http://www.openmath.org/`

Fig. 2.1. The status of markup standardization for mathematical formulae

OPENMATH was originally a development driven mainly by the Computer Algebra community in Europe trying to standardize the communication of mathematical objects between Computer Algebra Systems. The format has been discussed in a series of workshops and has been funded by a series of grants by the European Union. This process led to the OPENMATH 1 standard in June 1999 and eventually to the incorporation of the OPENMATH society as the institutional guardian of the OPENMATH standard. MATHML has developed out of the effort to include presentation primitives for mathematical notation (in TEX quality) into HTML, and was the first XML application to reach recommendation status[1] at the W3C [BDD⁺99].

---

[1] As such, MATHML played a great role as technology driver in the development of XML. This role gives MATHML a somewhat peculiar status at the W3C; it is the only "vertical" (application/domain-driven) XML application standardized by the W3C, which otherwise concentrates on "horizontal" (technology-driven) standards.

The competition and collaboration between these two approaches to representation of mathematical formulae and objects has led to a large overlap between the two developer communities. MATHML deals principally with the *presentation* of mathematical objects, while OPENMATH is solely concerned with their semantic meaning or *content*. While MATHML does have some limited facilities for dealing with content, it also allows semantic information encoded in OPENMATH to be embedded inside a MATHML structure. Thus the two technologies may be seen as highly compatible[2] and complementary (in aim).

2.1.1 MATHML

> MATHML is an XML application for describing mathematical *notation* and capturing both its *structure* and *content*. The goal of MATHML is to enable mathematics to be served, received, and processed on the World Wide Web, just as HTML has enabled this functionality for text.
>
> *from the MathML2 Recommendation* [ABC+03]

To reach this goal, MATHML offers two sub-languages: Presentation-MATHML for marking up the two-dimensional, visual appearance of mathematical formulae, and Content-MATHML as a markup infrastructure for the functional structure of mathematical formulae.

To mark up the visual appearance of formulae Presentation-MATHML represents mathematical formulae as a tree of layout primitives. For instance the expression $\frac{3}{x+2}$ would be represented as the layout tree in Figure 2.2. The layout primitives arrange "inner boxes" (given in black) and provide an outer box (given in gray here) for the next level of layout. In Figure 2.2 we see the general layout schemata for numbers (m:mn), identifiers (m:mi), operators (m:mo), bracketed groups (m:mfence), and fractions (m:mfrac); others include horizontal grouping (m:mrow), roots (m:mroot), scripts (m:msup, m:msub, m:msubsup), bars and arrows (m:munder, m:mover, m:munderover), and scoped CSS styling (m:mstyle). Mathematical symbols are taken from UNICODE and provided with special mnemonic entities by the MATHML DTD, e.g. ∑ for Σ.

Since the aim of MATHML is to do most of the formatting inside the browser, where resource considerations play a large role, it restricts itself to a fixed set of mathematical concepts – the K-14 fragment (Kindergarten to 14th grade; i.e. undergraduate college level) of mathematics. K-14 contains a large set of commonly used glyphs for mathematical symbols and very general and powerful presentation primitives, similar to those that make up the lower level

---

[2] e.g. MATHML is the preferred presentation format for OPENMATH objects and OPENMATH content dictionaries are the primary specification language for MATHML semantics.

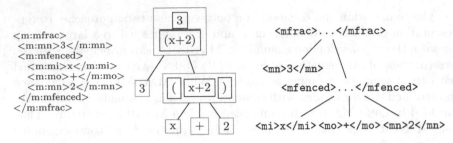

```
<m:mfrac>
 <m:mn>3</m:mn>
 <m:mfenced>
   <m:mi>x</m:mi>
   <m:mo>+</m:mo>
   <m:mn>2</m:mn>
 </m:mfenced>
</m:mfrac>
```

Fig. 2.2. The layout tree for the formula $\frac{3}{x+2}$

of TeX. However, it does not offer the programming language features of TeX[3] for the obvious computing resource considerations. Presentation-MathML is supported by current versions of the browsers Amaya [Vat], MS Internet Explorer [Cor] (via the MathPlayer plug-in [Sci]), and Mozilla [Org].

MathML also offers content markup for mathematical formulae, a sublanguage called **Content-**MathML to contrast it from the **Presentation-**MathML described above. Here, a mathematical formula is represented as a tree as well, but instead of marking up the visual appearance, we mark up the functional structure. For our example $\frac{3}{x+2}$ we obtain the tree in Figure 2.3, where we use @ as the function application operator (it interprets the first child as a function and applies it to the rest of the children as arguments).

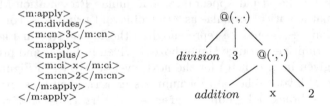

```
<m:apply>
 <m:divides/>
 <m:cn>3</m:cn>
 <m:apply>
   <m:plus/>
   <m:ci>x</m:ci>
   <m:cn>2</m:cn>
 </m:apply>
</m:apply>
```

Fig. 2.3. The functional structure of $\frac{3}{x+2}$

Content-MathML offers around 80 specialized elements for the most common K-14 functions and individuals. In Figure 2.3 we see function application (m:apply), content identifiers (m:ci), content numbers (m:cn) and the functions for division (m:divide) and addition (m:plus).

Finally, MathML offers a specialized m:semantics element that allows to annotate MathML formulae with alternative representations. This feature can be used to provide combined content- and presentation-MathML representations. Figure 2.4 shows an example of this for our expression $\frac{3}{x+2}$. The

---

[3] TeX contains a full, Turing-complete – if somewhat awkward – programming language that is mainly used to write style files. This is separated out by MathML to the CSS and XSLT style languages it inherits from XML.

outermost m:semantics element is used for mixing presentation and content markup. The first child of the m:semantics element contains Presentation-MATHML (this is used by the MATHML-aware browser), the subsequent m:annotation-xml element contains Content-MATHML markup for the same formula. Corresponding sub-expressions are co-referenced by cross-references: The presentation element carries an id attribute, which serves as the target for an xlink:href attribute in the content markup. This technique is called parallel markup, it allows to select logical sub-expressions by selecting layout sub-schemata in the browser, e.g. for copy and paste. Note that a m:semantics element can have more than one m:annotation-xml child, so that other content formats such as OPENMATH can also be incorporated.

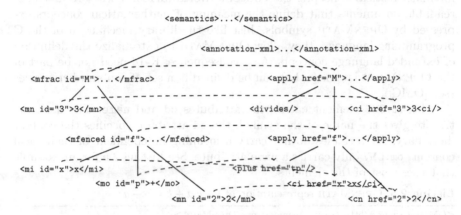

Fig. 2.4. Mixing presentation and Content-MATHML

2.1.2 OPENMATH

> [...] OPENMATH: a standard for the representation and communication of mathematical objects. [...]
> OPENMATH allows the *meaning* of an object to be encoded rather than just a visual representation. It is designed to allow the free exchange of mathematical objects between software systems and human beings. On the worldwide web it is designed to allow mathematical expressions embedded in web pages to be manipulated and computed with in a meaningful and correct way. It is designed to be machine-generatable and machine-readable, rather than written by hand.
> *from the* OPENMATH2 *Standard* [BCC+04]

Driven by the intention of representing the *meaning* of mathematical objects expressed in the quote above, the OPENMATH format is not primarily an XML application. Rather, OPENMATH defines an abstract (mathematical) object model for mathematical objects and specifies an XML encoding (and a binary[4] encoding) for that[5].

The central construct of OPENMATH is that of an OPENMATH **object** (realized by the element om:OMOBJ in the XML encoding), which has a tree-like representation made up of applications (om:OMA), binding structures (om:OMBIND using om:OMBVAR to specify the bound variables[6]), variables (om:OMV), and symbols (om:OMS).

The handling of symbols — which are used to represent the multitude of mathematical domain constants — is maybe the largest difference between OPENMATH and Content-MATHML. Instead of providing elements for all K-14 concepts, the OPENMATH standard adds an extension mechanism for mathematical concepts, the **content dictionaries**. These are machine-readable documents that define the meaning of mathematical concepts expressed by OPENMATH symbols. Just like the library mechanism of the C programming language, they allow OPENMATH to externalize the definition of extended language concepts. As a consequence, K-14 need not be part of the OPENMATH language, but can be defined in a set of content dictionaries (see [OMC]).

The om:OMS element carries the attributes cd and **name**. The **name** attribute gives the name of the symbol, the cd attribute specifies the content dictionary. As variables do not carry a meaning independent of their local content, om:OMV only carries a **name** attribute. See Listing 2.5 for an example that uses most of the elements.

Listing 2.5. OPENMATH representation of $\forall a, b. a + b = b + a$

```
   <OMOBJ xmlns="http://www.openmath.org/OpenMath">
     <OMBIND cdbase="http://www.openmath.org/cd">
       <OMS cd="quant1" name="forall"/>
       <OMBVAR><OMV name="a"/><OMV name="b"/></OMBVAR>
5      <OMA><OMS cd="relation" name="eq"/>
         <OMA><OMS cd="arith1" name="plus"/>
           <OMV name="a"/>
           <OMV name="b"/>
         </OMA>
10         <OMA><OMS cd="arith1" name="plus"/>
           <OMV name="b"/>
           <OMV name="a"/>
         </OMA>
       </OMA>
15     </OMBIND>
   </OMOBJ>
```

---

[4] The binary encoding allows to optimize encoding size and (more importantly) parsing time for large OPENMATH objects. The binary encoding for OPENMATH objects will not play a role for the OMDoc format, so we will not pursue this here.

[5] The MATHML specification is very vague on what the meaning of Content-MATHML fragments might be; we have to assume that its XML document object model [DOM] or the or its infoset [Cow04] must be.

[6] Binding structures are somewhat awkwardly realized via the m:apply element with an m:bvar child in Content-MATHML.

Listing 2.5 shows the XML encoding of the law of commutativity for addition (the formula $\forall a, b.a + b = b + a$) in OPENMATH. Note that as we have discussed above, this representation is not self-contained but relies on the availability of content dictionaries `quant1`, `relation1`, and `arith1`. Note that in this example they can be accessed via the URL specified in the `cdbase` attribute, but in general, the content dictionaries are only used for *identification of symbols*. In particular, in the classical OPENMATH model, content dictionaries are only viewed as a resource for system developers, who use them as a reference decide which symbol to use in an export/import facility for a computer algebra system. In the communication between mathematical software systems, they are no longer needed: If two systems agree on a set of content dictionaries, then they agree on the meaning of all OPEN-MATH objects that can be constructed using their symbols (the meaning of applications and bindings is known from the folklore).

The content dictionary architecture is the greatest strength of the OPEN-MATH format. It establishes an object model and XML encoding based on what we call "semantics by pointing". Two OPENMATH objects have the same meaning in this model, iff they have the same structure and all symbols point to the same content dictionaries[7].

In the standard encoding of OPENMATH content dictionary, the meaning of a symbol is specified by a set of

"formal mathematical properties" The `omcd:FMP` element contains an OPENMATH object that expresses the desired property.
"commented mathematical properties" The `omcd:CMP` element contains a natural language description of a desired property.

For instance, the specification in Listing 2.6 is part of the standard OPEN-MATH content dictionary `arith1.ocd` [OMC] for the elementary arithmetic operations.[8]

Listing 2.6. Part of the OPENMATH content dictionary `arith1`.

```
<CDDefinition>
  <Name>plus</Name>
  <CDDescription>
    The symbol representing an n−ary commutative function plus.
  </CDDescription>
  <CMP> for all a,b | a + b = b + a </CMP>
  <FMP>∀a, b.a + b = b + a</FMP>
</CDDefinition>
```

---

[7] Note that we can interpret the Content-MATHML model as a "semantics by pointing" model as well. Only that here the K-14 elements do not point to machine-readable content dictionaries, but at the (human-readable) MATHML specification, which specifies their meaning.

[8] The content of the `omcd:FMP` element is actually the OPENMATH object in the representation in Listing 2.5, we have abbreviated it here in the usual mathematical notation, and we will keep doing this in the remaining document: wherever an XML element in a figure contains mathematical notation, it stands for the corresponding OPENMATH element.

On the other hand, the content dictionary encoding defined in the OPEN-MATH standard (and the particular content dictionaries blessed by the OPEN-MATH society) are the greatest weakness of OPENMATH. The represent the knowledge in a very unstructured way — to name just a few problems:

- in the omcd:CMP, we can only make use of ASCII representation of formulae.
- The relation between a particular omcd:CMP and omcd:FMP elements is unclear.
- For properties like the distributivity of addition over multiplication it is unclear, whether we should express this in the definition of the symbol plus or the symbol times.
- Are all properties constitutive for the meaning of the symbol? Should they be verified for an implementation of a content dictionary?
- What is the relationship between content dictionaries? Are they translation-equivalent? Does one entail the other?

The OPENMATH2 standards acknowledges these problems and explicitly opens up the content dictionary format allowing other representations that meet certain minimal criteria relegating the standard encoding above to a reference implementation of the minimal model.

We will analyze the questions raised above from a general standpoint when discussing the remaining two levels of mathematical knowledge. This analysis constitutes the basic intuitions for the OMDoc format.

2.2   Mathematical Texts and Statements

The mathematical markup languages OPENMATH and MATHML we have discussed in the last section have dealt with mathematical objects and formulae. The formats either specify the semantics of the mathematical object involved in the standards document itself (MATHML) or in a fixed set of generally agreed-upon documents (OPENMATH content dictionaries). In both cases, the mathematical knowledge involved is relatively fixed. Even in the case of OPENMATH, which has an extensible library mechanism, the content dictionaries are not in themselves objects of communication (they are mainly background reference for the implementation of OPENMATH interfaces).

For the communication among mathematicians (rather than computation systems) this level of support is insufficient, because the mathematical knowledge expressed in definitions, theorems (stating properties of defined objects), their proofs, and even whole mathematical theories is the primary focus of mathematical communication. For content markup of mathematical knowledge, we have to turn implicit or presentational structuring devices in mathematical documents into explicit ones. For instance, **mathematical statements** like the ones in the document fragment in Figure 2.7 are delimited by keywords (e.g. **Definition**, **Lemma** and □) or by changes in text font.

Definition 3.2.5 (Monoid)
A monoid is a semigroup $S = (G, \circ)$ with an element $e \in G$, such that $e \circ x = x$ for all $x \in G$. e is called a left unit of S.

Lemma 3.2.6
A monoid has at most one left unit.
Proof: We assume that there is another left unit f ...
This contradicts our assumption, so we have proven the claim.   □

Fig. 2.7. A fragment of a traditional mathematical document

Of course, the content of a mathematical statement, e.g. the statement of an assertion that "addition is commutative" can be expressed by a Content-MATHML or OPENMATH formula like the one in Listing 2.5, but the information that this formula is a theorem that has a proof, cannot be directly expressed without extending the formalism. Even formalizations of mathematics like Russell and Whitehead's famous "Principia Mathematica" [WR10] treat this information on the meta-level. If we are willing to extend the mathematical formalism to include primitives for such information, we arrive at formalisms called **logical frameworks** (see [Pfe01] for an overview), where they are treated as the primary objects of study. The most prevalent approach here uses the "formulae as types" idea that delegates mathematical formulae to the status of types. Logical frameworks capture mathematical statements in formulae and as such can be expressed in Content-MATHML or OPENMATH. However, this approach relies on full formalization of the mathematical content, and cannot be directly used to capture mathematical practice. In particular, the gap between formal mathematics and informal (but rigorous) treatments of mathematics that rely on natural language as we find them in textbooks and journal articles is wide. The formalization process is so tedious, that it is seldom executed in practice (the "Principia Mathematica" and the MIZAR mathematical library [Comb] are solitary examples).

2.3 Large-Scale Structure and Context in Mathematics

The large-scale structure of mathematical knowledge is much less apparent than that for formulae and even statements. Experienced mathematicians are nonetheless aware of it, and use it for navigating the vast space of mathematical knowledge and to anchor their communication.

Much of this structure can be found in networks of **mathematical theories**: groups of mathematical statements, e.g. those in a monograph "Introduction to Group Theory" or a chapter or section in a textbook. The relations among such theories are described in the text, sometimes supported by mathematical statements called representation theorems. We can observe that mathematical texts can only be understood with respect to a particular mathematical context given by a theory which the reader can usually infer from the document. The context can be stated explicitly (e.g. by the title of a book) or implicitly (e.g. by the fact that the e-mail comes from a person that we know works on finite groups, and that she is talking about math).

If we make the structure of the context as explicit as the structure of the mathematical objects (we will speak of **context markup**), then mathematical software systems will be able to provide novel services that rely on this structure. We contend that without an explicit representation of context structure, tasks like semantics-based searching and navigation or object classification can only be performed by human mathematicians that can understand the implicitly given structure.

Mathematical theories have been studied by mathematicians and logicians in the search of a rigorous foundation for mathematical practice. They have been formalized as collections of symbol declarations — giving names to mathematical objects that are particular to the theory — and logical formulae, which state the laws governing the properties of the theory. A key research question was to determine conditions for the consistency of mathematical theories. In inconsistent theories all statements are vacuously valid[9], and therefore only consistent theories make interesting statements about mathematical objects.

It is one of the critical observations of meta-mathematics that theories can be extended without endangering consistency, if the added formulae can be proven from the formulae already in the theory (such formulae are called theorems). As a consequence, consistency of a theory can be determined by examining the **axiom**s (formulae without a proof) alone. Thus the role of proofs is twofold, they allow to push back the assumptions about the world to simpler and simpler axioms, and they allow to test the model by deriving consequences of these basic assumptions that can be tested against the data.

A second important observation is that new symbols together with axioms defining their properties can be added to a theory without endangering consistency, if they are of a certain restricted syntactical form. These **definitional form**s mirror the various types of mathematical **definition**s (e.g. equational, recursive, implicit definitions). This leads to the *"principle of conservative extension"*, which states that conservative extensions to theories (by theorems and definitions) are safe for mathematical theories, and that possible sources for inconsistencies can be narrowed down to small sets of axioms.

Even though all of this has theoretically been known to (meta)-mathematicians for almost a century, it has only been an explicit object of formal study and exploited by mathematical software systems in the last decades. Much of the meta-mathematics has been formally studied in the context of proof de-

---

[9] A statement is valid in a theory, iff it is true for all models of the theory. If there are none, it is vacuously valid.

velopment systems like AUTOMATH [dB80] NUPRL [CAB+86], HOL [GM93], MIZAR [Rud92] and ΩMEGA [BCF+97] which utilize strong logical systems that allow to express both mathematical statements and proofs as mathematical objects. Some systems like ISABELLE [PN90] and TWELF [Pfe91] even allow the specification of the logic language itself, in which the reasoning takes place. Such semi-automated theorem proving systems have been used to formalize substantial parts of mathematics and mechanically verify many theorems in the respective areas. These systems usually come with a library system that manages and structures the body of mathematical knowledge formalized in the system so far.

In software engineering, mathematical theories have been studied under the label of "(algebraic) specifications". Theories are used to specify the behavior of programs and software components. Under the pressure of industrial applications, the concept of a theory (specification) has been elaborated from a practical point of view to support the structured development of specifications, theory reuse, and modularization. Without this additional structure, real world specifications become unwieldy and unmanageable in practice. Just as in the case of the theorem proving systems, there is a whole zoo of specification languages, most of them tied to particular software systems. They differ in language primitives, theoretical expressivity, and the level of tool support.

Even though there have been standardization efforts, the most recent one being the CASL standard (Common Algebraic Specification Language; see [CoF04]) there have been no efforts of developing this into a general markup language for mathematics with attention to web communication and standards. The OMDOC format attempts to provide a content-oriented markup scheme that supports all the aspects and structure of mathematical knowledge we have discussed in this section. Before we define the language in the next chapter, we will briefly go over the consequences of adopting a markup language like OMDOC as a standard for web-based mathematics.

3

OMDoc: Open Mathematical Documents

Based on the analysis of the structure inherent in mathematical knowledge and existing content markup systems for mathematics we will now briefly introduce basic design assumptions and the development history of the OM-DOC format, situate it, and discuss possible applications.

3.1 A Brief History of the OMDOC Format

OMDOC initially developed from the quest for a solution of the problem of representing knowledge on the one hand and integrating external mathematical reasoning systems in the ΩMEGA project at Saarland University on the other. ΩMEGA [SBB+02] is a large-scale proof development environment that integrates various reasoning engines (automated theorem provers, decision procedures, computer algebra systems) via knowledge-based proof planning with the aim of creating a mathematical assistant system.

3.1.1 The Design Problem

One of the hard practical problems of building such systems is to represent, provision, and manage the relevant (factual, tactic, and intuitive) knowledge human mathematicians use in developing mathematical theories and proofs: Knowledge-based reasoning systems use explicit representations of this knowledge to automate the search for a proof, and before a system can be applied to a mathematical domain it must be formalized, the proof tactics of this domain must be identified, and the intuitions of when to use which tactic must be coaxed from practitioners. Ideally, as a valuable and expensive resource, this knowledge would be shared between mathematical assistant systems to be able to compare the relative strength of the systems and to enhance practical coverage. This poses the problem that the knowledge must be represented at a level that would accommodate the different systems' representational quirks and bridge between them.

Developing an agent-oriented framework for distributed reasoning via remote procedure calls to achieve system scalability (MATHWEB-SB [FK99, ZK02]; see Chapter 9 for an OMDOC-based reformulation) revealed that the underlying problem in integrating mathematical systems is a semantic one: all the reasoning systems make differing ontological assumptions that have to be reconciled to achieve a correct (i.e. meaning-preserving) integration. This integration problem is quite similar to the one at the knowledge level: if the knowledge ingrained in the system design could be explicitly described, then it would be possible to find applicable systems and deploy the necessary (syntactic) and (semantic) bridges automatically.

The approaches and solutions offered by the automated reasoning communities at that time were insular at best: They standardized character-level syntax standardizing on first-order logic [SSY94, HKW96], or explored bilateral system integrations overcoming deep ontological discrepancies between the systems [FH97].

At the same time, (ca 1998) the Computer Algebra Community was grappling with similar integration problems. The OPENMATH standard that was emerging shad solved the web-scalability problem in representing mathematical formulae by adopting the emerging XML framework as a syntactical basis and providing structural markup with explicit context references as a syntax-independent representation approach. First attempts by the author to influence OPENMATH standardization so that the format would allow mathematical knowledge representation (i.e. the statements and context level) were unsuccessful. The OPENMATH community had intensively discussed similar issues under the heading of "content dictionary inheritance" and "conformance specification", and had decided that they were too controversial for standardization.

3.1.2 Design Principles

The start of the development of OMDOC as a content-based representation format for mathematical knowledge was triggered by an e-mail by Alan Bundy to the author in 1998, where he lamented the fact that one of the great hindrances of knowledge-based reasoning is the fact that formalizing mathematical knowledge is very time-consuming and that it is very hard for young researchers to gain recognition for formalization work. This led to the idea of developing a global repository of formalized mathematics, which would eventually allow peer-reviewed publication of formalized mathematical knowledge, thus generating academic recognition for formalization work and eventually lead to the much enlarged corpus of formalized mathematics that is necessary for knowledge-based formal mathematical reasoning. Young researchers would contribute formalizations of mathematical knowledge in the form of mathematical documents that would be both formal and thus machine-readable, as well as human-readable, so that humans could find and understand them[1].

This idea brought the final ingredient to the design principles: in a nutshell, the OMDOC format was to

1. be *Ontologically uncommitted* (like the OPENMATH format), so that it could serve as a *integration format* for mathematical software systems.
2. provide a representation format for *mathematical documents* that combined *formal* and *informal* views of all the *mathematical knowledge* contained in them.
3. be based on *sound logic/representational principles* (as not to embarrass the author in front of his colleagues from automated reasoning)
4. be based on *structural/content markup* to guarantee both 1.) and 2.).

3.1.3 Development History

Version 1.0 of the OMDOC format was released on November 1st 2000 to give users a stable interface to base their documents and systems on. It was adopted by various projects in automated deduction, algebraic specification, and computer-supported education. The experience from these projects uncovered a multitude of small deficiencies and extension possibilities of the format, that have been subsequently discussed in the OMDOC community.

OMDOC 1.1 was released on December 29th 2001 as an attempt to roll the uncontroversial and non-disruptive part of the extensions and corrections into a consistent language format. The changes to version 1.0 were largely conservative, adding optional attributes or child elements. Nevertheless, some non-conservative changes were introduced, but only to less used parts of the format or in order to remedy design flaws and inconsistencies of version 1.0.

OMDOC 1.2 is the mature version in the OMDOC 1 series of specifications. It contains almost no large-scale changes to the document format, except that Content-MATHML is now allowed as a representation for mathematical objects. But many of the representational features have been finetuned and brought up to date with the maturing XML technology (e.g. ID attributes now follow the XML ID specification [MVW05], and the Dublin Core elements follow the official syntax [DUB03a]). The main development is that the OMDOC specification, the DTD, and schema are split into a system of interdependent modules that support independent development of certain language aspects and simpler specification and deployment of sub languages. Version 1.2 of OMDOC freezes the development so that version 2 can be started off on the modules.

---

[1] Here the strong influence of the MIZAR project under Andrzej Trybulec must be acknowledged, at that time, the project had already realized these two goals. They had even established the "Journal of Formalized Mathematics", where LATEX articles were generated from the automatically verified MIZAR source. However, the MIZAR mathematical language [Miz06] used a human-oriented syntax that defied outside parsing and web-integration, had a tightly integrated largely undocumented sort system, and made very strong ontological commitments.

3.2   Three Levels of Markup

To achieve content and context markup for mathematical knowledge, OM-
DOC uses three levels of modeling corresponding to the concerns raised previously. We have visualized this architecture in Figure 3.1.

Level of Representation	OMDOC Example
Theory Level: Development Graph – Inheritance via symbol-mapping – Theory inclusion via proof-obligations – Local (one-step) vs. global links	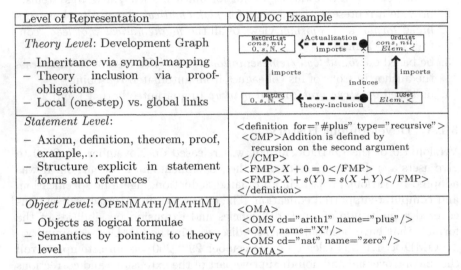

Statement Level:
 	
– Axiom, definition, theorem, proof, example,...
– Structure explicit in statement forms and references
	`<definition for="#plus" type="recursive">`
	`<CMP>`Addition is defined by
	    recursion on the second argument
	`</CMP>`
	`<FMP>`$X + 0 = 0$`</FMP>`
	`<FMP>`$X + s(Y) = s(X + Y)$`</FMP>`
	`</definition>`

Object Level: OPENMATH/MATHML
 	
– Objects as logical formulae
– Semantics by pointing to theory level
	`<OMA>`
	`<OMS cd="arith1" name="plus"/>`
	`<OMV name="X"/>`
	`<OMS cd="nat" name="zero"/>`
	`</OMA>`

Fig. 3.1. OMDOC in a nutshell (the three levels of modeling)

Building on the discussion in Chapter 2 we distinguish three levels of representation in OMDOC

Mathematical Theories (see Section 2.1) At this level, OMDOC supplies original markup for clustering sets of statements into theories, and for specifying relations between theories by morphisms. By using this scheme,
   mathematical knowledge can be structured into reusable chunks. Theories also serve as the primary notion of context in OMDOC, they are the
   natural target for the context aspect of formula and statement markup.
Mathematical Statements (see Section 2.2) OMDOC provides original markup infrastructure for making the structure of mathematical statements
   explicit. Again, we have content and context markup aspects. For instance the definition in the right hand side of the second row of Figure 3.1 contains an informal description of the definition as a first child
   and a formal description in the two recursive equations in the second and
   third children supported by the `type` attribute, which states that this is
   a recursive definition. The context markup in this example is simple: it
   states that this piece of markup pertains to a symbol declaration for the
   symbol `plus` in the current theory (presumably the theory `arith1`).

Mathematical Formulae (see Section 2.3) At the level of mathematical formulae, OMDOC uses the established standards OPENMATH [BCC+04] and Content-MATHML [ABC+03]. These provide content markup for the structure of mathematical formulae and context markup in the form of URI references in the symbol representations (see Chapter 13 for an introduction).

All levels are augmented by markup for various auxiliary information that is present in mathematical documents, e.g. notation declarations, exercises, experimental data, program code, etc.

3.3   Situating the OMDoc Format

The space of representation languages for mathematical knowledge reaches from the input languages of computer algebra systems (CAS) to presentation markup languages for mathematical vernacular like TEX/LATEX. We have organized some of the paradigmatic examples in a diagram mapping coverage (which kinds of mathematical knowledge can be expressed) against machine support (which services the respective software system can offer) in Figure 3.2.

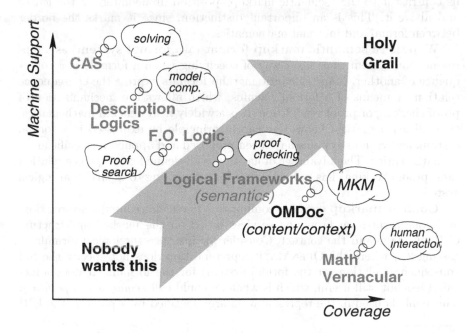

Fig. 3.2. Situating content markup: math. knowledge management

On the left hand side we see CAS like MATHEMATICA®[Wol02] or MAPLE™ [CGG⁺92] that are relatively restricted in the mathematical objects — they can deal with polynomials, group representations, differential equations only, but in this domain they can offer sophisticated services like equation solving, factorization, etc. More to the right we see systems like automated theorem provers, whose language — usually first-order logic — covers much more of mathematics, but that cannot perform computational services[2] like the CAS do.

In the lower right hand corner, we find languages like "mathematical vernacular", which is just the everyday mathematical language. Here coverage is essentially universal: we can use this language to write international treaties, math books, and love letters; but machine support is minimal, except for typesetting systems for mathematical formulae like TEX, or keyword search in the natural language part.

The distribution of the systems clusters around the diagonal stretching from low-coverage, high-support systems like CAS to wide-coverage, low-support natural language systems. This suggests that there is a trade-off between coverage and machine support. All of the representation languages occupy legitimate places in the space of representation languages, trying to find sweet-spots along this coverage/support trade-off. OMDOC tries to occupy the "content markup" position. To understand this position better, let us contrast it to the "semantic markup" position immediately to the left of and above it. This is an important distinction, since it marks the border between formal and informal mathematics.

We define a **semantic markup format** (aka **formal system**) as a representation system that has a way of specifying when a formula is a consequence of another. Many semantic markup formats express the consequence relation by means of a formal calculus, which allows the mechanization of proof checking or proof verification. It is a widely held belief in mathematics, that all mathematical knowledge can in principle be expressed in a formal system, and various systems have been proposed and applied to specific areas of mathematics. The advantage of having a well-defined consequence relation (and proof-checking) has to be paid for by committing to a particular logical system.

Content markup does not commit to a particular consequence relation, and concentrates on providing services based on the marked up structure of the content and the context. Consider for instance the logical formula in Listing 2.5, where the OPENMATH representation does not specify the full consequence relation (or the formal system) for the formula. It does something less but still useful, which is what we could call *semantics by pointing*: The symbols used in the representation are identified by a pointer (the URI

---

[2] Of course in principle, the systems could, since computation and theorem proving are inter-reducible, but in practice theorem provers get lost in the search spaces induced by computational tasks.

jointly specified in the cd and name attributes) to a defining document (in this case an OPENMATH content dictionary). Note that URI equality is a sufficient condition for two symbols to be equal, but not a necessary condition: Two symbols can be semantically equal without pointing to the same document, e.g. if the two defining documents are semantically marked up and the definitions are semantic consequences of each other.

In this sense, content markup offers a more generic markup service (for all formal systems; we do not have to commit ourselves) at the cost of being less precise (we for instance miss out on some symbol equalities). Thus, content markup is placed to the lower right of semantic markup in Figure 3.2. Note however, that content markup can easily be turned into semantic markup by adding a consequence relation, e.g. by pointing to defining documents that are marked up semantically. Unlike OPENMATH and Content-MATHML, the OMDOC format straddles the content/semantics border by closing the loop and providing a content markup format for both formulae and the defining documents. In particular, *an* OMDOC *document is semantic if all the documents it references are.*

As a consequence, OMDOC can serve as a migration format from formal to informal mathematics (and thus from representations that for human consumption to such that can be supported by machines). A document collection can be marked for content and context structure, making the structures and context references explicit in a first pass. Note that this pass may involve creating additional documents or identifying existing documents that serve as targets for the context references so that the document collection is self-contained. In a second (and possible semi-automatic) step, we can turn this self-contained document collection into a formal representation (semantic markup) by committing on consequence relations and adding the necessary detail to the referenced documents.

3.4   The Future: An Active Web of (Mathematical) Knowledge

It is a crucial – if relatively obvious – insight that true cooperation of mathematical services is only feasible if they have access to a joint corpus of mathematical knowledge. Moreover, having such a corpus would allow to develop added-value services like

- Cut and paste on the level of computation (take the output from a web search engine and paste it into a computer algebra system),
- Automatically proof checking published proofs,
- Math explanation (e.g. specializing a proof to an example that simplifies the proof in this special case),
- Semantic search for mathematical concepts (rather than keywords),

- Data mining for representation theorems (are there unnoticed groups out there?),
- Classification: Given a concrete mathematical structure, is there a general theory for it?

As the online mathematical knowledge is presently only machine-*readable*, but not machine-*understandable*, all of these services can currently only be performed by humans, limiting the accessibility and thus the potential value of the information. Services like this will transform the now passive and human-centered fragment of the Internet that deals with mathematical content, into an active (supported by semantic services) web of mathematical knowledge.

This promise of activating a web of knowledge is not limited to mathematics: the task of transforming the current presentation-oriented world-wide web into a "Semantic Web" [Lee98] has been identified as one of the main challenges by the world W3C. With the OMDoc format we pursue an alternative vision of a 'Semantic Web' for Mathematics. Like Tim Berners-Lee's vision we aim to make the Web (here mathematical knowledge) machine-understandable instead of merely machine-readable. However, instead of a top-down metadata-driven approach, which tries to approximate the content of documents by linking them to web ontologies (expressed in terminologic logics), we explore a bottom-up approach and focus on making explicit the intrinsic structure of the underlying scientific knowledge. A connection of documents to web ontologies is still possible, but a secondary effect.

The direct applications of OMDoc (apart from the general effect towards a Semantic Web) are not confined to mathematics proper either. The MathML working group in the W3C has led the way in many web technologies (presenting mathematics on the web taxes the current web technology to its limits); the endorsement of the MathML standard by the W3 Committee is an explicit testimony to this. We expect that the effort of creating an infrastructure for digital mathematical libraries will play a similar role, since mathematical knowledge is the most rigorous and condensed form of knowledge and will therefore pinpoint the problems and possibilities of the semantic web.

All modern sciences have a strongly mathematicised core and will benefit. The real market and application area for the techniques developed in this project lies with high-tech and engineering corporations that rely on huge formula databases. Currently, both the content markup as well as the added-value services alluded to above are very underdeveloped, limiting the usefulness of vital knowledge. The content-markup aspect needed for mining this information treasure is exactly what we are developing in OMDoc.

## Part II

An OMDoc Primer

This part of the book provides an easily approachable description of the OMDOC format by way of paradigmatic examples of OMDOC documents. The primer should be used alongside the formal descriptions of the language contained in Part III.

The intended audience for the primer are users who only need a casual exposure to the format, or authors that have a specific text category in mind. The examples presented here also serve as specifications of "best practice", to give the readers an intuition for how to encode various kinds of mathematical knowledge.

Each chapter of the OMDOC primer deals with a different category of mathematical document and introduces new features of the OMDOC format in the context of concrete examples.

Chapter 4: Mathematical Textbooks and Articles. discusses the markup process for an informal but rigorous mathematical texts. We will use a fragment of Bourbaki's "Algebra" as an example. The development marks up the content in four steps, from the document structure to a full formalization of the content that could be used by automated theorem provers. The first page of Bourbaki's "Algebra" serves as an example of the treatment of a rigorous presentation of pure mathematics, as it can be found in textbooks and articles.

Chapter 5 OpenMath Content Dictionaries. transforms an OPENMATH content dictionary into an OMDOC document. OPENMATH content dictionaries are semi-formal documents that serve as references for mathematical symbols in OPENMATH encoded formulae. As of OPENMATH2, OMDOC is an admissible OPENMATH content dictionary format. They are a good example for mathematical glossaries, and background references, both formal and informal.

Chapter 6 Structured and Parametrized Theories. shows the power of theory markup in OMDoc for theory reuse and modular specification. The example builds a theory of ordered lists of natural numbers from a generic theory of ordered lists and the theory of natural numbers which acts as a parameter in the actualization process.

Chapter 7 A Development Graph for Elementary Algebra. extends the range of theory-level structure by specifying the elementary algebraic hierarchy. The rich fabric of relations between these theories is made explicit in the form of theory morphisms, and put to use for proof reuse.

Chapter 8 Courseware and the Narrative/Content Distinction. covers markup for a fragment of a computer science course in the OMDoc format, dwelling on the difference between the narrative structure of the course and the background knowledge. Course materials like slides or writings on blackboards are usually much more informal than textbook presentations of mathematics. They also openly structure materials by didactic criteria and leave out important parts of the rigorous development, which the student is required to pick up from background materials like textbooks or the teacher's recitation.

Chapter 9 Communication with and between Mathematical Software Systems. uses an OMDoc fragment as content for communication protocols between mathematical software systems on the Internet. Since the communicating parties in this situation are machines, OMDoc fragments are embedded into other XML markup that serves as a protocol for the distribution layer.

Together these examples cover many of the mathematical documents involved in communicating mathematics. As the first two chapters build upon each other and introduce features of the OMDoc format, they should be read in succession. The remaining three chapters build on these, but are largely independent.

To keep the presentation of the examples readable, we will only present salient parts of the OMDoc representations in the discussion. The full text of the examples can be accessed at `https://svn.mathweb.org/repos/mathweb.org/branches/omdoc-1.2/examples/spec`.

4

Mathematical Textbooks and Articles

In this chapter we will work an example of a stepwise formalization of mathematical knowledge. This is the task of e.g. an editor of a mathematical textbook preparing it for web-based publication. We will use an informal, but rigorous text: a fragment of Bourbaki's Algebra [Bou74], which we show in Figure 4.1. We will mark it up in four stages, discussing the relevant OMDOC elements and the design decisions in the OMDOC format as we go along. Even though the text was actually written prior to the availability of the TeX/LaTeX system, we will take a LaTeX representation as the starting point of our markup experiment, since this is the prevalent source markup format in mathematics nowadays.

Section 4.1 discusses the minimal markup that is needed to turn an arbitrary document into a valid OMDOC document — albeit one, where the markup is worthless of course. It discusses the necessary XML infrastructure and adds some meta-data to be used e.g. for document retrieval or archiving purposes.

In Section 4.2 we mark up the top-level structure of the text and classify the paragraphs by their category as mathematical statements. This level of markup already allows us to annotate and extract some meta-data and would allow applications to slice the text into individual units, store it in databases like MBASE (see Section 26.4), or the In2Math knowledge base [Dah01, BB01], or assemble the text slices into individualized books e.g. covering only a sub-topic of the original work. However, all of the text itself, still contains the LaTeX markup for formulae, which is readable only by experienced humans, and is fixed in notation. Based on the segmentation and meta-data, suitable systems like the ACTIVEMATH system described in Section 26.8 can re-assemble the text in different orders.

In Section 4.3, we will map all mathematical objects in the text into OPENMATH or Content-MATHML objects. To do this, we have to decide which symbols we want to use for marking up the formulae, and how to structure the theories involved. This will not only give us the ability to generate specialized and user-adaptive notation for them (see Chapter 25), but

---

1. LAWS OF COMPOSITION

DEFINITION 1. *Let E be a set. A mapping of $E \times E$ is called a law of composition on E. The value $f(x, y)$ of f for an ordered pair $(x, y) \in E \times E$ is called the composition of x and y under this law. A set with a law of composition is called a magma.*

The composition of x and y is usually denoted by writing x and y in a definite order and separating them by a characteristic symbol of the law in question (a symbol which it may be agreed to omit). Among the symbols most often used are $+$ and $\cdot$, the usual convention being to omit the latter if desired; with these symbols the composition of x and y is written respectively as $x + y$, $x.y$ or xy. A law denoted by the symbol $+$ is usually called *addition* (the composition $x + y$ being called the *sum* of x and y) and we say that it is *written additively*; a law denoted by the symbol . is usually called *multiplication* (the composition $x.y = xy$ being called the *product* for x and y) and we say that it is *written multiplicatively*.

In the general arguments of paragraphs 1 to 3 of this chapter we shall generally use the symbols $\top$ and $\bot$ to denote arbitrary laws of composition.

By an abuse of language, a mapping of a *subset* of $E \times E$ into E is sometimes called a law of composition *not everywhere defined* on E.

Examples. (1) The mappings $(X, Y) \mapsto X \cup Y$ and $(X, Y) \mapsto X \cap Y$ are laws of composition on the set of subsets of a set E.

(2) On the set $\mathbf{N}$ of natural numbers addition, multiplication, and exponentiation are laws of composition (the compositions of $x \in \mathbf{N}$ and $y \in \mathbf{N}$ under these laws being denoted respectively by $x + y$, xy, or $x.y$ and x^y) (*Set Theory*, III, §3, no. 4).

(3) Let E be a set; the mapping $(X, Y) \mapsto X \circ Y$ is a law of composition on the set of subsets of $E \times E$ (*Set Theory*, II, §3, no. 3, Definition 6); the mapping $(f, g) \mapsto f \circ g$ is a law of composition on the set of mappings from E into E (*Set Theory*, II, §5, no. 2).

Fig. 4.1. A fragment from Bourbaki's algebra [Bou74]

also to copy and paste them to symbolic math software systems. Furthermore, an assembly into texts can now be guided by the semantic theory structure, not only by the mathematical text categories or meta-data.

Finally, in Section 4.4 we will fully formalize the mathematical knowledge. This involves a transformation of the mathematical vernacular in the statements into some logical formalism. The main benefit of this is that we can verify the mathematical contents in theorem proving environments like NUPRL [CAB+86], HOL [GM93], MIZAR [Rud92] and OMEGA [BCF+97].

4.1 Minimal OMDoc Markup

It actually takes very little change to an existing document to make it a valid OMDOC document. We only need to wrap the text into the appropriate XML document tags. In Listing 4.2, we have done this and also added meta-data.

Actually, since the **metadata** and the document type declaration are optional in OMDoc, just wrapping the original text with lines 1, 4, 7, 31, 32, and 36 to 38 is the simplest way to create an OMDoc document.

Listing 4.2. The outer part of the document

```
<?xml version="1.0" encoding="utf−8"?>
<!DOCTYPE omdoc PUBLIC "−//OMDoc//DTD OMDoc Basic V1.2//EN"
              "http://www.mathweb.org/omdoc/dtd/omdoc−basic.dtd" []>

5   <omdoc xml:id="algebra1.omdoc" version="1.2" modules="@basic"
          xmlns:dc="http://purl.org/dc/elements/1.1/"
          xmlns:cc="http://creativecommons.org/ns"
          xmlns="http://www.mathweb.org/omdoc">
      <metadata>
10      <dc:title>Laws of Composition</dc:title>
        <dc:creator role="trl">Michael Kohlhase</dc:creator>
        <dc:date action="created">2002−01−03T07:03:00</dc:date>
        <dc:date action="updated">2002−11−23T18:17:00</dc:date>
        <dc:description>
15        A first migration step for a fragment of Bourbaki's Algebra
        </dc:description>
        <dc:source>
          Nicolas Bourbaki, Algebra, Springer Verlag 1989, ISBN 0−387−19373−1
        </dc:source>
20      <dc:type>Text</dc:type>
        <dc:format>application/omdoc+xml</dc:format>
        <dc:rights>Copyright (c) 2005 Michael Kohlhase</dc:rights>
        <cc:license>
          <cc:permissions reproduction="permitted" distribution="permitted"
25                       derivative_works="permitted"/>
          <cc:prohibitions commercial_use="permitted"/>
          <cc:requirements notice="required" copyleft="required" attribution="required"/>
        </cc:license>
      </metadata>
30
      <omtext xml:id="all">
        <CMP xml:lang="en">
        {\sc Definition 1.} Let E be a set. A mapping E × E is called a law of
        ...
35      mappings from E into E ({\emph{Set Theory}}, II, §5, no. 2).
        </CMP>
      </omtext>
    </omdoc>
```

We will now explain the general features of the OMDoc representation in detail by line numbers. The references point to the relevant sections in the OMDoc specification; details and normative rules for using the elements in questions can be found there.

line	Description	ref.
1	This document is an XML 1.0 file that is encoded in the UTF-8 encoding.	

2,3	The parser is told to use a document type definition for validation. The string `omdoc` specifies the name of the root element, the identifier `PUBLIC` specifies that the DTD (we use the "OMDOC basic" DTD; see Subsection 22.3.1), which can be identified by the public identifier in the first string and looked up in an XML catalog or (if that fails) can be found at the URL specified in the second string. A DTD declaration is not strictly needed for an OMDOC document, but is recommended, since the DTD supplies default values for some attributes.	24.1 p. 228
4	In general, XML files can contain as much whitespace as they want between elements, here we have used it for structuring the document.	
5	Start tag of the root element of the document. It declares the version (OMDOC 1.2) via the `version`, and an identifier of the document using the `xml:id` attribute. The optional `modules` specifies the sub-language used in this document. This is used when no DTD is present (see Subsection 22.3.1).	11.1 p. 90
6,7	the namespace prefix declarations for the Dublin Core, Creative Commons, and OPENMATH namespaces. They declare the prefixes `dc:`, `cc:`, and `om:`, and bind them to the specified URIs. We will need the OPENMATH namespace only in the third markup step described in Section 4.3, but spurious namespace prefix declarations are not a problem in the XML world.	10 p. 83
8	the namespace declaration for the document; if not prefixed, all elements live in the OMDOC namespace.	10.1 p. 83
9–29	The metadata for the whole document in Dublin Core format	11.2 p. 92
10	The title of the document	12.1 p. 98
11	The document creator, here in the role of a translator	12.2 p. 101
12	The date and time of first creation of the document in ISO 8601 norm format.	12.1 p. 99
13	The date and time of the last update to the document in ISO 8601 norm format.	12.1 p. 99
14–16	A short description of the contents of the document	12.1 p. 99
17–19	Here we acknowledge that the OMDOC document is just a translation from an earlier work.	12.1 p. 100
20	The type of the document, this can be `Dataset` (un-ordered mathematical knowledge) or `Text` (arranged for human consumption).	12.1 p. 100
21	The format/MIME type [FB96] of the document, for OMDOC, this is `application/omdoc+xml`.	12.1 p. 100
22	The copyright resides with the creator of the OMDOC document	12.1 p. 100
23–28	The creator licenses the document to the world under certain conditions as specified in the Creative Commons license specified in this element.	12.3 p. 102

24,25	The cc:permissions element gives the world the permission to reproduce and distribute it freely. Furthermore the license grants the public the right to make derivative works under certain conditions.	12.3 p. 103
26	The cc:prohibitions can be used to prohibit certain uses of the document, but this one is unencumbered.	12.3 p. 103
27	The cc:requirements states conditions under which the license is granted. In our case the licensee is required to keep the copyright notice and license notices intact during distribution, to give credit to the copyright holder, and that any derivative works derived from this document must be licensed under the same terms as this document (the copyleft clause).	12.3 p. 103
31-37	The omtext element is used to mark up text fragments. Here, we have simply used a single omtext to classify the whole text in the fragment as unspecific "text".	14.3 p. 124
32-36	The CMP element holds the actual text in a multilingual group. Its xml:lang specifies the language. If the document is used with a DTD or an XML schema (as we are) this attribute is redundant, since the default value given by the DTD or schema is en. More keywords in other languages can be given by adding more CMP elements.	14.1 p. 122
33–35	The text of the LaTeX fragment we are migrating. For simplicity we do not change the text, and leave that to later stages of the migration.	
38	The closing tag of the root omdoc element. There may not be text after this in the file.	11.1 p. 90

4.2 Marking up the text structure and statements

In the next step, we analyze and mark up the structure of the text of the further, and embed the paragraphs into markup for mathematical statements or text segments. Instead of lines 19–25 in Listing 4.2, we will now have the representation in Listing 4.3.

Listing 4.3. The segmented text

```
<omtext xml:id="magma.def" type="definition">
  <CMP>Let <legacy format="TeX">E</legacy> be a set ... called a magma.</CMP>
</omtext>

<omtext xml:id="t1">
  <CMP>The composition of <legacy format="TeX">x</legacy> ... multiplicatively.</CMP>
</omtext>
<omtext xml:id="t2">
  <CMP>In the general ... composition.</CMP>
</omtext>
<omtext xml:id="t3">
  <CMP>By an abuse ... on <legacy format="TeX">E.</legacy></CMP>
</omtext>

<omgroup xml:id="magma−ex" type="enumeration">
    <metadata><dc:title>Examples</dc:title></metadata>

    <omtext type="example" xml:id="e1.magma">
```

```
     <CMP>
20      The mappings <legacy format="TeX">(X,Y)</legacy>
        ... subsets of a set <legacy format="TeX">E</legacy>.
     </CMP>
     </omtext>
     <omtext type="example" xml:id="e2.magma">
25      <CMP>
        On the set <legacy format="TeX">N</legacy> ... III, §3, no. 4).
        </CMP>
     </omtext>
     <omtext type="example" xml:id="e3.magma">
30      <CMP>
        Let <legacy format="TeX">E</legacy> be a set; ... II, §5, no. 2).
        </CMP>
     </omtext>
     </omgroup>
```

In summary, we have sliced the text into `omtext` fragments and individually classified them by their mathematical role. The formulae inside have been encapsulated into `legacy` elements that specify their format for further processing. The higher-level structure has been captured in OMDoc grouping elements and the document as well as some of the slices have been annotated by metadata.

line	Description	ref.
1	The `omtext` element classifies the text fragment as a `definition`, other types for mathematical statements include `axiom`, `example`, `theorem`, and `lemma`. Note that the numbering of the original text is lost, but can be re-created in the text presentation process. The optional `xml:id` attribute specifies a document-unique identifier that can be used for reference later.	14.3 p. 124
2	A multilingual group of `CMP` elements that hold the text (in our case, there is only the English default). Here the TeX formulae have been marked up with `legacy` elements characterizing them as such. This might simplify a later automatic transformation to OPENMATH or Content-MATHML.	13.5 p. 120
4–13	We have classified every paragraph in the original as a separate `omtext` element, which does not carry a `type` since it does not fit any other mathematical category at the moment.	14.3 p. 124
15	The three examples in the original in Figure 4.1 are grouped into an enumeration. We use the OMDoc `omgroup` element for this. The optional attribute `xml:id` can be used for referencing later. We have chosen `enumeration` for the `type` attribute to specify the numbering of the examples in the original.	11.4 p. 93
16	We can use the `metadata` of the `omgroup` element to accommodate the title "Examples" in the original. We could enter more metadata at this level.	14.1 p. 122
18	The `type` attribute of this `omtext` element classifies this text fragment as an example.	14.3 p. 124

4.3   Marking up the Formulae

After we have marked up the top-level structure of the text to expose the content, the next step will be to mark up the formulae in the text to content mathematical form. Up to now, the formulae were still in TEX notation, which can be read by TEX/LATEX for presentation to the human user, but not used by symbolic mathematics software. For this purpose, we will re-represent the formulae as OPENMATH objects or Content-MATHML, making their functional structure explicit.

So let us start turning the TEX formulae in the text into OPENMATH objects. Here we use the hypothetical `mbc.mathweb.org` as repository for theory collections.

Listing 4.4. The definition of a magma with OPENMATH objects

```
    <!DOCTYPE omdoc PUBLIC "−//OMDoc//DTD OMDoc CD V1.2//EN"
                  "http://www.mathweb.org/omdoc/dtd/omdoc−cd.dtd"
             [<!ENTITY % om.prefixed "INCLUDE">]>
5   <theory xml:id="magmas">
     <imports from="background.omdoc#products"/>
     <imports from="http://mbc.mathweb.org/omstd/relation1.omdoc#relation1"/>

     <symbol name="magma">
10      <metadata><dc:description>Magma</dc:description></metadata>
     </symbol>
     <symbol name="law_of_composition"/>

     <definition xml:id="magma.def" for="#magma #law_of_composition">
15     <CMP>
        Let <om:OMOBJ><om:OMV name="E"/></om:OMOBJ> be a set. A mapping of
        <om:OMOBJ>
          <om:OMA><om:OMS cd="products" name="Cartesian−product"/>
          <om:OMV name="E"/><om:OMV name="E"/>
20        </om:OMA>
        </om:OMOBJ> is called a
        <term cd="magmas" name="magma" role="definiens">law of composition</term>
        on <om:OMOBJ><om:OMV name="E"/></om:OMOBJ>. The value
        <om:OMOBJ>
25        <om:OMA><om:OMV name="f"/>
            <om:OMV name="x"/><om:OMV name="y"/>
          </om:OMA>
        </om:OMOBJ>
        of <om:OMOBJ><om:OMV name="f"/></om:OMOBJ> for an ordered pair
30      <om:OMOBJ>
          <om:OMA><om:OMS cd="sets" name="in"/>
          <om:OMA><om:OMS cd="products" name="pair"/>
            <om:OMV name="x"/><om:OMV name="y"/>
          </om:OMA>
35        <om:OMA><om:OMS cd="products" name="Cartesian−product"/>
            <om:OMV name="E"/><om:OMV name="E"/>
          </om:OMA>
        </om:OMA>
        </om:OMOBJ> is called the
40      <term cd="magmas" name="law_of_composition"
                      role="definiens−applied">composition</term>
        of <om:OMOBJ><om:OMV name="x"/></om:OMOBJ> and
        <om:OMOBJ><om:OMV name="y"/></om:OMOBJ> under this law.
        A set with a law of composition is called a
45      <term cd="magmas" name="magma" role="definiens">magma</term>.
       </CMP>
     </definition>
     ...
    </theory>
50  ...
```

Of course all the other mathematical statements in the documents have to be treated in the same way.

line	Description	ref.
1–4	The `omdoc-basic` document type definition is no longer sufficient for our purposes, since we introduce new symbols that can be used in other documents. The DTD for OMDoc content dictionaries (see Chapter 5), which allows this. Correspondingly, we would specify the value `cd` for the attribute `module`. The part in line 4 is the internal subset of the DTD, which sets a parameter entity for the modularized DTD to instruct it to accept OPENMATH elements in their namespace prefixed form. Of course a suitable namespace prefix declaration is needed as well.	22.3.2 p. 218
5	The start tag of a theory. We need this, since symbols and definitions can only appear inside `theory` elements.	15.6 p. 149
6,7	We need to import the theory `products` to be able to use symbols from it in the definition below. The value of the `from` is a relative URI reference to a `theory` element much like the one in line 5. The other `imports` element imports the theory `relation1` from the OPENMATH standard content dictionaries[1]. Note that we do not need to import the theory `sets` here, since this is already imported by the theory `products`.	15.6.1 p. 150
9–11	A symbol declaration: For every definition, OMDoc requires the declaration of one or more `symbol` elements for the concept that is to be defined. The `name` attribute is used to identify it. The `dc:description` element allows to supply a multilingual (via the `xml:lang` attribute) group of keywords for the declared symbol	15.2.1 p. 136
12	Upon closer inspection it turns out that the definition in Listing 4.4 actually defines three concepts: "law of composition", "composition", and "magma". Note that "composition" is just another name for the value under the law of composition, therefore we do not need to declare a symbol for this. Thus we only declare one for "law of composition".	15.2.1 p. 136
14	A definition: the `definition` element carries a `name` attribute for reference within the theory. We need to reference the two symbols defined here in the `for` attribute of the `definition` element; it takes a whitespace-separated list of `name` attributes of `symbol` elements in the same theory as values.	15.2.4 p.139
16	We use an OPENMATH object for the set E. It is an `om:OMOBJ` element with an `om:OMV` daughter, whose `name` attribute specifies the object to be a variable with name E. We have chosen to represent the set E as a variable instead of a constant (via an `om:OMS` element) in the theory, since it seems to be local to the definition. We will discuss this further in the next section, where we talk about formalization.	13.1.1 p. 108

---

[1] The originals are available at `http://www.openmath.org/cd`; see Chapter 5 for a discussion of the differences of the original OPENMATH format and the OMDoc format used here.

17–21	This `om:OMOBJ` represents the Cartesian product $E \times E$ of the set E with itself. It is an application (via an `om:OMA` element) of the symbol for the binary Cartesian product relation to E.	13.1.1 p. 108
18	The symbol for the Cartesian product constructor is represented as an `om:OMS` element. The `cd` attribute specifies the theory that defines the symbol, and the `name` points to the `symbol` element in it that declares this symbol. The value of the `cd` attribute is a theory identifier. Note that this theory has to be imported into the current theory, to be legally used.	13.1.1 p. 108
22	We use the `term` element to characterize the defined terms in the text of the definition. Its `role` attribute can used to mark the text fragment as a `definiens`, i.e. a concept that is under definition.	14.5 p. 128
24–28	This object stands for $f(x, y)$	
30–39	This object represents $(x, y) \in E \times E$. Note that we make use of the symbol for the elementhood relation from the OPEN-MATH core content dictionary `set1` and of the pairconstructor from the theory of products from the Bourbaki collection there.	

The rest of the representation in Listing 4.4 is analogous. Thus we have treated the first definition in Figure 4.1. The next two paragraphs contain notation conventions that help the human reader to understand the text. They are annotated as `omtext` elements. The third paragraph is really a definition (even if the wording is a bit bashful), so we mark it up as one in the style of Listing 4.4 above.

Finally, we come to the examples at the end of our fragment. In the markup shown in Listing 4.5 we have decided to construct a new theory for these examples since the examples use concepts and symbols that are independent of the theory of magmas. Otherwise, we would have to add the `imports` element to the theory in Listing 4.4, which would have misrepresented the actual dependencies. Note that the new theory has to import the theory `magmas` together with the theories from which examples are taken, so their symbols can be used in the examples.

Listing 4.5. Examples for magmas with OPENMATH objects

```
   <theory xml:id="magmas−examples">
     <metadata><dc:title>Examples</dc:title></metadata>

     <imports from="http://mbc.mathweb.org/omstd/fns1.omdoc##fns1"/>
 5   <imports from="background.omdoc#nat"/>
     <imports from="background.omdoc#functions"/>
     <imports from="#magmas"/>

     <omgroup xml:id="magma−ex" type="enumeration">
10     <metadata><dc:title>Examples</dc:title></metadata>

       <example xml:id="e1.magma" for="#law_of_composition" type="for">
         <CMP>The mappings
           <om:OMOBJ>
15             <om:OMBIND><om:OMS cd="fns1" name="lambda"/>
               <om:OMBVAR>
                 <om:OMV name="X"/><om:OMV name="Y"/>
```

```
          </om:OMBVAR>
          <om:OMA><om:OMS cd="functions" name="pattern-defined"/>
20          <om:OMA><om:OMS cd="products" name="pair"/>
            <om:OMV name="X"/>
            <om:OMV name="Y"/>
          </om:OMA>
          <om:OMA><om:OMS cd="sets" name="union"/>
25          <om:OMV name="X"/>
            <om:OMV name="Y"/>
          </om:OMA>
        </om:OMA>
      </om:OMBIND>
30    </om:OMOBJ> and
      <om:OMOBJ>
        <om:OMBIND><om:OMS cd="fns1" name="lambda"/>
          <om:OMBVAR>
            <om:OMV name="X"/><om:OMV name="Y"/>
35        </om:OMBVAR>
          <om:OMA><om:OMS cd="functions" name="pattern-defined"/>
          <om:OMA><om:OMS cd="products" name="pair"/>
            <om:OMV name="X"/>
            <om:OMV name="Y"/>
40        </om:OMA>
          <om:OMA><om:OMS cd="sets" name="intersection"/>
            <om:OMV name="X"/>
            <om:OMV name="Y"/>
          </om:OMA>
45        </om:OMA>
      </om:OMBIND>
    </om:OMOBJ>
    are <term cd="magmas" name="law_of_composition">laws of composition</term>
    on the set of subsets of a set
50  <om:OMOBJ><om:OMS cd="magmas" name="E"/></om:OMOBJ>.
    </CMP>
  </example>

  <example xml:id="e2.magma" for="#law_of_composition" type="for">
55  <CMP>
    On the set <om:OMOBJ><om:OMS cd="nat" name="Nat"/></om:OMOBJ>
    of <term cd="nats" name="nats">natural numbers</term>,
    <term cd="nats" name="plus">addition</term>,
    <term cd="nats" name="times">multiplication</term>, and
60  <term cd="nats" name="power">exponentiation</term> are ...
    </CMP>
  </example>
  </omgroup>
</theory>
```

The `example` element in line 13 is used for mathematical examples of a special form in OMDOC: objects that have or fail to have a specific property. In our case, the two given mappings have the property of being a law of composition. This structural property is made explicit by the `for` attribute that points to the concept that these examples illustrate, in this case, the symbol `law_of_composition`. The `type` attribute has the values `for` and `against`. In our case `for` applies, `against` would for counterexamples. The content of an `example` is a multilingual `CMP` group. For examples of other kinds — e.g. usage examples, OMDOC does not supply specific markup, so we have to fall back to using an `omtext` element with type `example` as above.

In our text fragment, where the examples are at the end of the section that deals with magmas, creating an independent theory for the examples (or even multiple theories, if examples from different fields are involved) seems appropriate. In other cases, where examples are integrated into the text, we can equivalently embed theories into other theories. Then we would have the following structure:

Listing 4.6. Examples embedded into a theory

```
<theory xml:id="magmas">
  <imports xml:id="imp3" from="background.omdoc#products"/>
  <imports from="http://mbc.mathweb.org/omstd/relation1.omdoc#relation1"/>
  ...
5 <theory xml:id="magmas−examples"
    <imports xml:id="imp4">
      from="http://www.mathweb.org/omdoc/examples/omstd/fns1.omdoc#fns1"/>
    <imports xml:id="imp5" from="background.omdoc#nat"/>
    <imports xml:id="imp6" from="background.omdoc#functions"/>
10   ...
  </theory>
  ...
</theory>
```

Note that the embedded theory (`magmas-examples`) has access to all the symbols in the embedding theory (`magmas`), so it does not have to import it. However, the symbols imported into the embedded theory are only visible in it, and do not get imported into the embedding theory.

4.4 Full Formalization

The final step in the migration of the text fragment involves a transformation of the mathematical vernacular in the statements into some logical formalism. The main benefit of this is that we can verify the mathematical contents in theorem proving environments. We will start out by dividing the first definition into two parts. The first one defines the symbol `law_of_composition` (see Listing 4.7), and the second one `magma` (see Listing 4.8).

Listing 4.7. The formal definition of a law of composition

```
<symbol name="law_of_composition">
  <metadata><dc:description>A law of composition on a set.</dc:description></metadata>
</symbol>
<definition xml:id="magma.def" for="#law_of_composition" type="simple">
5  <CMP>
    Let <om:OMOBJ><om:OMV name="E"/></om:OMOBJ> be a set. A mapping of
    <om:OMOBJ><om:OMR href="#comp.1"/></om:OMOBJ>
    is called a <term cd="magmas" name="law_of_composition"
                role="definiens">law of composition </term>
10  on <om:OMOBJ><om:OMV name="E"/></om:OMOBJ>.
  </CMP>
  <om:OMOBJ>
    <om:OMBIND>
      <om:OMS cd="fns1" name="lambda"/>
15    <om:OMBVAR>
        <om:OMV name="E"/><om:OMV name="F"/>
      </om:OMBVAR>
      <om:OMA><om:OMS cd="pl0" name="and"/>
      <om:OMA><om:OMS cd="sets" name="set"/>
20      <om:OMV name="E"/>
      </om:OMA>
```

```
        <om:OMA>
          <om:OMS cd="functions" name="function"/>
          <om:OMA id="comp.1">
25          <om:OMS cd="products" name="Cartesian−product"/>
            <om:OMV name="E"/>
            <om:OMV name="E"/>
          </om:OMA>
          <om:OMV name="E"/>
30        </om:OMA>
        </om:OMA>
      </om:OMBIND>
    </om:OMOBJ>
  </definition>
```

The main difference of this definition to the one in the section above is the om:OMOBJ element, which now accompanies the CMP element. It contains a formal definition of the property of being a law of composition in the form of a λ-term $\lambda E, F.set(E) \wedge F : E \times E \rightarrow E^2$. The value simple of the type attribute in the definition element signifies that the content of the om:OMOBJ element can be substituted for the symbol law_of_composition, wherever it occurs. So if we have law_of_composition(A, B) somewhere this can be reduced to $(\lambda E, F.set(E) \wedge F : E \times E \rightarrow E)(A, B)$ which in turn reduces[3] to $set(A) \wedge B : A \times A \rightarrow A$ or in other words law_of_composition(A, B) is true, iff A is a set and B is a function from $A \times A$ to A. This definition is directly used in the second formal definition, which we depict in Listing 4.8.

Listing 4.8. The formal definition of a magma

```
    <definition xml:id="magma.def" for="#magma" type="implicit">
    <CMP> A set with a law of composition is called a
      <term cd="magmas" name="magma" role="definiens">magma</term>.
    </CMP>
5   <FMP>
      <om:OMOBJ>
        <om:OMBIND><om:OMS cd="pl1" name="forall"/>
          <om:OMBVAR><om:OMV name="M"/></om:OMBVAR>
          <om:OMA><om:OMS cd="pl0" name="iff"/>
10          <om:OMA><om:OMS cd="magmas" name="magma"/>
            <om:OMV name="M"/>
          </om:OMA>
          <om:OMBIND>
            <om:OMS cd="pl1" name="exists"/>
15          <om:OMBVAR>
              <om:OMV name="E"/><om:OMV name="C"/>
            </om:OMBVAR>
            <om:OMA><om:OMS cd="pl0" name="and"/>
              <om:OMA><om:OMS cd="relation1" name="eq"/>
20            <om:OMV name="M"/>
                <om:OMA><om:OMS cd="products" name="Cartesian−product"/>
                  <om:OMV name="E"/>
                  <om:OMV name="C"/>
```

---

[2] We actually need to import the theories pl1 for first-order logic (it imports the theory pl0) to legally use the logical symbols here. Since we did not show the theory element, we assume it to contain the relevant imports elements.

[3] We use the λ-calculus as a formalization framework here: If we apply a λ-term of the form $\lambda X.A$ to an argument B, then the result is obtained by binding all the formal parameters X to the actual parameter B, i.e. the result is the value of A, where all the occurrences of X have been replaced by B. See [Bar80, And02] for an introduction.

```
25        </om:OMA>
          </om:OMA>
          <om:OMA><om:OMS cd="magmas" name="law_of_composition"/>
          <om:OMV name="E"/>
          <om:OMV name="F"/>
          </om:OMA>
30        </om:OMA>
        </om:OMBIND>
      </om:OMA>
    </om:OMBIND>
    </om:OMOBJ>
35    </FMP>
    </definition>
```

Here, the **type** attribute on the **definition** element has the value **implicit**, which signifies that the content of the **FMP** element should be understood as a logical formula that is made true by exactly one object: the property of being a magma. This formula can be written as

$$\forall M.magma(M) \Leftrightarrow \exists E, F.M = (E, F) \land law\_of\_composition(E, F)$$

in other words: M is a magma, iff it is a pair (E, F), where F is a law of composition over E.

Finally, the examples get a formal part as well. This mainly consists of formally representing the object that serves as the example, and making the way it does explicit. The first is done simply by adding the object to the example as a sibling node to the CMP. Note that we are making use of the OpenMath reference mechanism here that allows to copy subformulae by linking them with an **om:OMR** element that stands for a copy of the object pointed to by the **href** attribute (see Section 13.1), which makes this very simple. Also note that we had to split the example into two, since OMDoc only allows one example per **example** element. However, the **example** contains two **om:OMOBJ** elements, since the property of being a law of composition is binary.

The way this object is an example is made explicit by adding an assertion that makes the claim of the example formal (in our case that for every set E, the function $(X, Y) \mapsto X \cup Y$ is a law of composition on the set of subsets of E). The assertion is referenced by the **assertion** attribute in the **example** element.

Listing 4.9. A formalized magma example

```
    <example xml:id="e11.magma" for="#law_of_composition"
            type="for" assertion="e11.magma.ass">
      <CMP> The mapping <om:OMOBJ><om:OMR href="#e11.magma.1"/></om:OMOBJ> is
      a law of composition on the set of subsets of a set
5     <om:OMOBJ><om:OMS cd="magmas" name="E"/></om:OMOBJ>.
      </CMP>
      <om:OMOBJ>
        <om:OMA id="e11.magma.2"><om:OMS cd="sets" name="subset"/>
        <om:OMV name="E"/>
10      </om:OMA>
      </om:OMOBJ>
      <om:OMOBJ>
        <om:OMBIND id="e11.magma.1">
          <om:OMS cd="fns1" name="lambda"/>
15        <om:OMBVAR><om:OMV name="X"/><om:OMV name="Y"/></om:OMBVAR>
          <om:OMA>
            <om:OMS cd="functions" name="pattern-defined"/>
            <om:OMA><om:OMS cd="products" name="pair"/>
            <om:OMV name="X"/>
20          <om:OMV name="Y"/>
            </om:OMA>
            <om:OMA><om:OMS cd="sets" name="union"/>
            <om:OMV name="X"/>
            <om:OMV name="Y"/>
25          </om:OMA>
          </om:OMA>
        </om:OMBIND>
      </om:OMOBJ>
    </example>
30
    <assertion xml:id="e11.magma.ass">
      <FMP>
        <om:OMOBJ>
          <om:OMBIND>
35          <om:OMS cd="pl1" name="forall"/>
            <om:OMBVAR><om:OMV name="E"/></om:OMBVAR>
            <om:OMA>
              <om:OMS cd="magmas" name="law_of_composition"/>
              <om:OMR href="#e11.magma.2"/>
40            <om:OMR href="#e11.magma.1"/>
            </om:OMA>
          </om:OMBIND>
        </om:OMOBJ>
      </FMP>
45  </assertion>
```

5

OpenMath Content Dictionaries

Content Dictionaries are structured documents used by the OPENMATH standard [BCC+04] to codify knowledge about mathematical symbols and concepts used in the representation of mathematical formulae. They differ from the mathematical documents discussed in the last chapter in that they are less geared towards introduction of a particular domain, but act as a reference/-glossary document for implementing and specifying mathematical software systems. Content Dictionaries are important for the OMDoc format, since the OMDoc architecture, and in particular the integration of OPENMATH builds on the equivalence of OPENMATH content dictionaries and OMDoc theories.

Concretely, we will look at the content dictionary `arith1.ocd` which defines the OPENMATH symbols `abs`, `divide`, `gcd`, `lcm`, `minus`, `plus`, `power`, `product`, `root`, `sum`, `times`, `unary_minus` (see [OMC] for the original). We will discuss the transformation of the parts listed below into OMDoc and see from this process that the OPENMATH content dictionary format is (isomorphic to) a subset of the OMDoc format. In fact, the OPENMATH2 standard only presents the content dictionary format used here as one of many encodings and specifies abstract conditions on content dictionaries that the OMDoc encoding below also meets. Thus OMDoc is a valid content dictionary encoding.

Listing 5.1. Part of the OPENMATH content dictionary `arith1.ocd`

```
    <CD>
      <CDName> arith1 </CDName>
      <CDURL> http://www.openmath.org/cd/arith1.ocd </CDURL>
      <CDReviewDate> 2003−04−01 </CDReviewDate>
 5    <CDStatus> official </CDStatus>
      <CDDate> 2001−03−12 </CDDate>
      <CDVersion> 2 </CDVersion>
      <CDRevision> 0 </CDRevision>
      <dc:description>
10      This CD defines symbols for common arithmetic functions.
      </dc:description>

      <CDDefinition>
      <Name> lcm </Name>
15    <Description>
```

The symbol to represent the n−ary function to return the least common
multiple of its arguments.
</Description>

20   <CMP> lcm(a,b) = a*b/gcd(a,b) </CMP>
<FMP>... </FMP>

<CMP>
for all integers a,b |
25   There does not exist a c>0 such that c/a is an Integer and c/b is an
Integer and lcm(a,b) > c.
</CMP>
<FMP>...</FMP>
...
30   </CD>

Generally, OPENMATH content dictionaries are represented as mathematical
theories in OMDOC. These act as containers for sets of symbol declarations
and knowledge about them, and are marked by **theory** elements. The result
of the transformation of the content dictionary in Listing 5.1 is the OMDOC
document in Listing 5.2.

The first 25 lines in Listing 5.1 contain administrative information and
metadata of the content dictionary, which is mostly incorporated into the
metadata of the **theory** element. The translation adds further metadata to
the **omdoc** element that were left implicit in the original, or are external to
the document itself. These data comprise information about the translation
process, the creator, and the terms of usage, and the source, from which this
document is derived (the content of the **omcd:CDURL** element is recycled in
Dublin Core metadata element **dc:source** in line 12.

The remaining administrative data is specific to the content dictionary
per se, and therefore belongs to the **theory** element. In particular, the
omcd:CDName goes to the **xml:id** attribute on the **theory** element in line
36. The **dc:description** element is directly used in the **metadata** in line 38.
The remaining information is encapsulated into the **cd*** attributes.

Note that we have used the OMDOC sub-language "OMDOC Content
Dictionaries" described in Subsection 22.3.2 since it suffices in this case, this
is indicated by the **modules** attribute on the **omdoc** element.

Listing 5.2. The OPENMATH content dictionary **arith1** in OMDOC form

---

```
<?xml version="1.0" encoding="utf−8"?>
<omdoc xml:id="arith1.omdoc" modules="@cd"
       xmlns:dc="http://purl.org/dc/elements/1.1/">

5    <metadata>
      <dc:title>The OpenMath Content Dictionary arith1.ocd in OMDoc Form</dc:title>
      <dc:creator role="trl">Michael Kohlhase</dc:creator>
      <dc:creator role="ant">The OpenMath Society</dc:creator>
      <dc:date action="updated"> 2004−01−17T09:04:03Z </dc:date>
10     <dc:source>
       Derived from the OpenMath CD http://www.openmath.org/cd/arith1.ocd.
      </dc:source>
      <dc:type>Text</dc:type>
      <dc:format>application/omdoc+xml</dc:format>
15     <dc:rights>Copyright (c) 2000 Michael Kohlhase;
       This OMDoc content dictionary is released under the OpenMath license:
```

```
          http://www.openmath.org/cdfiles/license.html
        </dc:rights>
      </metadata>
20
    <theory xml:id="arith1"
            cdstatus="official" cdreviewdate="2003−04−01" cdversion="2" cdrevision="0">
      <metadata>
        <dc:title>Common Arithmetic Functions</dc:title>
25      <dc:description>This CD defines symbols for common arithmetic functions.</dc:description>
        <dc:date action="updated"> 2001−03−12 </dc:date>
      </metadata>
      <imports from="#sts"/>

30    <symbol name="lcm">
        <metadata>
          <dc:description>The symbol to represent the n−ary function to return the least common
            multiple of its arguments.
          </dc:description>
35        <dc:description xml:lang="de">
            Das Symbol für das kleinste gemeinsame Vielfache (als n-äre Funktion).
          </dc:description>
          <dc:subject>lcm, least common mean</dc:subject>
          <dc:subject xml:lang="de">kgV, kleinstes gemeinsames Vielfaches</dc:subject>
40        </metadata>
        <type system="sts">
          <OMOBJ>
            <OMA><OMS name="mapsto" cd="sts"/>
            <OMA><OMS name="nassoc" cd="sts"/><OMV name="SemiGroup"/></OMA>
45          <OMV name="SemiGroup"/>
            </OMA>
          </OMOBJ>
        </type>
      </symbol>
50
    <presentation xml:id="pr_lcm" for="#lcm">
      <use format="default">lcm</use>
      <use format="default" xml:lang="de">kgV</use>
      <use format="cmml" element="lcm"/>
55    </presentation>

    <definition xml:id="lcm−def" for="#lcm" type="pattern">
      <CMP>We define <OMOBJ><OMR href="#lcm−def.O"/></OMOBJ>
        as <OMOBJ><OMR href="#lcm−def.1"/></OMOBJ></CMP>
60    <CMP xml:lang="de">
        Wir definieren <OMOBJ><OMR href="#lcm−def.O"/></OMOBJ>
        als <OMOBJ><OMR href="#lcm−def.1"/></OMOBJ></CMP>
      <requation>
        <OMOBJ>
65        <OMA id="lcm−def.O">
            <OMS cd="arith1" name="lcm"/>
            <OMV name="a"/><OMV name="b"/>
          </OMA>
        </OMOBJ>
70      <OMOBJ>
          <OMA id="lcm−def.1">
            <OMS cd="arith1" name="divide"/>
            <OMA><OMS cd="arith1" name="times"/>
            <OMV name="a"/>
75          <OMV name="b"/>
            </OMA>
            <OMA><OMS cd="arith1" name="gcd"/>
            <OMV name="a"/>
            <OMV name="b"/>
80          </OMA>
          </OMA>
        </OMOBJ>
```

```
        </requation>
        </definition>
85
        <theory>
          <imports from="#relation1"/>
          <imports from="#quant1"/>
          <imports from="#logic1"/>
90
          <assertion xml:id="lcm-prop-3" type="lemma">
            <CMP>For all integers <OMOBJ><OMV name="a"/></OMOBJ>,
              <OMOBJ><OMV name="b"/></OMOBJ> there is no
              <OMOBJ><OMR href="#lcm-prop-3.1"/></OMOBJ> such that
95            <OMOBJ><OMR href="#lcm-prop-3.2"/></OMOBJ> and
              <OMOBJ><OMR href="#lcm-prop-3.3"/></OMOBJ> and
              <OMOBJ><OMR href="#lcm-prop-3.4"/></OMOBJ>.
            </CMP>
            <CMP xml:lang="de">Für alle ganzen Zahlen
100            <OMOBJ><OMV name="a"/></OMOBJ>,
              <OMOBJ><OMV name="b"/></OMOBJ>
              gibt es kein <OMOBJ><OMR href="#lcm-prop-3.1"/></OMOBJ> mit
              <OMOBJ><OMR href="#lcm-prop-3.2"/></OMOBJ> und
              <OMOBJ><OMR href="#lcm-prop-3.3"/></OMOBJ> und
105            <OMOBJ><OMR href="#lcm-prop-3.4"/></OMOBJ>.
            </CMP>
            <FMP>
              <OMOBJ><OMBIND><OMS cd="quant1" name="forall"/>
                <OMBVAR><OMV name="a"/><OMV name="b"/></OMBVAR>
110              <OMA><OMS cd="logic1" name="implies"/>
                  <OMA>...</OMA>
                  <OMA><OMS cd="logic1" name="not"/>
                    <OMBIND><OMS cd="quant1" name="exists"/>
                      <OMBVAR><OMV name="c"/></OMBVAR>
115                    <OMA><OMS cd="logic1" name="and"/>
                        <OMA id="lcm-prop-3.1">...</OMA>
                        <OMA id="lcm-prop-3.2">...</OMA>
                        <OMA id="lcm-prop-3.3">...</OMA>
                        <OMA id="lcm-prop-3.4">...</OMA>
120                    </OMA>
                    </OMBIND>
                  </OMA>
                </OMA>
              </OMBIND>
125            </OMOBJ>
            </FMP>
          </assertion>
          ...
        </theory>
130      ...
        </theory>
```

One important difference between the original and the OMDOC version of the OPENMATH content dictionary is that the latter is intended for machine manipulation, and we can transform it into other formats. For instance, the human-oriented presentation of the OMDOC version might look something like the following[1]:

----

[1] These presentation was produced by the style sheets discussed in Section 25.3.

The OpenMath Content Dictionary arith1.ocd in OMDoc Form
Michael Kohlhase, The OpenMath Society
January 17. 2004
This CD defines symbols for common arithmetic functions.

Concept 1. lcm (lcm, least common mean)
Type (sts): $SemiGroup^* \to SemiGroup$
The symbol to represent the n-ary function to return the least common multiple of its arguments.

Definition 2. (lcm-def)
We define $lcm(a, b)$ as $\frac{a \cdot b}{gcd(a,b)}$

Lemma 3. *For all integers a, b there is no $c > 0$ such that $(a|c)$ and $(b|c)$ and $c < lcm(a,b)$.*

Fig. 5.3. A human-oriented presentation of the OMDoc CD

The OpenMath Content Dictionary arith1.ocd in OMDoc form
Michael Kohlhase, The OpenMath Society
17. Januar 2004
This CD defines symbols for common arithmetic functions.

Konzept 1. lcm (kgV, kleinstes gemeinsames Vielfaches)
Typ (sts): $SemiGroup^* \to SemiGroup$
Das Symbol für das kleinste gemeinsame Vielfache (als n-äre Funktion).

Definition 2. (lcm-def)
Wir definieren $kgV(a, b)$ als $\frac{a \cdot b}{ggT(a,b)}$

Lemma 3. *Für alle ganzen Zahlen a, b gibt es kein $c > 0$ mit $(a|c)$ und $(b|c)$ und $c < kgV(a,b)$.*

Fig. 5.4. A human-oriented presentation in German

6

Structured and Parametrized Theories

In Chapter 5 we have seen a simple use of theories in OPENMATH content dictionaries. There, theories have been used to reference OPENMATH symbols and to govern their visibility. In this chapter we will cover an extended example showing the structured definition of multiple mathematical theories, modularizing and re-using parts of specifications and theories. Concretely, we will consider a structured specification of lists of natural numbers. This example has been used as a paradigmatic example for many specification formats ranging from CASL (Common Abstract Specification Language [CoF04]) standard to the PVS theorem prover [ORS92], since it uses most language elements without becoming too unwieldy to present.

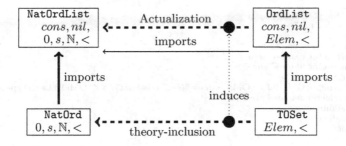

Fig. 6.1. A structured specification of lists (of natural numbers)

In this example, we specify a theory OrdList of lists that is generic in the elements (which is assumed to be a totally ordered set, since we want to talk about ordered lists). Then we will instantiate OrdList by applying it to the theory NatOrd of natural numbers to obtain the intended theory NatOrdList of lists of natural numbers. The advantage of this approach is that we can re-use the generic theory OrdList to apply it to other element theories like that of "characters" to obtain a theory of lists of characters. In algebraic specification languages, we speak of **parametric theories**. Here, the theory OrdList has a formal parameter (the theory TOSet) that can be instantiated

later with concrete values to get a **theory instance** (in our example the
theory NatOrdList). We call this process theory **actualization**.

We begin the extended example with the theories in the lower half of
Figure 6.1. The first is a (mock up of a) theory of totally ordered sets. Then
we build up the theory of natural numbers as an **abstract data type** (see
Chapter 16 for an introduction to abstract data types in OMDOC and a
more elaborate definition of ℕ). The sortdef element posits that the set of
natural numbers is given as the **sort NatOrd**, with the constructors **zero** and
succ. Intuitively, a sort represents an inductively defined set, i.e. it contains
exactly those objects that can be represented by the constructors only, for
instance the number three is represented as $s(s(s(0)))$, where s stands for
the successor function (given as the constructor **succ**) and 0 for the number
zero (represented by the constructor **zero**). Note that the theory **nat** does
not have any explicitly represented axioms. They are implicitly given by the
abstract data type structure, in our case, they correspond to the five Peano
Axioms (see Figure 15.1). Finally, the argument elements also introduce one
partial inverse to the constructor functions per argument; in our case the
predecessor function.

```
    <theory xml:id="TOSet">
      <symbol name="set"/>
      <symbol name="ord"/>
      <axiom xml:id="toset"><CMP>ord is a total order on set.</CMP></axiom>
 5  </theory>

    <theory xml:id="nat">
      <adt>
        <sortdef name="Nat">
10        <constructor name="zero"/>
          <constructor name="succ">
            <argument>
              <type><OMOBJ><OMS name="Nat" cd="nat"/></OMOBJ></type>
              <selector name="pred"/>
15          </argument>
          </constructor>
        </sortdef>
      </adt>
    </theory>
20
    <theory xml:id="NatOrd">
      <imports from="#nat"/>
      <imports from="#TOSet"/>
      <symbol name="leq"/>
25    <definition xml:id="leq.def" for="#leq" type="implicit"
             existence="#leq.ex" uniqueness="#leq.uniq">
        <FMP>∀x.0 ≤ x ∧ ∀x, y.x ≤ y ⇒ s(x) ≤ s(y)</FMP>
      </definition>
      <assertion xml:id="leq.ex"><CMP>≤ exists.</CMP></assertion>
30    <assertion xml:id="leq.unique"><CMP>≤ is unique</CMP></assertion>
      <assertion xml:id="leq.TO"><CMP>≤ is a total order on Nat.</CMP></assertion>
    </theory>
```

Finally we have extended the natural numbers by an ordering function ≤
(symbol leq) which we show to be a total ordering function in assertion
leq.TO. Note that to state the assertion, we had to import the notion of a
total ordering from theory TOSet. We can directly use this result to establish
a **theory inclusion** between TOSet as the **source theory** and NatOrd as the

target theory. A theory inclusion is a formula mapping between two theories, such that the translations of all axioms in the source theory are provable in the target theory. In our case, the mapping is given by the recursive function given in the `morphism` element in Listing 6 that maps the respective base sets and the ordering relations to each other. The `obligation` element just states that translation of the only theory-constitutive (see Subsection 15.2.4) element of the source theory (the axiom `toset`) has been proven in the target theory, as witnessed by the assertion `leq.TO`[1].

```
<theory−inclusion xml:id="elem−nat−incl" to="#NatOrd" from="#TOSet">
   <morphism xml:id="elem−nat" type="pattern">
      <requation>
         <OMOBJ><OMS cd="TOSet" name="set"/></OMOBJ>
5        <OMOBJ><OMS cd="NatOrd" name="Nat"/></OMOBJ>
      </requation>
      <requation>
         <OMOBJ><OMS cd="TOSet" name="ord"/></OMOBJ>
         <OMOBJ><OMS cd="NatOrd" name="leq"/></OMOBJ>
10    </requation>
   </morphism>
   <obligation induced−by="#toset" assertion="#leq.TO"/>
</theory−inclusion>
```

We continue our example by building a generic theory `OrdList` of ordered lists. This is given as the abstract data type generated by the symbols `cons` (construct a list from an element and a rest list) and `nil` (the empty list) together with a defined symbol `ordered`: a predicate for ordered lists. Note that this symbol cannot be given in the abstract data type, since it is not a constructor symbol. Note that `OrdList` imports theory `TOSet` for the base set of the lists and the ordering relation $\leq$.

```
<theory xml:id="OrdList">
   <imports from="#TOSet"/>
   <adt xml:id="list−adt">
      <sortdef name="lists">
5        <constructor name="cons">
            <argument>
               <type><OMOBJ><OMS name="set" cd="TOSet"/></OMOBJ></type>
               <selector name="head"/>
            </argument>
10          <argument>
               <type><OMOBJ><OMS name="lists" cd="OrdList"/></OMOBJ></type>
               <selector name="rest"/>
            </argument>
         </constructor>
15       <constructor name="nil"/>
      </sortdef>
   </adt>

   <symbol name="ordered"/>
20 <definition xml:id="ordered−def" for="#ordered" type="informal">
      <CMP>A list l is called ordered, iff head(l) $\leq$ z for all elements z $\in$ rest(l) and
      rest(l) is ordered.</CMP>
   </definition>
</theory>
```

---

[1] Note that as always, OMDOC only cares about the structural aspects of this: The OMDOC model only insists that there is the statement of an assertion, whether the author chooses to prove it or indeed whether the statement is true at all is left to other levels of modeling.

The theory NatOrdList of lists of natural numbers is built up by importing from the theories NatOrd and OrdList. Note that the attribute type of the imports element nat-list.im-elt is set to local, since we only want to import the local axioms of the theory OrdList and not the whole theory OrdList (which would include the axioms from TOSet; see Section 18.3 for a discussion). In particular the symbols set and ord are not imported into theory NatOrdList: the theory TOSet is considered as a **formal parameter theory**, which is actualized to the **actual parameter theory** with this construction. The effect of the actualization comes from the morphism elem-nat in the import of OrdList that renames the symbol set (from theory TOSet) with Nat (from theory NatOrd). The actualization from OrdList to NatOrdList only makes sense, if the parameter theory NatOrd also has a suitable ordering function. This can be ensured using the OMDOC inclusion element.

```
<theory xml:id="NatOrdList">
  <imports xml:id="natordlist.im−natord" from="#NatOrd"/>
  <imports xml:id="natordlist.im−elt" from="#OrdList" type="local">
    <morphism base="#elem−nat"/>
  </imports>
  <inclusion via="elem−nat−incl"/>
</theory>
```

The benefit of this inclusion requirement is twofold: If the theory inclusion from TOSet to NatOrd cannot be verified, then the theory NatOrdList is considered to be undefined, and we can use the development graph techniques presented in Section 18.5 to obtain a theory inclusion from OrdList to NatOrdList: We first establish an axiom inclusion from theory TOSet to NatOrdList by observing that this is induced by composing the theory inclusion from TOSet to NatOrd with the theory inclusion given by the imports from NatOrd to NatOrdList. This gives us a **decomposition** situation: every theory that the source theory OrdList inherits from has an axiom inclusion to the target theory NatOrdList, so the local axioms of those theories are provable in the target theory. Since we have covered all of the inherited ones, we actually have a theory inclusion from the source- to the target theory.

```
<axiom−inclusion xml:id="toset−natordlist−incl" from="#TOSet" to="#NatOrdList">
  <morphism base="#elem−nat"/>
  <path−just local="#elem−nat−incl" globals="#natordlist.im−natord"/>
</axiom−inclusion>

<theory−inclusion from="#OrdList" to="#NatOrdList">
  <morphism base="#elem−nat"/>
  <decomposition links="#toset−natordlist−incl #elem−nat−incl"/>
</theory−inclusion>
```

This concludes our example, since we have seen that the theory OrdList is indeed included in NatOrdList via renaming.

Note that with this construction we could simply extend the graph by actualizations for other theories, e.g. to get lists of characters, as long as we can prove theory inclusions from TOSet to them.

A Development Graph for Elementary Algebra

We will now use the technique presented in the last chapter for the elementary algebraic hierarchy. Figure 7.1 gives an overview of the situation. We will build up theories for semigroups, monoids, groups, and rings and a set of theory inclusions from these theories to themselves given by the converse of the operation.

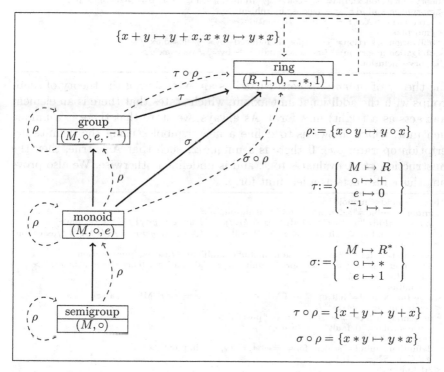

Fig. 7.1. A development graph for elementary algebra

We start off with the theory for *semigroup*s. It introduces two symbols, the base set M and the operation $\circ$ on M together with two axioms that state that M is closed under $\circ$ and that $\circ$ is associative on M. We have a structural theory inclusion from this theory to itself that uses the fact that M together with the converse $\sigma(\circ)$ of $\circ$ is also a semigroup: the obligation for the axioms can be justified by themselves (for the closure axiom we have $\sigma(\forall x, y \in M . x \circ y \in M) = \forall y, x \in M . x \circ y \in M$, which is logically equivalent to the axiom.)

```
<theory xml:id="semigroup">
  <symbol name="base−set"/>
  <presentation for="#base−set"><use format="default">M</use></presentation>
  <symbol name="op"/>
  <presentation for="#op"><use format="default">∘</use></presentation>
  <axiom xml:id="closed.ax"><FMP>∀x, y ∈ M.x ∘ y ∈ M</FMP></axiom>
  <axiom xml:id="assoc.ax">
      <FMP>∀x, y, z ∈ M.(x ∘ y) ∘ z = x ∘ (y ∘ z)</FMP>
  </axiom>
</theory>

<theory−inclusion xml:id="sg−conv−sg" from="#semigroup" to="#semigroup">
  <morphism xml:id="sg−conv−sg.morphism">
      <requation>X ∘ Y ⤳ Y ∘ X</requation>
  </morphism>
  <obligation assertion="conv.closed" induced−by="#closed.ax"/>
  <obligation assertion="#assoc.ax" induced−by="#assoc.ax"/>
</theory−inclusion>
```

The theory of *monoid*s is constructed as an extension of the theory of semigroups with the additional unit axiom, which states that there is an element that acts as a (right) unit for $\circ$. As always, we state that there is a unique such unit, which allows us to define a new symbol e using the definite description operator $\tau x.$: If there is a unique x, such that **A** is true, then the construction $\tau x.$**A** evaluates to x, and is undefined otherwise. We also prove that this e also acts as a left unit for $\circ$.

```
<theory xml:id="monoid">
  <imports xml:id="sg2mon" from="#semigroup"/>
  <axiom xml:id="unit.ax"><FMP>∃x ∈ M.∀y ∈ M.y ∘ x = y</FMP></axiom>
  <assertion xml:id="unit.unique"><FMP>∃¹x ∈ M.∀y ∈ M.y ∘ x = y</FMP></assertion>
  <symbol name="unit"/>
  <presentation for="#unit"><use format="default">e</use></presentation>
  <definition xml:id="unit.def" for="#unit" type="simple" existence="#unit.unique">
      τx ∈ M.∀y ∈ M.y ∘ x = y
  </definition>
  <assertion xml:id="left.unit"><FMP>∀x ∈ M.e ∘ x = x</FMP></assertion>
  <symbol name="setstar"/>
  <presentation for="#setstar" fixity="postfix">
      <use format="default">*</use>
  </presentation>
  <definition xml:id="ss.def" for="#setstar" type="implicit">
      ∀S ⊆ M.S* = S\{e}
  </definition>
</theory>
```

Building on this, we first establish an axiom-selfinclusion from the theory of monoids to itself. We can make this into a theory selfinclusion using the theory-selfinclusion for semigroups as the local part of a path justification (recall that theory inclusions are axiom inclusions by construction) and the definitional theory inclusion induced by the import from semigroups to monoids as the global path.

```
<axiom−inclusion xml:id="mon−conv−mon.local" from="#monoid" to="#monoid">
  <morphism base="#sg−conv−sg.morphism"/>
  <obligation assertion="#left.unit" induced−by="#unit.ax"/>
</axiom−inclusion>

<axiom−inclusion xml:id="sg−conv−mon" from="#semigroup" to="#monoid">
  <morphism base="#sg−conv−sg.morphism"/>
  <path−just local="#sg−conv−sg" globals="#sg2mon"/>
</axiom−inclusion>
<theory−inclusion xml:id="mon−conv−mon.global" from="#monoid" to="#monoid">
  <morphism base="#sg−conv−sg.morphism"/>
  <decomposition links="#sg−conv−sg #sg−conv−mon"/>
</theory−inclusion>
```

Note that all of these axiom inclusions have the same morphism (denoted by ρ in Figure 7.1), in OMDoc we can share this structure using the **base** on the **morphism** element. This normally points to a morphism that is the base for extension, but if the **morphism** element is empty, then this just means that the morphisms are identical.

For groups, the situation is very similar: We first build a theory of groups by adding an axiom claiming the existence of inverses and constructing a new function $\cdot^{-1}$ from that via a definite description.

```
<theory xml:id="group">
  <imports xml:id="mon2grp" from="#monoid"/>
  <axiom xml:id="inv.ax"><FMP>∀x ∈ M.∃y ∈ M.x ∘ y = e</FMP></axiom>
  <symbol name="inv"/>
  <presentation for="#inv" role="applied">
    <use format="default" lbrack="" rbrack="" fixity="postfix">−1</use>
  </presentation>
  <definition xml:id="inv.def" for="#inv" type="pattern">
    <requation>x−1 ↝ τy.x ∘ y = e</value></requation>
  </definition>
  <assertion xml:id="conv.inv"><FMP>∀x ∈ M.∃y ∈ M.y ∘ x = e</FMP></assertion>
</theory>
```

Again, we have to establish a couple of axiom inclusions to justify the theory inclusion of interest. Note that we have one more than in the case for monoids, since we are one level higher in the inheritance structure, also, the local chains are one element longer.

```
<axiom−inclusion xml:id="grp−conv−grp.local" from="#group" to="#group">
  <morphism base="#sg−conv−sg.morphism"/>
  <obligation assertion="conv.inv" induced−by="#inv.ax"/>
</axiom−inclusion>
<axiom−inclusion xml:id="sg−conv−grp" from="#semigroup" to="#group">
  <morphism base="#sg−conv−sg.morphism"/>
  <path−just local="#sg−conv−sg" globals="#mon2grp #sg2mon"/>
</axiom−inclusion>
<axiom−inclusion xml:id="mon−conv−grp" from="#monoid" to="#group">
  <morphism base="#sg−conv−sg.morphism"/>
  <path−just local="#mon−conv−mon.local" globals="#mon2grp"/>
</axiom−inclusion>
<theory−inclusion xml:id="grp−conv−grp" from="#group" to="#group">
  <morphism base="#sg−conv−sg.morphism"/>
  <decomposition links="#sg−conv−grp #mon−conv−grp #grp−conv−grp.local"/>
</theory−inclusion>
```

Finally, we extend the whole setup to a theory of rings. Note that we have a dual import from **group** and **monoid** with different morphisms (they are represented by σ and τ in Figure 7.1). These rename all of the imported symbols apart (interpreting them as additive and multiplicative) except of the punctuated set constructor $\cdot^*$, which is imported from the additive group structure only. We avoid a name clash with the operator that would have been imported from the multiplicative structure by specifying that this is not imported using the **hiding** on the **morphism** in the respective **imports** element[1].

```
   <theory xml:id="ring">
     <symbol name="R"/>
     <presentation for="#R"><use format="default">R</use></presentation>
     <symbol name="zero"/>
5    <presentation for="#plus" role="applied">
       <use format="default">+</use>
     </presentation>
     <symbol name="one"/>
     <presentation for="#zero"><use format="default">0</use></presentation>
10   <symbol name="times"/>
     <presentation for="#negative" role="applied">
       <use format="default">-</use>
     </presentation>
     <symbol name="times"/>
15   <presentation for="#times" role="applied">
       <use format="default">*</use>
     </presentation>
     <symbol name="one"/>
     <presentation for="#one"><use format="default">1</use></presentation>
20   <imports xml:id="add.import" from="#group">
       <morphism>M ↦ R, x ∘ y ↦ x * y, e ↦ 1, ·⁻¹ ↦ -</morphism>
     </imports>
     <imports xml:id="mult.import" from="#monoid">
       <morphism hiding="setstar">M ↦ M*, x ∘ y ↦ x * y, e ↦ 1</morphism>
25   </imports>
     <axiom xml:id="dist.ax"><FMP>x * (y + z) = (x * y) + (x * z)</FMP></axiom>
     <assertion xml:id="dist.conv"><FMP>(z + y) * x = (z * x) + (y * x)</FMP></assertion>
   </theory>
```

Again, we have to establish some axiom inclusions to justify the theory self-inclusion we are after in the example. Note that in the rings case, things are more complicated, since we have a dual import in the theory of **rings**. Let us first establish the additive part.

```
   <axiom−inclusion xml:id="sg−conv−rg.add" from="#semigroup" to="#ring">
     <morphism base="#sg−conv−sg.morphism #add.import"/>
     <path−just local="#sg−conv−sg" globals="#sg2mon #mon2grp #add.import "/>
   </axiom−inclusion>
5  <axiom−inclusion xml:id="mon−conv−rg.add" from="#monoid" to="#group">
     <morphism base="#sg−conv−sg.morphism #add.import"/>
     <path−just local="#mon−conv−mon.local" globals="#mon2grp #add.import"/>
   </axiom−inclusion>
   <axiom−inclusion xml:id="grp−conv−rg.add" from="#group" to="#group">
10   <morphism base="#sg−conv−sg.morphism #add.import"/>
     <path−just local="#grp−conv−grp.local" globals="#add.import"/>
   </axiom−inclusion>
```

---

[1] An alternative (probably better) to this would have been to explicitly include the operators in the morphisms, creating new operators for them in the theory of **rings**. But the present construction allows us to exemplify the **hiding**, which has not been covered in an example otherwise.

The multiplicative part is totally analogous, we will elide it to conserve space. Using both parts, we can finally get to the local axiom self-inclusion and extend it to the intended theory inclusion justified by the axiom inclusions established above.

```
   <axiom−inclusion xml:id="rg−conv−rg.local" from="#ring" to="#ring">
     <morphism xml:id="rg−conv−rg.morphism">x + y ↦ y + x, x * y ↦ y * x</morphism>
     <obligation assertion="#dist.conv" induced−by="#dist.ax"/>
   </axiom−inclusion>
5  <theory−inclusion xml:id="rg−conv−rg" from="#ring" to="#ring">
     <morphism base="#rg−conv−rg.morphism"/>
     <decomposition links="#rg−conv−rg.local
                    #sg−conv−rg.add #mon−conv−rg.add #grp−conv−rg.add
                    #sg−conv−rg.mult #mon−conv−rg.mult #grp−conv−rg.mult"/>
10 </theory−inclusion>
```

This concludes our example. It could be extended to higher constructs in algebra like fields, magmas, or vector spaces easily enough using the same methods, but we have seen the key features already.

8

Courseware and the Narrative/Content Distinction

In this chapter we will look at another type of mathematical document: courseware; in this particular case a piece from an introductory course "Fundamentals of Computer Science" (Course 15-211 at Carnegie Mellon University). The OMDoc documents produced from such courseware can be used as input documents for ACTIVEMATH (see Section 26.8) and can be produced e.g. by CPOINT (see Section 26.14).

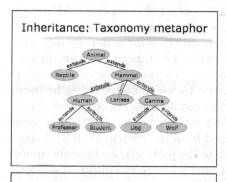

Inheritance: Taxonomy metaphor

Classes vs instances

- Every *object* is an *instance* of a *class.*
- The characteristics of an object are defined by its class.
- An object *inherits* characteristics from all of its *superclasses*

Classes vs instances

- Example:
  - *Danny Sleator* is an *instance* of the *Professor* class.
  - He is therefore also an instance of the *Human, Mammal,* and *Animal* classes.
  - Sometimes we say that Danny Sleator *"is a"* Professor (or Human or Mammal...)
  - Danny also *"has a"* wife and son, who are also instances of the Human class

In Java:

- `public class Animal { _ }` *Implicitly extends class Object*
- `public class Mammal extends Animal { _ }`
- `public class Human extends Mammal { _ }`
- `public class Professor extends Human { _ }`
- `public class MyClass {`
- `  Public static main () {`
- `  Professor danny = new Professor();`
- `  }`

Fig. 8.1. Three slides from 15-211

We have chosen a fragment that is relatively far from conventional mathematical texts to present the possibility of semantic markup in OMDoc even under such circumstances. We will highlight the use of OMDoc theories for such an application. Furthermore, we will take seriously the difference between marking up the knowledge (implicitly) contained in the slides and the slide presentation as a structured document. As a consequence, we will capture the slides in *two* documents:

- a *knowledge-centered document*, which contains the knowledge conveyed in the course organized by its inherent logical structure
- a *narrative-structured document* references the knowledge items and adds rhetorical and didactic structure of a slide presentation.

This separation of concerns into two documents is good practice in marking up mathematical texts: It allows to make explicit the structure inherent in the respective domain and at the same time the structure of the presentation that is driven by didactic needs. We call knowledge-structured documents **content OMDocs** and narrative-structured ones **narrative OMDocs**. The separation also simplifies management of academic content: The content OMDoc of course will usually be shared between individual installments of the course, it will be added to, corrected, cross-referenced, and kept up to date by different authors. It will eventually embody the institutional memory of an organization like a university or a group of teachers. The accompanying narrative OMDocs will capture the different didactic tastes and approaches by individual teachers and can be adapted for the installments of the course. Since the narrative OMDocs are relatively light-weight structures (they are largely void of original content, which is referenced from the content OMDoc) constructing or tailoring a course to the needs of the particular audience becomes a simpler endeavor of choosing a path through a large repository of marked up knowledge embodied in the content OMDoc rather than re-authoring[1] the content with a new slant.

Let us look at the four slides in Figure 8.1. The first slide shows a graphic of a simple taxonomy of animals, the second one introduces first concepts from object-oriented programming, the third one gives examples for these interpreting the class hierarchy introduced in the first slide, finally the fourth slide gives code concrete snippets as examples for the concepts introduced in the first three ones.

We will first discuss content OMDoc and then the narrative OMDoc in Section 8.2.

---

[1] Since much of the re-authoring is done by copy and paste in the current model, it propagates errors in the course materials rather than corrections.

8.1   A Knowledge-Centered View

In this section, we will take a look at how we can make the knowledge that is contained in the slides in Figure 8.1 and its structure explicit so that a knowledge management system like MBASE (see Section 26.4) or knowledge presentation system like ACTIVEMATH (see Section 26.8) can take advantage of it. We will restrict ourselves to knowledge that is explicitly represented in the slides in some form, even though the knowledge document would probably acquire more and more knowledge in the form of examples, graphics, variant definitions, and explanatory text as it is re-used in many courses.

The first slide introduces a theory, which we call **animals-tax**; see Listing 8.2. It declares primitive symbols for all the concepts[2] (the ovals), and for all the links introduced in the graphic it has **axiom** elements stating that the parent node in the tree extends the child node. The axiom uses the symbol for concept extension from a theory **kr** for knowledge representation which we import in the theory and which we assume in the background materials for the course.

Listing 8.2. The OMDOC representation for slide 1 from Figure 8.1

```
   <theory xml:id="animals−tax">
    <imports xml:id="tax_imports_taxonomy" from="#taxonomies"/>
    <imports xml:id="tax_imports_kr" from="#kr"/>
    <symbol name="human">
5     <type system="stlc"><OMOBJ><OMS cd="kr" name="concept"/></OMOBJ></type>
    </symbol>
    <symbol name="mammal">
      <type system="stlc"><OMOBJ><OMS cd="kr" name="concept"/></OMOBJ></type>
    </symbol>
10  ...
    <axiom xml:id="mammal−ext−human">
      <CMP>Humans are Animals.</CMP>
      <FMP>
       <OMOBJ>
15       <OMA><OMS cd="kr" name="extends"/>
         <OMS cd="animal−taxonomy" name="mammal"/>
         <OMS cd="animal−taxonomy" name="human"/>
       </OMA>
      </OMOBJ>
20     </FMP>
    </axiom>
    ...
   </theory>

25  <private xml:id="tax−image" for="#animals−tax" reformulates="#animals−tax">
    <data format="image/jpeg" href="animals−taxonomy.jpg"/>
    <data format="application/postscript" href="animals−taxonomy.ps"/>
   </private>
```

The **private** element contains the reference to the image in various formats. Its **reformulates** attribute hints that the image contained in this element can be used to illustrate the theory above (in fact, it will be the only thing used from this theory in the narrative OMDOC in Listing 8.7.)

---

[2] The type information in the symbols is not strictly included in the slides, but may represent the fact that the instructor said that the ovals represent "concepts".

The second slide introduces some basic concepts in object oriented programming. These give rise to the five primitive symbols of the theory. Note that this theory is basic, it does not import any other. The three text blocks are marked up as axioms, using the attribute `for` to specify the symbols involved in these axioms. The value of the `for` attribute is a whitespace-separated list of URI references to `symbol` elements.

Listing 8.3. The OMDoc representation for slide 2 from Figure 8.1

```
<theory xml:id="cvi">
  <symbol name="object" xml:id="cvi.object"/>
  <symbol name="instance" xml:id="cvi.instance"/>
  <symbol name="class" xml:id="cvi.class"/>
5  <symbol name="inherits" xml:id="cvi.inherits"/>
  <symbol name="superclass" xml:id="cvi.superclass"/>

  <axiom xml:id="ax1" for="#cvi.object #cvi.instance #cvi.class">
    <CMP>Every <phrase style="font-style:italic;color:blue">object</phrase>
10   is an <phrase style="font-style:italic; color:red">instance</phrase>
     of a <phrase style="font-style:italic; color:blue">class</phrase>.
    </CMP>
  </axiom>

15  <axiom xml:id="ax2" for="#cvi.class">
    <CMP>The characteristics of an object are defined by its class.</CMP>
  </axiom>

  <axiom xml:id="ax3" for="#cvi.inherits #cvi.superclass">
20   <CMP> An object <phrase style="font-style:italic;color:blue">inherits</phrase>
     characteristics from all of its
       <phrase style="font-style:italic; color:red">superclasses</phrase>.</CMP>
    </axiom>
</theory>
```

For the third slide it is not entirely obvious which of the OMDoc elements we want to use for markup. The intention of the slide is obviously to give some examples for the concepts introduced in the second slide in terms of the taxonomy presented in the first slide in Figure 8.1. However, the OMDoc `example` element seems to be too specific to directly capture the contents (see p. 146). What is immediately obvious is that the slide introduces some new knowledge and symbols, so we have to have a separate theory for this slide. The first item in the list headed by the word Example is a piece of new knowledge, it is therefore not an example at all, but an axiom[3]. The second item in the list is a statement that can be deduced from the knowledge we already have at our disposal from theories `animals-tax` and `cvi`. Therefore, the new theory `cvi-examples` in Listing 8.4 imports these two. Furthermore, it introduces the new symbol `danny` for "Danny Sleator" which is clarified in the `axiom` element with `xml:id="ax1"`. Finally, the third item in the list does not have the function of an example either, it introduces a new concept, the "is a" relation[4]. So we arrive at the theory in Listing 8.4. Note that this

---

[3] We could say that the function of being an example has moved up from mathematical statements to mathematical theories; we will not pursue this here.

[4] Actually, this text block introduces a new concept "by reference to examples", which is not a formal definition at all. We will neglect this for the moment.

markup treats the last text block on the third slide without semantic function in the theory – it points out that there are other relations among humans – and leaves it for the narrative-structured OMDOC in Section 8.2[5].

Listing 8.4. The OMDOC representation for slide 3 from Figure 8.1

```
<theory xml:id="cvi−examples">
  <imports from="#animals−tax"/><imports from="#cvi"/>

  <symbol name="danny" xml:id="cvi−examples.danny">
5   <metadata><dc:description>Danny Sleator</dc:description></metadata>
  </symbol>

  <axiom xml:id="danny−professor" for="#cvi.class #cvi.instance #cvi−examples.danny">
    <CMP><phrase style="font−style:italic;color:blue">Danny Sleator</phrase>
10    is an <phrase style="font−style:italic; color:red">instance</phrase>
      of the <phrase style="font−style:italic; color:blue">Professor</phrase>
      class.
    </CMP>
  </axiom>
15
  <assertion xml:id="dannys−classes" type="theorem">
    <CMP>He is therefore also an instance of the
      <phrase style="font−style:italic; color:blue">Human</phrase>,
      <phrase style="font−style:italic; color:blue">Mammal</phrase>,
20    <phrase style="font−style:italic; color:blue">Animal</phrase> classes.
    </CMP>
  </assertion>

  <symbol name="is_a" scope="global">
25    <metadata><dc:subject>'is a' relation</dc:subject></metadata>
  </symbol>

  <definition xml:id="is_a−def" for="#is_a" type="informal">
    <CMP>Sometimes we say that Danny Sleator
30      &#x201C;<phrase style="font−style:italic;color:red">is a</phrase>&#x201D;
      Professor (or Human or Mammal&#x2026;)
    </CMP>
  </definition>
</theory>
```

An alternative, more semantic way to mark up the `assertion` element in the theory above would be to split it into multiple `assertion` and `example` elements, as in Listing 8.5, where we have also added formal content. We have split the assertion `dannys-classes` into three — we have only shown one of them in Listing 8.5 — separate assertions about class instances, and used them to justify the explicit examples. These are given as OMDOC `example` elements. The `for` attribute of an `example` element points to the concepts that are exemplified here (in this case the symbols for the concepts "instance", "class" from the theory `cvi` and the concept "mammal" from the animal taxonomy). The `type` specifies that this is not a counter-example, and the `assertion` points to the justifying assertion. In this particular case, the reasoning behind the example is pretty straightforward (therefore it has been omitted in the slides), but we will make it explicit to show the mechanisms

---

[5] Of course this design decision is debatable, and depends on the intuitions of the author. We have mainly treated the text this way to show the possibilities of semantic markup

involved. The `assertion` element just re-states the assertion implicit in the example, we refrain from giving the formal statement in an `FMP` child here to save space. The `proofs` can be used to point to set of proofs for this assertion, in this case only the one given in Listing 8.5. We use the OMDOC `proof` element to mark up this proof. It contains a series of `derive` proof steps. In our case, the argument is very simple, we can see that Danny Sleator is an instance of the human class, using the knowledge that

1. Danny is a professor (from the axiom in the `cvi-examples` theory)
2. An object inherits all the characteristics from its superclasses (from the axiom `ax3` in the `cvi` theory)
3. The human class is a superclass of the professor class (from the axiom `human-extends-professor` in the `animal-taxonomy` theory).

The use of this knowledge in the proof step is made explicit by the `premise` children of the `derive` element.

The information in the proof could for instance be used to generate very detailed explanations for students who need help understanding the content of the original slides in Figure 8.1.

Listing 8.5. An alternative representation using `example` elements

```
     . . .
     <example xml:id="danny−mammal" type="for" assertion="dannys−mammal−thm"
              for="#cvi.instance #cvi.class #animal−taxonomy.mammal">
       <CMP>Danny Sleator is an instance of the
5        <phrase style="font−style:italic; color:blue">Mammal</phrase> class.
       </CMP>
       <OMOBJ><OMS cd="cvi−examples" name="danny"/></OMOBJ>
     </example>

10   <assertion xml:id="dannys−mammal−thm" type="theorem" proofs="danny−mammal−pf">
       <CMP>Danny Sleator is an instance of the Human class.</CMP>
     </assertion>

     <proof xml:id="danny−human−pf" for="#dannys−mammal−thm">
15     <derive xml:id="d1">
         <CMP>Danny Sleator is an instance of the human class.</CMP>
         <method>
           <premise xref="#danny−professor"/>
           <premise xref="#cvi.ax3"/>
20         <premise xref="#animal−tax.human−extends−professor"/>
         </method>
       </derive>
       <derive xml:id="concl">
         <CMP>Therefore he is an instance of the human class.</CMP>
25       <method>
           <premise xref="#d1"/>
           <premise xref="#cvi.ax3"/>
           <premise xref="#animal−tax.mammal−extends−human"/>
         </method>
30     </derive>
     </proof>
     . . .
```

The last slide contains a set of Java code fragments that are related to the material before. We have marked them up in the `code` elements in Listing 8.6. The actual code is encapsulated in a `data` element, whose `format` specifies the format the data is in. The program text is encapsulated in a CDATA section to suspend the XML parser (there might be characters like < or & in there which offend it). The `code` elements allow to document the input, output, and side-effects in `input`, `output`, `effect` elements as children of the `code` elements. Since the code fragments in question do not have input or output, we have only described the side-effect (class declaration and class extension). As the code elements do not introduce any new symbols, definitions or axioms, we do not have to place them in a theory. The second `code` element also carries a `requires` attribute, which specifies that to execute this code snippet, we need the previous one. An application can use this information to make sure that that one is loaded before executing this code fragment.

Listing 8.6. OMDoc representation of program code

```
<code xml:id="cvic−code1">
  <data format="Java"><![CDATA[public class Animal {... }]]></data>
  <effect><CMP>class declaration</CMP></effect>
</code>

<code xml:id="cvic−code2" requires="cvic−code1" >
  <data format="Java"><![CDATA[public class Mammal extends Animal {...}]]></data>
  <effect><CMP>class extension</CMP></effect>
</code>
...
```

8.2 A Narrative-Structured View

In this section we present an OMDoc document that captures the structure of the slide show as a document. It references the knowledge items from the theories presented in the last section and adds rhetorical and didactic structure of a slide presentation.

The individual slides are represented as `omgroup` elements with `type slide`.

The representation of the first slide in Figure 8.1 is rather straightforward: we use the `dc:title` element in `metadata` to represent the slide title. Its `class` attribute references a CSS `class` definition in a style file. To represent the image with the taxonomy tree we use an `omtext` element with an `omlet` element.

The second slide marks up the list structure of the slide with the `omgroup` element (the value `itemize` identifies it as an itemizes list). The items in the list are given by `ref` elements, whose `xref` attribute points to the axioms in the knowledge-structured document (see Listing 8.3). The effect of this markup is shared between the document: the content of the axioms are copied over from the knowledge-structured document, when the narrative-structured is presented to the user. However, the `ref` element cascades its `style` attribute (and the `class` attribute, if present) with the `style` and class attributes of the target element, essentially adding style directives during the copying process. In our example, this adds positioning information and specifies a particular image for the list bullet type.

Listing 8.7. The narrative OMDoc for Figure 8.1

```
...
  <omgroup xml:id="slide-847" type="slide">
    <metadata>
      <dc:title class="15-211-title">Inheritance: Taxonomy metaphor</dc:title>
5   </metadata>

    <omtext xml:id="the-tax">
      <CMP>
        <omlet data="#tax-image" style="width:540;height:366"
10          action="display" show="embed"/>
      </CMP>
    </omtext>
  </omgroup>

15  <omgroup xml:id="slide-848" type="slide">
    <metadata><dc:title class="15-211-title">Classes vs. instances</dc:title></metadata>
    <omgroup type="itemize" style="list-style-type:url(square.gif)">
      <ref style="position:30% 10%" xml:id="obj" xref="slide1_content.omdoc#ax1"/>
      <ref style="position:55% 10%" xml:id="class" xref="slide1_content.omdoc#ax2"/>
20      <ref style="position:80% 10%" xml:id="inh" xref="slide1_content.omdoc#ax3"/>
    </omgroup>
  </omgroup>

    <omgroup xml:id="slide-849" type="slide">
25    <metadata><dc:title class="15-211-title">Classes vs. instances</dc:title></metadata>
    <omgroup type="itemize" style="list-style-type:url(square.gif)">
      <omtext style="position:30% 10%" xml:id="ex"><CMP>Example:</CMP></omtext>
      <omgroup type="itemize" style="list-style-type:url(triangle.gif)">
        <ref style="position:400% 15%"
30          xml:id="danny" xref="slide1_content.omdoc#danny-professor"/>
        <ref style="position:55% 15%"
            xml:id="inst" xref="slide1_content.omdoc#dannys-classes"/>
        <ref style="position:70% 15%" xml:id="is_a" xref="slide1_content.omdoc#is-a-def"/>
      </omgroup>
35      <omtext style="position:83% 10%" xml:id="has_a">
        <CMP>
          Danny also &#x201C;<phrase style="font-style:italic;color:red">has
          a</phrase>&#x201D; wife and son, who are also instances of the Human class
        </CMP>
40      </omtext>
    </omgroup>
  </omgroup>

    <omgroup xml:id="slide-850" type="slide">
45    <metadata><dc:title class="15-211-title">In Java</dc:title></metadata>
    <omgroup type="itemize">
      <omtext xml:id="slide-850.t1" style="position:80% 10%;color:red">
        <CMP>Implicitly extends class object</CMP>
      </omtext>
50      <omtext xml:id="slide-850.t2">
        <CMP><omlet data="#cvic-code1" action="display" show="embed"/></CMP>
      </omtext>
      <omtext xml:id="slide-850.t3">
        <CMP><omlet data="#cvic-code2" action="display" show="embed"/></CMP>
55    </omtext>
    </omgroup>
  </omgroup>
  ...
```

8.3   Choreographing Narrative and Content OMDoc

The interplay between the narrative and content OMDoc above was relatively simple. The content OMDoc contained three theories that were linearized according to the dependency relation. This is often sufficient, but more complex rhetoric/didactic figures are also possible. For instance, when we introduce a new concept, we often first introduce a naive reduced approximation $\mathcal{N}$ of the real theory $\mathcal{F}$, only to show an example $\mathcal{E}_{\mathcal{N}}$ of where this is insufficient. Then we propose a first (straw-man) solution $\mathcal{S}$, and show an example $\mathcal{E}_{\mathcal{S}}$ of why this does not work. Based on the information we gleaned from this failed attempt, we build the eventual version $\mathcal{F}$ of the concept or theory and demonstrate that this works on $\mathcal{E}_{\mathcal{F}}$.

Let us visualize the narrative- and content structure in Figure 8.8. The structure with the solid lines and boxes at the bottom of the diagram represents the content structure, where the boxes $\mathcal{N}$, $\mathcal{E}_{\mathcal{N}}$, $\mathcal{S}$, $\mathcal{E}_{\mathcal{S}}$, $\mathcal{F}$, and $\mathcal{E}_{\mathcal{F}}$ signify theories for the content of the respective concepts and examples, much in the way we had them in Section 8.1. The arrows represent the theory inheritance structure, e.g. Theory $\mathcal{F}$ imports theory $\mathcal{N}$.

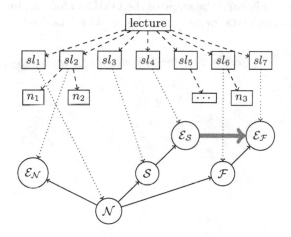

Fig. 8.8. An introduction of a concept via a straw-man theory

The top part of the diagram with the dashed lines stands for the narrative structure, where the arrows mark up the document structure. For instance, the slides sl_i are grouped into a lecture. The dashed lines between the two documents visualize **ref** elements with pointers into the content structure. In the example in Figure 8.8, the second slide of "lecture" presents the first example: the text fragment n_1 links the content $\mathcal{E}_{\mathcal{N}}$, which is referenced from the content structure, to slide 1. The fragment n_2 might say something like "this did not work in the current situation, so we have to extend the conceptualization...".

Just as for content-based systems on the formula level, there are now MKM systems that generate presentation markup from content markup, based on general presentation principles, also on this level. For instance, the ACTIVEMATH system [MBG+03] generates a simple narrative structure (the presentation; called a personalized book) from the underlying content structure (given in OMDOC) and a user model.

8.4 Summary

As we have seen, the narrative and content fulfill different, but legitimate content markup needs, that can coincide (as in the main example in this chapter), but need not (as in the example in the last section). In the simple case, where the dependency and narrative structure largely coincide, systems like the ACTIVEMATH system described in Section 26.8 can generate narrative OMDOCs from content OMDOCs automatically. To generate more complex rhetoric/didactic figures, we would have to have more explicit markup for relations like "can act as a straw-man for". Providing standardized markup for such relations is beyond the scope of the OMDOC format, but could easily be expressed as metadata, or as external, e.g. RDF-based relations.

9

Communication with and Between Mathematical Software Systems

OMDOC can be used as content language for communication protocols between mathematical software systems on the Internet. The ability to specify the context and meaning of the mathematical objects makes the OMDOC format ideally suited for this task.

In this chapter we will discuss a message interface in a fictitious software system MATHWEB-WS[1], which connects a wide-range of reasoning systems (*mathematical services*), such as automated theorem provers, automated proof assistants, computer algebra systems, model generators, constraint solvers, human interaction units, and automated concept formation systems, by a common *mathematical software bus*. Reasoning systems integrated in MATHWEB-WS can therefore offer new services to the pool of services, and can in turn use all services offered by other systems.

On the protocol level, MATHWEB-WS uses SOAP remote procedure calls with the HTTP binding [GHMN03] (see [Mit03] for an introduction to SOAP) interface that allows client applications to request service objects and to use their service methods. For instance, a client can simply request a service object for the automated theorem prover SPASS [Wei97] via the HTTP GET request in Listing 9.1 to a MATHWEB-WS broker node.

Listing 9.1. Discovering automated theorem provers (request)

```
GET /ws.mathweb.org/broker/getService?name=SPASS HTTP/1.1
Host: ws.mathweb.org
Accept: application/soap+xml
```

---

[1] "MATHWEB **W**eb Services"; The examples discussed in this chapter are inspired by the MATHWEB-SB [FK99, ZK02] ("MATHWEB **S**oftware **B**us") service infrastructure, which offers similar functionality based on the XML-RPC protocol (an XML encoding of Remote Procedure Calls (RPC) [Coma]). We use the SOAP-based formulation, since SOAP (**S**imple **O**bject **A**ccess **P**rotocol) is the relevant W3C standard and we can show the embedding of OMDOC fragments into other XML namespaces. In XML-RPC, the XML representations of the content language OMDOC would be transported as base-64-encoded strings, not as embedded XML fragments.

As a result, the client receives a SOAP message like the one in Listing 9.2 containing information about various instances of services embodying the SPASS prover known to the broker service.

Listing 9.2. Discovering automated theorem provers (response)

```
  HTTP/1.1 200 OK
  Content−Type: application/soap+xml
  Content−Length: 990

5 <?xml version='1.0'?>
  <env:Envelope xmlns:env="http://www.w3.org/2003/05/soap−envelope">
    <env:Body>
      <ws:prover env:encodingStyle="http://www.w3.org/2003/05/soap−encoding"
          xmlns:ws="http://www.mathweb.org/ws−fictional">
10      <ws:name>SPASS</ws:name>
        <ws:version>2.1</ws:version>
        <ws:URL>http://spass.mpi−sb.mpg.de/webspass/soap</ws:URL>
        <ws:uptime>P3D5H6M45S</ws:uptime>
        <ws:sysinfo>
15        <ws:ostype>SunOS 5.6</ws:ostype>
          <ws:mips>3825</ws:mips>
        </ws:sysinfo>
      </ws:prover>
      <ws:prover env:encodingStyle="http://www.w3.org/2003/05/soap−encoding"
20        xmlns:ws="http://www.mathweb.org/ws−fictional">
        <ws:name>SPASS</ws:name>
        <ws:version>2.0</ws:version>
        <ws:URL>http://asuka.mt.cs.cmu.edu/atp/spass/soap</ws:URL>
        <ws:uptime>P5M2D15H56M5S</ws:uptime>
25      <ws:sysinfo>
          <ws:ostype>linux−2.4.20</ws:ostype>
          <ws:mips>1468</ws:mips>
        </ws:sysinfo>
      <ws:prover>
30    </env:Body>
  </end:Envelope>
```

The client can then select one of the provers (say the first one, because it runs on the faster machine) and post theorem proving requests like the one in Listing 9.3[2] to the URL which uniquely identifies the service object in the Internet (this was part of the information given by the broker; see line 11 in Listing 9.2).

Listing 9.3. A SOAP RPC call to SPASS

```
  POST http://spass.mpi−sb.mpg.de/webspass/soap HTTP/1.1
  Host: http://spass.mpi−sb.mpg.de/webspass/soap
  Content−Type: application/soap+xml;
  Content−Length: 1123

5 <?xml version='1.0'?>
  <env:Envelope xmlns:env="http://www.w3.org/2003/05/soap−envelope">
    <env:Body>
      <ws:prove env:encodingStyle="http://www.w3.org/2003/05/soap−encoding"
10        xmlns:ws="http://www.mathweb.org/ws−fictional">
        <omdoc:assertion xml:id="peter−hates−somebody" type="conjecture"
            xmlns:omdoc="http://www.mathweb.org/omdoc"
            theory="http://mbase.mathweb.org:8080/RPC2#lovelife">
          <omdoc:CMP>Peter hates somebody</omdoc:CMP>
```

---

[2] We have made the namespaces involved explicit with prefixes in the examples, to show the mixing of different XML languages.

```
15        <omdoc:FMP>
            <om:OMOBJ xmlns:om="http://www.openmath.org/OpenMath">
              <om:OMBIND>
                <om:OMS cd="quant1" name="exists"/>
                <om:OMBVAR><om:OMV name="X"/></om:OMBVAR>
20              <om:OMA>
                  <om:OMS cd="lovelife" name="hate"/>
                  <om:OMS cd="lovelife" name="peter"/>
                  <om:OMV name="X"/>
                </om:OMA>
25            </om:OMBIND>
            </om:OMOBJ>
          </omdoc:FMP>
        </omdoc:assertion>
        <ws:replyWith><ws:state>proof</ws:state></ws:replyWith>
30      <ws:timeout>20</ws:timeout>
      </ws:prove>
    </env:Body>
  </env:Envelope>
```

This SOAP remote procedure call uses a generic method "prove" that can
be understood by the first-order theorem provers on MATHWEB-SB, and in
particular the SPASS system. This method is encoded as a ws:prove element;
its children describe the proof problem and are interpreted by the SOAP RPC
node as a parameter list for the method invocation. The first parameter is an
OMDOC representation of the assertion to be proven. The other parameters
instruct the theorem prover service to reply with the proof (instead of e.g.
just a yes/no answer) and gives it a time limit of 20 seconds to find it.

Note that OMDOC fragments can be seamlessly integrated into an XML
message format like SOAP. A SOAP implementation in the client's implementation language simplifies this process drastically since it abstracts from
HTTP protocol details and offers SOAP nodes using data structures of the
host language. As a consequence, developing MATHWEB clients is quite simple in such languages. Last but not least, both MS Internet Explorer and
the open source WWW browser FIREFOX now allow to perform SOAP calls
within JavaScript. This opens new opportunities for building user interfaces
based on web browsers.

Note furthermore that the example in Listing 9.3 depends on the information given in the theory lovelife referenced in the theory attribute in the
assertion element (see Section 15.6 for a discussion of the theory structure
in OMDOC). In our instance, this theory might contain formalizations (in
first-order logic) of the information that Peter hates everybody that Mary
loves and that Mary loves Peter, which would allow SPASS to prove the assertion. To get the information, the MATHWEB-WS service based on SPASS
would first have to retrieve the relevant information from a knowledge base
like the MBASE system described in Section 26.4 and pass it to the SPASS theorem prover as background information. As MBASE is also a MATHWEB-WS
server, this can be done by sending the query in Listing 9.4 to the MBASE
service at http://mbase.mathweb.org:8080.

Listing 9.4. Requesting a theory from MBASE

```
GET /mbase.mathweb.org:8080/soap/getTheory?name=lovelife HTTP/1.1
Host: mbase.mathweb.org:8080
Accept: application/soap+xml
```

The answer would be of the form given in Listing 9.5. Here, the SOAP envelope contains the OMDOC representation of the requested theory (irrespective of what the internal representation of MBASE was).

Listing 9.5. The background theory for message 9.3

```
   HTTP/1.1 200 OK
   Content−Type: application/soap+xml
   Content−Length: 602

5  <?xml version='1.0'?>
   <env:Envelope xmlns:env="http://www.w3.org/2003/05/soap−envelope">
     <env:Body>
       <theory xml:id="lovelife" xmlns="http://www.mathweb.org/omdoc">
         <symbol name="peter"/><symbol name="mary"/>
10        <symbol name="love"/><symbol name="hate"/>
         <axiom xml:id="opposite">
           <CMP>Peter hates everybody Mary loves</CMP>
           <FMP>∀x.loves(mary, x) ⇒ hates(peter, x)</FMP>
         </axiom>
15        <axiom xml:id="mary−loves−peter">
           <CMP>Mary loves Peter</CMP>
           <FMP>loves(mary, peter)</FMP>
         </axiom>
       </theory>
20     </env:Body>
   </env:Envelope>
```

This information is sufficient to prove the theorem in Listing 9.3; and the SPASS service might reply to the request with the message in Listing 9.6 which contains an OMDOC representation of a proof (see Chapter 17 for details). Note that the `for` attribute in the `proof` element points to the original assertion from Listing 9.3.

Listing 9.6. A proof that Peter hates someone

```
   HTTP/1.1 200 OK
   Content−Type: application/soap+xml
   Content−Length: 588

5  <?xml version='1.0'?>
   <env:Envelope xmlns:env="http://www.w3.org/2003/05/soap−envelope">
     <env:Body>
       <proof xml:id="p347" for="#peter−hates−somebody"
              xmlns="http://www.mathweb.org/omdoc">
10       <derive xml:id="d1">
           <FMP>hates(peter, peter)</FMP>
           <method xref="nd.omdoc#ND.chain">
             <premise xref="#lovelife.mary−loves−peter"/>
             <premise xref="#lovelife.opposite"/>
15         </method>
         </derive>
         <derive xml:id="concl">
           <method xref="nd.omdoc#ND.ExI"><premise xref="#d1"/></method>
         </derive>
20       </proof>
     </env:Body>
   </env:Envelope>
```

The proof has two steps: The first one is represented in the **derive** element, which states that "Peter hates Peter". This fact is derived from the two axioms in the theory **lovelife** in Listing 9.5 (the **premise** elements point to them) by the "chaining rule" of the natural deduction calculus. This inference rule is represented by a symbol in the theory **ND** and referred to by the **xref** attribute in the **method** element. The second proof step is given in the second **derive** element and concludes the proof. Since the assertion of the conclusion is the statement of the proven assertion, we do not have a separate **FMP** element that states this here. The sole premise of this proof step is the previous one. For details on the representation of proofs in OMDOC see Chapter 17.

Note that the SPASS theorem prover does not in itself give proofs in the natural deduction calculus, so the SPASS service that provided this answer presumably enlisted the help of another MATHWEB-WS service like the TRAMP system [Mei00] that transforms resolution proofs (the native format of the SPASS prover) to natural deduction proofs.

## Part III

The OMDoc Document Format

The OMDOC (Open Mathematical Documents) format is a content markup scheme for (collections of) mathematical documents including articles, textbooks, interactive books, and courses. OMDOC also serves as the content language for agent communication of mathematical services on a mathematical software bus.

This part of the book is the specification of version 1.2 of the OMDOC format, the final and mature release of OMDOC version 1. It defines the OMDOC language features and their meaning. The content of this part is normative for the OMDOC format; an OMDOC document is valid as an OMDOC document, iff it meets all the constraints imposed here. OMDOC applications will normally presuppose valid OMDOC documents and only exhibit the intended behavior on such.

OMDoc as a Modular Format

A modular approach to design is generally accepted as best practice in the development of any type of complex application. It separates the application's functionality into a number of "building blocks" or "modules", which are subsequently combined according to specific rules to form the entire application. This approach offers numerous advantages: The increased conceptual clarity allows developers to share ideas and code, and it encourages reuse by creating well-defined modules that perform a particular task. Modularization also reduces complexity by decomposition of the application's functionality and thus decreases debugging time by localizing errors due to design changes. Finally, flexibility and maintainability of the application are increased because single modules can be upgraded or replaced independently of others.

The OMDoc vocabulary has been split by thematic role, which we will briefly overview in Figure 10.1 before we go into the specifics of the respective modules in Chapters 13 to 21. To avoid repetition, we will introduce some attributes already in this chapter that are shared by elements from all modules. In Chapter 22 we will discuss the OMDoc document model and possible sub-languages of OMDoc that only make use of parts of the functionality (Section 22.3).

The first four modules in Figure 10.1 are required (mathematical documents without them do not really make sense), the other ones are optional. The document-structuring elements in module DOC have an attribute `modules` that allows to specify which of the modules are used in a particular document (see Chapter 11 and Section 22.3).

10.1 The OMDoc Namespaces

The namespace for the OMDoc format is the URI `http://www.mathweb.org/omdoc`. Note that the OMDoc namespace does not reflect the versions, this is done in the `version` attribute on the document root element `omdoc` (see Chapter 11). As a consequence, the OMDoc vocabulary identified by this

Module	Title	Required?	Chapter
MOBJ	Mathematical Objects	yes	Chapter 13
Formulae are a central part of mathematical documents; this module integrates the content-oriented representation formats OPENMATH *and* MATHML *into* OMDOC			
MTXT	Mathematical Text	yes	Chapter 14
Mathematical vernacular, i.e. natural language with embedded formulae			
DOC	Document Infrastructure	yes	Chapter 11
A basic infrastructure for assembling pieces of mathematical knowledge into functional documents and referencing their parts			
DC	Dublin Core Metadata	yes	Sections 12.1 and 12.2
Contains bibliographical **"data about data"***, which can be used to annotate many* OMDoc *elements by descriptive and administrative information that facilitates navigation and organization*			
CC	Creative Commons Metadata	yes	Section 12.3
Licenses for text use			
RT	Rich Text Structure	no	Section 14.6
Rich text structure in mathematical vernacular (lists, paragraphs, tables, . . .)			
ST	Mathematical Statements	no	Chapter 15
Markup for mathematical forms like theorems, axioms, definitions, and examples that can be used to specify or define properties of given mathematical objects and theories to group mathematical statements and provide a notion of context.			
PF	Proofs and proof objects	no	Chapter 17
Structure of proofs and argumentations at various levels of details and formality			
ADT	Abstract Data Types	no	Chapter 16
Definition schemata for sets that are built up inductively from constructor symbols			
CTH	Complex Theories	no	Chapter 18
Theory morphisms; they can be used to structure mathematical theories			
DG	Development Graphs	no	Section 18.5
Infrastructure for managing theory inclusions, change management			
EXT	Applets, Code, and Data	no	Chapter 20
Markup for applets, program code, and data (e.g. images, measurements, . . .)			
PRES	Presentation Information	no	Chapter 19
Limited functionality for specifying presentation and notation information for local typographic conventions that cannot be determined by general principles alone			
QUIZ	Infrastructure for Assessments	no	Chapter 21
Markup for exercises integrated into the OMDOC *document model*			

Fig. 10.1. The OMDoc Modules

namespace is not static, it can change with each new OMDOC version. However, if it does, the changes will be documented in later versions of the specification: the latest released version can be found at [Kohb].

In an OMDOC document, the OMDOC namespace must be specified either using a namespace declaration of the form `xmlns="http://www.mathweb.org/omdoc"` on the `omdoc` element or by prefixing the local names of the OMDOC elements by a namespace prefix (OMDOC customarily use the prefixes `omdoc:` or `o:`) that is declared by a namespace prefix declaration of the form `xmlns:o="http://www.mathweb.org/omdoc"` on some element dominating the OMDOC element in question (see Section 1.3 for an introduction). OMDOC also uses the following namespaces[1]:

Format	namespace URI	see
Dublin Core	http://purl.org/dc/elements/1.1/	Sections 12.1 and 12.2
Creative Commons	http://creativecommons.org/ns	Section 12.3
MATHML	http://www.w3.org/1998/Math/MathML	Section 13.2
OPENMATH	http://www.openmath.org/OpenMath	Section 13.1
XSLT	http://www.w3.org/1999/XSL/Transform	Chapter 19

Thus a typical document root of an OMDOC document looks as follows:

```
<?xml version="1.0" encoding="utf−8"?>
<omdoc xml:id="test.omdoc" version="1.2"
  xmlns="http://www.mathweb.org/omdoc"
  xmlns:cc="http://creativecommons.org/ns"
  xmlns:dc="http://purl.org/dc/elements/1.1/"
  xmlns:om="http://www.openmath.org/OpenMath"
  xmlns:m="http://www.w3.org/1998/Math/MathML">
...
</omdoc>
```

10.2   Common Attributes in OMDoc

Generally, the OMDOC format allows any attributes from foreign (i.e. non-OMDOC) namespaces on the OMDOC elements. This is a commonly found feature that makes the XML encoding of the OMDOC format extensible. Note that the attributes defined in this specification are in the default (empty) namespace: they do not carry a namespace prefix. So any attribute of the form `na:xxx` is allowed as long as it is in the scope of a suitable namespace prefix declaration.

Many OMDOC elements have optional `xml:id` attributes that can be used as identifiers to reference them. These attributes are of type ID, they must be unique in the document which is important, since many XML applications offer functionality for referencing and retrieving elements by ID-type attributes. Note that unlike other ID-attributes, in this special case it is the name `xml:id` [MVW05] that defines the referencing and uniqueness

---

[1] In this specification we will use the namespace prefixes above on all the elements we reference in text unless they are in the OMDOC namespace.

functionality, not the type declaration in the DTD or XML schema (see Subsection 1.3.2 for a discussion).

Note that in the OMDoc format proper, all ID type attributes are of the form `xml:id`. However in the older OPENMATH and MATHML standards, they still have the form `id`. The latter are only recognized to be of type `ID`, if a document type or XMLschema is present. Therefore it depends on the application context, whether a DTD should be supplied with the OMDoc document.

For many occasions (e.g. for printing OMDoc documents), authors want to control a wide variety of aspects of the presentation. OMDoc is a content-oriented format, and as such only supplies an infrastructure to mark up content-relevant information in OMDoc elements. To address this dilemma XML offers an interface to Cascading Style Sheets (CSS) [Bos98], which allow to specify presentational traits like text color, font variant, positioning, padding, or frames of layout boxes, and even aural aspects of the text.

To make use of CSS, most OMDoc elements (all that have `xml:id` attributes[2]) have `style` attributes that can be used to specify CSS directives for them. In the OMDoc fragment in Listing 10.2 we have used the `style` attribute to specify that the text content of the `omtext` element should be formatted in a centered box whose width is 80% of the surrounding box (probably the page box), and that has a 2 pixel wide solid frame of the specified RGB color. Generally CSS directives are of the form `A:V`, where `A` is the name of the aspect, and `V` is the value, several CSS directives can be combined in one `style` attribute as a semicolon-separated list (see [Bos98] and the emerging CSS 3 standard).

Listing 10.2. Basic CSS directives in a `style` attribute

```
   <?xml version="1.0" encoding="utf-8"?>
   <?xml-stylesheet type="text/css" href="http://example.org/style.css"?>
   <omdoc xml:id="stylish">
       ...
5      <omtext xml:id="t1" style="width:80%;align:center;border:2px #006699 solid">
         <CMP>Here comes something
            <phrase style="font-weight:bold;color:green" class="emphasize">stylish</phrase>!
         </CMP>
       </omtext>
10     ...
   </omdoc>
```

Note that many CSS properties of parent elements are inherited by the children, if they are not explicitly specified in the child. We could for instance have set the font family of all the children of the `omtext` element by adding a directive `font-family:sans-serif` there and then override it by a directive for the property `font-family` in one of the children.

Frequently recurring groups of CSS directives can be given symbolic names in CSS style sheets, which can be referenced by the `class` attribute.

---

[2] The treatment of the CSS attributes has changed from OMDoc 1.1, see the discussion on page 320.

In Listing 10.2 we have made use of this with the class `emphasize`, which we assume to be defined in the style sheet `style.css` associated with the document in the "style sheet processing instruction" in the prolog[3] of the XML document (see [Cla99a] for details). Note that an OMDoc element can have both `class` and `style` attributes, in this case, precedence is determined by the rules for CSS style sheets as specified in [Bos98]. In our example in Listing 10.2 the directives in the `style` attribute take precedence over the CSS directives in the style sheet referenced by the `class` attribute on the `phrase` element. As a consequence, the word "stylish" would appear in green, bold italics.

---

[3] i.e. at the very beginning of the XML document before the document type declaration

Document Infrastructure (Module DOC)

Mathematical knowledge is largely communicated by way of a specialized set of documents (e.g. e-mails, letters, pre-prints, journal articles, and textbooks). These employ special notational conventions and visual representations to convey the mathematical knowledge reliably and efficiently.

When marking up mathematical knowledge, one always has the choice whether to mark up the structure of the document itself, or the structure of the mathematical knowledge that is conveyed in the document. Even though in most documents, the document structure is designed to help convey the structure of the knowledge, the two structures need not be the same. To frame the discussion we will distinguish two aspects of mathematical documents. In the *knowledge-centered view* we organize the mathematical knowledge by its function, and do not care about a way to present it to human recipients. In the *narrative-centered view* we are interested in the structure of the argument that is used to convey the mathematical knowledge to a human user.

We will call a document **knowledge-structured** and **narrative-structured**, based on which of the two aspects is prevalent in the organization of the material. Narrative-structured documents in mathematics are generally directed at human consumption (even without being in presentation markup). They have a general narrative structure: text interleaving with formal elements like assertions, proofs, ... Generally, the order of presentation plays a role in their effectiveness as a means of communication. Typical examples of this class are course materials or introductory textbooks. Knowledge-structured documents are generally directed at machine consumption or for referencing. They do not have a linear narrative spine and can be accessed randomly and even re-ordered without information loss. Typical examples of these are formula collections, OPENMATH content dictionaries, technical specifications, etc.

The distinction between knowledge-structured and narrative-structured documents is reminiscent of the presentation vs. content distinction discussed in Section 2.1, but now it is on the level of document structure. Note that mathematical documents are often in both categories: a mathematical textbook can be read from front to end, but it can also be used as a reference, accessing it by the index and the table of contents. The way humans work with knowledge also involves a change of state. When we are taught or explore a mathematical domain, we have a linear/narrative path through the material, from which we abstract more and more, finally settling for a semantic representation that is relatively independent from the path we acquired it by. Systems like ACTIVEMATH (see Section 26.8) use the OMDOC format in exactly that way playing on the difference between the two classes and generating narrative-structured representations from knowledge-structured ones on the fly.

So, maybe the best way to think about this is that the question whether a document is narrative- or knowledge-structured is not a property of the document itself, but a property of the application processing this document.

OMDOC provides markup infrastructure for both aspects. In this chapter, we will discuss the infrastructure for the narrative aspect — for a working example we refer the reader to Chapter 8. We will look at markup elements for knowledge-structured documents in Section 15.6.

Even though the infrastructure for narrative aspects of mathematical documents is somewhat presentation-oriented, we will concentrate on content-markup for it. In particular, we will not concern ourselves with questions like font families, sizes, alignment, or positioning of text fragments. Like in most other XML applications, this kind of information can be specified in the CSS `style` and `class` attributes described in Section 10.2.

11.1   The Document Root

The XML root element of the OMDOC format is the <u>omdoc</u> element, it contains all other elements described here. We call an OMDOC element a **top-level element**, if it can appear as a direct child of the `omdoc` element.

The `omdoc` element (and the `omgroup` element introduced below as well) has an optional attribute `xml:id` that can be used to reference the whole document. The `version` attribute is used to specify the version of the OMDOC format the file conforms to. It is fixed to the string `1.2` by this specification. This will prevent validation with a different version of the DTD or schema, or processing with an application using a different version of the OMDOC specification. The (optional) attribute `modules` allows to specify the OM-DOC modules that are used in this document. The value of this attribute is a whitespace-separated list of module identifiers (e.g. MOBJ the left column in Figure 10.1), OMDOC sub-language identifiers (see Figure 22.5), or URI references for externally given OMDOC modules or sub-language identifiers.[1]

---

[1] Allowing these external module references keeps the OMDOC format extensible. Like in the case with namespace URIs OMDOC do not mandate that these URI references reference an actual resource. They merely act as identifiers for the modules.

The intention is that if present, the `modules` specifies the list of all the modules used in the document (fragment). If a `modules` attribute is present, then it is an error, if the content of this element contains elements from a module that is not specified; spurious module declarations in the `modules` attributes are allowed.

The `omdoc` element acts as an implicit grouping element, just as the `omgroup` element to be introduced in Section 11.4. Both have an optional `type` attribute; we will discuss its values and meaning in Section 11.4.

Here and in the following we will use tables as the one in Figure 11.1 to give an overview over the respective OMDOC elements described in a chapter or section. The first column gives the element name, the second and third columns specify the required and optional attributes. We will use the fourth column labeled "DC" to indicate whether an OMDOC element can have a `metadata` child, which will be described in the next section. Finally the fifth column describes the content model — i.e. the allowable children — of the element. For this, we will use a form of Bachus Naur Form notation also used in the DTD: `#PCDATA` stands for "parsed character data", i.e. text intermixed with legal OMDOC elements.) A synopsis of all elements is provided in Appendix B.

Element	Attributes		D	Content
	Required	Optional	C	
omdoc	version, xmlns	xml:id, type, class, style, version, modules	+	(⟪top-level⟫)*
omgroup		xml:id, modules, type, class, style	+	(⟪top-level⟫)*
metadata		xml:id, inherits, class, style	−	⟪MDelt⟫*
ref	xref	type, class, style	−	
ignore		type, comment	−	ANY
where ⟪top-level⟫ stands for top-level OMDOC elements, and ⟪MDelt⟫ for those introduced in Chapter 12				

Fig. 11.1. OMDOC Elements for Specifying Document Structure.

11.2 Metadata

The World Wide Web was originally built for human consumption, and although everything on it is machine-readable, most of it is not machine-understandable. The accepted solution is to provide metadata (data about data) to describe the documents on the web in a machine-understandable format that can be processed automatically. Metadata commonly specifies aspects of a document like title, authorship, language usage, and administrative aspects like modification dates, distribution rights, and identifiers.

In general, metadata can either be embedded in the respective document, or be stated in a separate one. The first facilitates maintenance and control (metadata is always at your fingertips, and it can only be manipulated by the document's authors), the second one enables inference and distribution. OMDoc allows to embed metadata into the document, from where it can be harvested for external metadata formats, such as the XML resource description format (RDF [LS99]). We use one of the best-known metadata schemata for documents – the *Dublin Core* (cf. Sections 12.1 and 12.2). The purpose of annotating metadata in OMDoc is to facilitate the administration of documents, e.g. digital rights management, and to generate input for metadata-based tools, e.g. RDF-based navigation and indexing of document collections. Unlike most other document formats OMDoc allows to add metadata at many levels, also making use of the metadata for document-internal markup purposes to ensure consistency.

The `metadata` element contains elements for various metadata formats including bibliographic data from the Dublin Core vocabulary (as mentioned above), licensing information from the Creative Commons Initiative (see Section 12.3), as well as information for OPENMATH content dictionary management. Application-specific metadata elements can be specified by adding corresponding OMDoc modules that extend the content model of the `metadata` element.

The OMDoc `metadata` element can be used to provide information about the document as a whole (as the first child of the `omdoc` element), as well as about specific fragments of the document, and even about the top-level mathematical elements in OMDoc. This reinterpretation of bibliographic metadata as general data about knowledge items allows us to extract document fragments and re-assemble them to new aggregates without losing information about authorship, source, etc.

11.3 Document Comments

Many content markup formats rely on commenting the source for human understanding; in fact source comments are considered a vital part of document markup. However, as XML comments (i.e. anything between "`<!--`" and "`-->`" in a document) need not even be read by some XML parsers, we cannot guarantee that they will survive any XML manipulation of the OMDoc source.

Therefore, anything that would normally go into comments should be modeled with an `omtext` element (`type comment`, if it is a text-level comment; see Section 14.3) or with the `ignore` element for persistent comments, i.e. comments that survive processing. The content of the `ignore` element can be any well-formed OMDoc, it can occur as an OMDoc top-level element or inside mathematical texts (see Chapter 14). This element should be used if the author wants to comment the OMDoc representation, but the end user should not see their content in a final presentation of the document, so that OMDoc text elements are not suitable, e.g. in

```
<ignore type="todo" comment="this does not make sense yet, rework">
  <assertion xml:id="heureka">...</assertion>
</ignore>
```

Of course, `ignore` elements can be nested, e.g. if we want to mark up the comment text (a pure string as used in the example above is not enough to express the mathematics). This might lead to markup like

```
<ignore type="todo" comment="rework">
  <ignore type="todo-comment">
    <CMP>This does not make sense yet, in particular, the equation
      <OMOBJ>...</OMOBJ> cannot be true, think of <OMOBJ>...</OMOBJ>
    </CMP>
  </ignore>
  <assertion xml:id="heureka">...</assertion>
</ignore>
```

Another good use of the `ignore` element is to use it as an analogon to the in-place error markup in OPENMATH objects (see Subsection 13.1.2). In this case, we use the `type` attribute to specify the kind of error and the content for the faulty OMDOC fragment. Note that since the whole object must be a valid OMDOC object (or at least licensed by a DTD or schema), the content itself must be a well-formed OMDOC fragment. As a consequence, the `ignore` element can only be used for "mathematical errors" like sibling `CMP` or `FMP` elements that do not have the same meaning as in Listing 11.2. XML-well-formedness and validity errors will have to be handled by the XML tools involved.

Listing 11.2. Marking up mathematical errors using `ignore`

```
<ignore type="CMP-lang-error"
        comment="multilingual CMPs are not translations of each other">
  <assertion xml:id="ass1">
    <CMP>The proof is trivial</CMP>
    <CMP xml:lang="de">Der Beweis ist extrem schwer</CMP>
  </assertion>
</ignore>
```

For another use of the `ignore` element, see Figure 11.3 in Section 11.5.

11.4 Document Structure

Like other documents mathematical ones are often divided into units like chapters, sections, and paragraphs by tags and nesting information. OMDOC makes these document relations explicit by using the `omgroup` element with an optional attribute `type`. It can take the values[2]

---

[2] Version 1.1 of OMDOC also allowed values `dataset` and `labeled-dataset` for marking up tables. These values are deprecated in Version 1.2 of OMDOC, since we provide tables in module RT; see Section 14.6 for details. Furthermore, Version 1.1 of OMDOC allowed the value `narrative`, which was synonymous with `sequence`.

sequence for a succession of paragraphs. This is the default, and the normal way narrative texts are built up from paragraphs, mathematical statements, figures, etc. Thus, if no **type** is given the type **sequence** is assumed.

itemize for unordered lists. The children of this type of **omgroup** will usually be presented to the user as indented paragraphs preceded by a bullet symbol. Since the choice of this symbol is purely presentational, OMDoc use the CSS **style** or **class** attributes on the children to specify the presentation of the bullet symbols (see Section 10.2).

enumeration for ordered lists. The children of this type of **omgroup** are usually presented like unordered lists, only that they are preceded by a running number of some kind (e.g. "1.", "2."...or "a)", "b)"...; again the **style** or **class** attributes apply).

sectioning The children of this type of **omgroup** will be interpreted as sections. This means that the children will be usually numbered hierarchically, and their metadata will be interpreted as section heading information. For instance the **metadata/dc:title** information (see Section 12.1 for details) will be used as the section title. Note that OMDoc does not provide direct markup for particular hierarchical levels like "chapter", "section", or "paragraph", but assumes that these are determined by the application that presents the content to the human or specified using the CSS attributes.

Other values for the **type** attribute are also admissible, they should be URI references to documents explaining their intension.

We consider the **omdoc** element as an implicit **omgroup**, in order to allow plugging together the content of different OMDoc documents as **omgroups** in a larger document. Therefore, all the attributes of the **omdoc** element also appear on **omgroup** elements and behave exactly like those.

11.5   Sharing and Referring to Document Parts

As the document structure need not be a tree in hypertext documents, **omgroup** elements also allow empty **ref** elements whose **xref** attribute can be used to reference OMDoc elements defined elsewhere. The optional **xml:id** (its value must be document-unique) attribute identifies it and can be used for building reference labels for the included parts. Even though this attribute is optional, it is highly recommended to supply it. The **type** attribute can be used to describe the reference type. Currently OMDoc supports two values: **include** (the default) for in-text replacement and **cite** for a proper reference. The first kind of reference requires the OMDoc application to process the document as if the **ref** element were replaced with the OMDoc fragment specified in the **xref**. The processing of the type **cite** is application specific.

It is recommended to generate an appropriate label and (optionally) supply
a hyper-reference. There may be more supported values for type in time.

Let R be a ref element of type include. We call the element the URI
in the xref points to its target unless it is an omdoc element; in this case,
the target is an omgroup element which has the same children as the original
omdoc element[3].

We call the process of replacing a ref element by its target in a document
ref-reduction, and the document resulting from the process of systematically and recursively reducing all the ref elements the **ref-normal form** of
the source document. Note that ref-normalization may not always be possible, e.g. if the ref-targets do not exist or are inaccessible — or worse yet,
if the relation given by the ref elements is cyclic. Moreover, even if it is
possible to ref-normalize, this may not lead to a valid OMDoc document,
e.g. since ID type attributes that were unique in the target documents are
no longer in the ref-reduced one. We will call a document **ref-reducible**,
iff its ref-normal form exists, and **ref-valid**, iff the ref normal form exists
and is a valid OMDoc document.

Note that it may make sense to use documents that are not ref-valid for
narrative-centered documents, such as courseware or slides for talks that only
allude to, but do not fully specify the knowledge structure of the mathematical knowledge involved. For instance the slides discussed in Section 8.2 do not
contain the theory elements that would be needed to make the documents
ref-valid.

The ref elements also allow to "flatten" the tree structure in a document
into a list of leaves and relation declarations (see Figure 11.3 for an example).
It also makes it possible to have more than one view on a document using
omgroup structures that reference a shared set of OMDoc elements. Note
that we have embedded the ref-targets of the top-level omgroup element
into an ignore comment, so that an OMDoc transformation (e.g. to text
form) does not encounter the same content twice.

While the OMDoc approach to specifying document structure is a much
more flexible (database-like) approach to representing structured documents[4]
than the tree model, it puts a much heavier load on a system for presenting

---

[3] This transformation is necessary, since OMDoc does not allow to nest omdoc
elements, which would be the case if we allowed verbatim replacement for omdoc
elements. As we have stated above, the omdoc has an implicit omgroup element,
and thus behaves like one.

[4] The simple tree model is sufficient for simple markup of existing mathematical
texts and to replay them verbatim in a browser, but is insufficient e.g. for generating individualized presentations at multiple levels of abstractions from the
representation. The OMDoc text model — if taken to its extreme — allows to
specify the respective role and contributions of smaller text units, even down to
the sub-sentence level, and to make the structure of mathematical texts machine-
understandable. Thus, an advanced presentation engine like the ACTIVEMATH
system [SBC+00] can — for instance — extract document fragments based on
the preferences of the respective user.

```
<omgroup xml:id="text"
       type="sequence">
  <omtext xml:id="t1">$T_1$</omtext>
  <omgroup xml:id="enum"
         type="enumeration">
    <omtext xml:id="t2">$T_2$</omtext>
    <omtext xml:id="t3">$T_3$</omtext>
  </omgroup>
  <omtext xml:id="t4">$T_4$</omtext>
</omgroup>
```

$\leftrightarrow$

```
<omgroup xml:id="text" type="sequence">
  <ref xref="#t1"/>
  <ref xref="#enum"/>
  <ref xref="#t4"/>
</omgroup>

<ignore type="targets"
        comment="already referenced">
  <omtext xml:id="t1">$T_1$</omtext>
  <omtext xml:id="t2">$T_2$</omtext>
  <omtext xml:id="t3">$T_3$</omtext>
  <omtext xml:id="t4">$T_4$</omtext>

  <omgroup xml:id="enum"
         type="enumeration">
    <ref xref="#t2"/>
    <ref xref="#t3"/>
  </omgroup>
</ignore>
```

Fig. 11.3. Flattening a tree structure

the text to humans. In essence the presentation system must be able to recover the left representation from the right one in Figure 11.3. Generally, any OMDoc element defines a fragment of the OMDoc it is contained in: everything between the start and end tags and (recursively) those elements that are reached from it by following the cross-references specified in **ref** elements. In particular, the text fragment corresponding to the element with `xml:id="text"` in the right OMDoc of Figure 11.3 is just the one on the left.

In Section 10.2 we have introduced the CSS attributes `style` and `class`, which are present on all OMDoc elements. In the case of the **ref** element, there is a problem, since the content of these can be incompatible. In general, the rule for determining the style information for an element is that we treat the replacement element as if it were a child of the **ref** element, and then determine the values of the CSS properties of the **ref** element by inheritance.

Metadata (Modules DC and CC)

Metadata is "data about data" — in the case of OMDOC data about documents, such as titles, authorship, language usage, or administrative aspects like modification dates, distribution rights, and identifiers. To accommodate such data, OMDOC offers the **metadata** element in many places. The most commonly used metadata standard is the Dublin Core vocabulary, which is supported in some form by most formats. OMDOC uses this vocabulary for compatibility with other metadata applications and extends it for document management purposes in OMDOC. Most importantly OMDOC extends the use of metadata from documents to other (even mathematical) elements and document fragments to ensure a fine-grained authorship and rights management.

Element	Attributes		Content
	Req.	Optional	
dc:creator		xml:id, class, style, role	ANY
dc:contributor		xml:id, class, style, role	ANY
dc:title		xml:lang	⟪*math vernacular*⟫
dc:subject		xml:lang	⟪*math vernacular*⟫
dc:description		xml:lang	⟪*math vernacular*⟫
dc:publisher		xml:id, class, style	ANY
dc:date		action, who	ISO 8601
dc:type			fixed: "Dataset" or "Text"
dc:format			fixed: "application/omdoc+xml"
dc:identifier		scheme	ANY
dc:source			ANY
dc:language			ISO 639
dc:relation			ANY
dc:rights			ANY
for ⟪*math vernacular*⟫ see Section 14.1			

Fig. 12.1. Dublin core metadata in OMDOC

In the following we will describe the variant of Dublin Core metadata elements used in OMDoc[1]. Here, the `metadata` element can contain any number of instances of any Dublin Core elements described below in any order. In fact, multiple instances of the same element type (multiple `dc:creator` elements for example) can be interspersed with other elements without change of meaning. OMDoc extends the Dublin Core framework with a set of roles (from the MARC relator set [MAR03]) on the authorship elements and with a rights management system based on the Creative Commons Initiative.

12.1   The Dublin Core Elements (Module DC)

The descriptions in this section are adapted from [DUB03a], and augmented for the application in OMDoc where necessary. All these elements live in the Dublin Core namespace `http://purl.org/dc/elements/1.1/`, for which we traditionally use the namespace prefix `dc:`.

`dc:title` The title of the element — note that OMDoc metadata can be specified at multiple levels, not only at the document level, in particular, the Dublin Core `dc:title` element can be given to assign a title to a theorem, e.g. the "Substitution Value Theorem".

The `dc:title` element can contain mathematical vernacular, i.e. the same content as the `CMP` defined in Section 14.1. Also like the `CMP` element, the `dc:title` element has an `dc:lang` attribute that specifies the language of the content. Multiple `dc:title` elements inside a `metadata` element are assumed to be translations of each other.

`dc:creator` A primary creator or author of the publication. Additional contributors whose contributions are secondary to those listed in `dc:creator` elements should be named in `dc:contributor` elements. Documents with multiple co-authors should provide multiple `dc:creator` elements, each containing one author. The order of `dc:creator` elements is presumed to define the order in which the creators' names should be presented.

As markup for names across cultures is still un-standardized, OMDoc recommends that the content of a `dc:creator` element consists in a single name (as it would be presented to the user). The `dc:creator` element has an optional attribute `dc:id` so that it can be cross-referenced and a `role` attribute to further classify the concrete contribution to the element. We will discuss its values in Section 12.2.

`dc:contributor` A party whose contribution to the publication is secondary to those named in `dc:creator` elements. Apart from the significance of contribution, the semantics of the `dc:contributor` is identical to that of `dc:creator`, it has the same restriction content and carries the same

---

[1] Note that OMDoc 1.2 systematically changes the Dublin Core XML tags to synchronize with the tag syntax recommended by the Dublin Core Initiative. The tags were capitalized in OMDoc1.1

attributes plus a `dc:lang` attribute that specifies the target language in case the contribution is a translation.

`dc:subject` This element contains an arbitrary phrase or keyword, the attribute `dc:lang` is used for the language. Multiple instances of the `dc:subject` element are supported per `dc:lang` for multiple keywords.

`dc:description` A text describing the containing element's content; the attribute `dc:lang` is used for the language. As description of mathematical objects or OMDoc fragments may contain formulae, the content of this element is of the form "mathematical text" described in Chapter 14. The `dc:description` element is only recommended for `omdoc` elements that do not have a `CMP` group (see Section 14.1), or if the description is significantly shorter than the one in the `CMPs` (then it can be used as an abstract).

`dc:publisher` The entity for making the document available in its present form, such as a publishing house, a university department, or a corporate entity. The `dc:publisher` element only applies if the `metadata` is a direct child of the root element (`omdoc`) of a document.

`dc:date` The date and time a certain action was performed on the element that contains this. The content is in the format defined by XML Schema data type `dateTime` (see [XML04] for a discussion), which is based on the ISO 8601 norm for dates and times.

Concretely, the format is $\langle\!\langle YYYY \rangle\!\rangle$-$\langle\!\langle MM \rangle\!\rangle$-$\langle\!\langle DD \rangle\!\rangle$T$\langle\!\langle hh \rangle\!\rangle$:$\langle\!\langle mm \rangle\!\rangle$:$\langle\!\langle ss \rangle\!\rangle$ where $\langle\!\langle YYYY \rangle\!\rangle$ represents the year, $\langle\!\langle MM \rangle\!\rangle$ the month, and $\langle\!\langle DD \rangle\!\rangle$ the day, preceded by an optional leading "-" sign to indicate a negative number. If the sign is omitted, "+" is assumed. The letter "T" is the date/time separator and $\langle\!\langle hh \rangle\!\rangle$, $\langle\!\langle mm \rangle\!\rangle$, $\langle\!\langle ss \rangle\!\rangle$ represent hour, minutes, and seconds respectively. Additional digits can be used to increase the precision of fractional seconds if desired, i.e the format $\langle\!\langle ss \rangle\!\rangle.\langle\!\langle sss... \rangle\!\rangle$ with any number of digits after the decimal point is supported. The `dc:date` element has the attributes `action` and `who` to specify who did what: The value of `who` is a reference to a `dc:creator` or `dc:contributor` element and `dc` is a keyword for the action undertaken. Recommended values include the short forms `updated`, `created`, `imported`, `frozen`, `review-on`, `normed` with the obvious meanings. Other actions may be specified by URIs pointing to documents that explain the action.

`dc:type` Dublin Core defines a vocabulary for the document types in [DUB03b]. The best fit values for OMDoc are

`Dataset` defined as "*information encoded in a defined structure (for example lists, tables, and databases), intended to be useful for direct machine processing.*"

`Text` defined as "*a resource whose content is primarily words for reading. For example – books, letters, dissertations, poems, newspapers, articles, archives of mailing lists. Note that facsimiles or images of texts are still of the genre text.*"

> Collection defined as *"an aggregation of items. The term collection means that the resource is described as a group; its parts may be separately described and navigated"*.
>
> The more appropriate should be selected for the element that contains the dc:type. If it consists mainly of formal mathematical formulae, then Dataset is better, if it is mainly given as text, then Text should be used. More specifically, in OMDOC the value Dataset signals that the order of children in the parent of the metadata is not relevant to the meaning. This is the case for instance in formal developments of mathematical theories, such as the specifications in Chapter 18.

dc:format The physical or digital manifestation of the resource. Dublin Core suggests using MIME types [FB96]. Following [MSLK01] we fix the content of the dc:format element to be the string application/omdoc+xml as the MIME type for OMDOC.

dc:identifier A string or number used to uniquely identify the element. The dc:identifier element should only be used for public identifiers like ISBN or ISSN numbers. The numbering scheme can be specified in the scheme attribute.

dc:source Information regarding a prior resource from which the publication was derived. We recommend using either a URI or a scientific reference including identifiers like ISBN numbers for the content of the dc:source element.

dc:relation Relation of this document to others. The content model of the dc:relation element is not specified in the OMDOC format.

dc:language If there is a primary language of the document or element, this can be specified here. The content of the dc:language element must be an ISO 639 norm two-letter language specifier, like en $\widehat{=}$ English, de $\widehat{=}$ German, fr $\widehat{=}$ French, nl $\widehat{=}$ Dutch, ...

dc:rights Information about rights held in and over the document or element content or a reference to such a statement. Typically, a dc:rights element will contain a rights management statement, or reference a service providing such information. dc:rights information often encompasses Intellectual Property rights (IPR), Copyright, and various other property rights. If the dc:rights element is absent (and no dc:rights information is inherited), no assumptions can be made about the status of these and other rights with respect to the document or element.

OMDOC supplies specialized elements for the Creative Commons licenses to support the sharing of mathematical content. We will discuss them in Section 12.3.

Note that Dublin Core also defines a Coverage element that specifies the place or time which the publication's contents addresses. This does not seem appropriate for the mathematical content of OMDOC, which is largely independent of time and geography.

12.2 Roles in Dublin Core Elements

Because the Dublin Core metadata fields for dc:creator and dc:contributor do not distinguish roles of specific parties (such as author, editor, and illustrator), we will follow the Open eBook specification [Gro99] and use an optional role attribute for this purpose, which is adapted for OMDOC from the MARC relator code list [MAR03].

aut (author) Use for a person or corporate body chiefly responsible for the intellectual content of an element. This term may also be used when more than one person or body bears such responsibility.

ant (scientific/bibliographic antecedent) Use for the author responsible for a work upon which the element is based.

clb (collaborator) Use for a person or corporate body that takes a limited part in the elaboration of a work of another author or that brings complements (e.g., appendices, notes) to the work of another author.

edt (editor) Use for a person who prepares a document not primarily his/her own for publication, such as by elucidating text, adding introductory or other critical matter, or technically directing an editorial staff.

ths (thesis advisor) Use for the person under whose supervision a degree candidate develops and presents a thesis, memoir, or text of a dissertation.

trc (transcriber) Use for a person who prepares a handwritten or typewritten copy from original material, including from dictated or orally recorded material. This is also the role (on the dc:creator element) for someone who prepares the OMDOC version of some mathematical content.

trl (translator) Use for a person who renders a text from one language into another, or from an older form of a language into the modern form. The target language can be specified by dc:lang.

As OMDOC documents are often used to formalize existing mathematical texts for use in mechanized reasoning and computation systems, it is sometimes subtle to specify authorship. We will discuss some typical examples to give a guiding intuition. Listing 12.2 shows metadata for a situation where editor R gives the sources (e.g. in LATEX) of an element written by author A to secretary S for conversion into OMDOC format.

Listing 12.2. A document with editor (edt) and transcriber (trc)

```
<metadata>
    <dc:title>The Joy of Jordan C* Triples</dc:title>
    <dc:creator role="aut">A</dc:creator>
    <dc:contributor role="edt">R</dc:contributor>
5   <dc:contributor role="trc">S</dc:contributor>
</metadata>
```

In Listing 12.3 researcher R formalizes the theory of natural numbers following the standard textbook B (written by author A). In this case we recommend the first declaration for the whole document and the second one for specific math elements, e.g. a definition inspired by or adapted from one in book B.

Listing 12.3. A formalization with scientific antecedent (`ant`)

```
<omdoc xml:id="NNat" version="1.2" xmlns:dc="http://purl.org/dc/elements/1.1/">
  <metadata><dc:title>Natural Numbers</dc:title></metadata>
  ...
  <theory xml:id="NNat.thy">
5    <metadata>
      <dc:title>Natural Numbers</dc:title>
      <dc:creator role="aut">R</dc:creator>
      <dc:contributor role="ant">A</dc:contributor>
      <dc:source>B</dc:source>
10   </metadata>
     ...
  </theory>
  ...
</omdoc>
```

12.3 Managing Rights by Creative Commons Licenses (Module CC)

The Dublin Core vocabulary provides the `dc:rights` element for information about rights held in and over the document or element content, but leaves the content model unspecified. While it is legally sufficient to describe this information in natural language, a content markup format like OMDOC should support a machine-understandable format. As one of the purposes of the OMDOC format is to support the sharing and re-use of mathematical content, OMDOC provides markup for rights management via the Creative Commons (CC) licenses. Digital rights management (DRM) and licensing of intellectual property has become a hotly debated topic in the last years. We feel that the Creative Commons licenses that encourage sharing of content and enhance the (scientific) public domain while giving authors some control over their intellectual property establish a good middle ground. Specifying rights is important, since in the absence of an explicit or implicit (via inheritance) `dc:rights` element no assumptions can be made about the status of the document or fragment. Therefore OMDOC adds another child to the `metadata` element. This <u>`cc:license`</u> element is a symbolic representation of the Creative Commons legal framework, adapted to the OMDOC setting: The Creative Commons Metadata Initiative specifies various ways of embedding CC metadata into documents and electronic artefacts like pictures or MP3 recordings. As OMDOC is a source format, from which various presentation formats are generated, we need a content representation of the CC metadata from which the end-user representations for the respective formats can be generated.

The Creative Commons Metadata Initiative [Crea] divides the license characteristics in three types: **permissions, prohibitions** and **requirements**, which are represented by the three elements, which can occur as children of the `cc:license` element. The `cc:license` element has two optional argument:

Element	Attributes		Content
	Req.	Optional	
cc:license		jurisdiction	permissions, prohibitions, requirements
cc:permissions		reproduction, distribution, derivative_works	EMPTY
cc:prohibitions		commercial_use	EMPTY
cc:requirements		notice, copyleft, attribution	EMPTY

Fig. 12.4. The OMDOC elements for creative commons metadata

jurisdiction which allows to specify the country in whose jurisdiction the license will be enforced[2]. It's value is one of the top-level domain codes of the "Internet Assigned Names Authority (IANA)" [IAN]. If this attribute is absent, then the original US version of the license is assumed.

version which allows to specify the version of the license. If the attribute is not present, then the newest released version is assumed (version 2.0 at the time of writing this book)

The following three empty elements can occur as children of the cc:license element; their attribute specify the rights bestowed on the user by the license. All these elements have the namespace http://creativecommons.org/ns, for which we traditionally use the namespace prefix cc:.

— cc:permissions are the rights granted by the license, to model them the element has three attributes, which can have the values permitted (the permission is granted by the license) and prohibited (the permission isn't):

Attribute	Permission	Default
reproduction	the work may be reproduced	permitted
distribution	the work may be distributed, publicly displayed, and publicly performed	permitted
derivative_works	derivative works may be created and reproduced	permitted

— cc:prohibitions are the things the license prohibits.

Attribute	Prohibition	Default
commercial_use	stating that rights may be exercised for commercial purposes.	permitted

— cc:requirements are restrictions imposed by the license.

Attribute	Requirement	Default
notice	copyright and license notices must be kept intact	required
attribution	credit must be given to copyright holder and/or author	required
copyleft	derivative works, if authorized, must be licensed under the same terms as the work	required

---

[2] The Creative Commons Initiative is currently in the process of adapting their licenses to jurisdictions other than the USA, where the licenses originated. See [Crec] for details and to check for progress.

This vocabulary is directly modeled after the Creative Commons Metadata [Creb] which defines the meaning, and provides an RDF [LS99] based implementation. As we have discussed in Section 11.2, OMDOC follows an approach that specifies metadata in the document itself; thus we have provided the elements described here. In contrast to many other situations in OMDOC, the rights model is not extensible, since only the current model is backed by legal licenses provided by the creative commons initiative.

Listing 12.5 specifies a license grant using the Creative Commons "share-alike" license: The copyright is retained by the author, who licenses the content to the world, allowing others to reproduce and distribute it without restrictions as long as the copyright notice is kept intact. Furthermore, it allows others to create derivative works based on the content as long as it attributes the original work of the author and licenses the derived work under the identical license (i.e. the Creative Commons "share-alike" as well).

Listing 12.5. A creative commons license

```
<metadata>
  <dc:rights>Copyright (c) 2004 Michael Kohlhase</dc:rights>
  <license jurisdiction ="de" xmlns="http://creativecommons.org/ns">
    <permissions reproduction="permitted" distribution="permitted"
                 derivative_works="permitted"/>
    <prohibitions commercial_use="permitted"/>
    <requirements notice="required" copyleft="required" attribution="required"/>
  </license>
</metadata>
```

12.4   Inheritance of Metadata

The metadata elements can be added to many of the OMDOC elements, including grouping elements that can contain others that contain metadata. To avoid duplication, OMDOC assumes a priority-union semantics for the Dublin Core elements dc:creator, dc:contributor, dc:date, dc:type, dc:format, dc:source, dc:language, and dc:rights. A Dublin Core element, e.g. dc:creator that is missing in lower metadata declaration (i.e. there is no element of the same name) is inherited from the upper ones. So in Figure 12.6, the two boxes are equivalent, since the metadata in theory th1 and in definition d1 is inherited from the main declaration in the top-level omdoc element. If there is a metadata element of the same name present, the closer one takes precedence.

```
<omdoc xml:id="o1">                          <omdoc xml:id="o1">
 <metadata>                                  <metadata>
  <dc:creator>MiKo</dc:creator>               <dc:creator>MiKo</dc:creator>
 </metadata>                                 </metadata>

 <theory xml:id="th1">                       <theory xml:id="th1">
                                              <metadata>
                                               <dc:creator>MiKo</dc:creator>
                                              </metadata>

  <symbol name="s1"/>                          <symbol name="s1"/>
  <definition for="#s1" xml:id="d1"/>          <definition for="#s1" xml:id="d1">
                                                <metadata>
                                                 <dc:creator>MiKo</dc:creator>
                                                </metadata>
                                               </definition>
 </theory>                                   </theory>

 <theory xml:id="th2">                       <theory xml:id="th2">
  <metadata>                                   <metadata>
   <dc:creator>Paul</dc:creator>               <dc:creator>Paul</dc:creator>
  </metadata>                                  </metadata>
  <symbol name="s2"/>                          <symbol name="s2"/>
  <definition for="#s2" xml:id="d1">           <definition for="#s2" xml:id="d1">
   <metadata>                                   <metadata>
    <dc:creator>MiKo</dc:creator>               <dc:creator>MiKo</dc:creator>
   </metadata>                                  </metadata>
  </definition>                               </definition>
 </theory>                                   </theory>
</omdoc>                                     </omdoc>
```

⟷

Fig. 12.6. Inheritance of metadata

Mathematical Objects (Module MOBJ)

A distinguishing feature of mathematics is its ability to represent and manipulate ideas and objects in symbolic form as mathematical formulae. OMDoc uses the OpenMath and Content-MathML formats to represent mathematical formulae and objects. Therefore, the OpenMath standard [BCC+04] and the MathML 2.0 recommendation (second edition) [ABC+03] are part of this specification. We will review OpenMath objects (top-level element om:OMOBJ) in Section 13.1 and Content-MathML (top-level element m:math) in Section 13.2, and specify an OMDoc element for entering mathematical formulae (element legacy) in Section 13.5.

Element	Attributes		Content
	Required	Optional	
OMOBJ	id	class, style	See Figure 13.2
m:math		id, xlink:href	See Figure 13.8
legacy	format	xml:id, formalism	#PCDATA

Fig. 13.1. Mathematical objects in OMDoc

The recapitulation in the next two sections is not normative, please consult Section 2.1 for a general introduction and history and the OpenMath standard and the MathML 2.0 Recommendation for details and clarifications.

13.1 OpenMath

OpenMath is a markup language for mathematical formulae that concentrates on the meaning of formulae building on an extremely simple kernel (markup primitive for syntactical forms of content formulae), and adds an extension mechanism for mathematical concepts, the **content dictionaries**. These are machine-readable documents that define the meaning of mathematical concepts expressed by OpenMath symbols. The current released version of the OpenMath standard is OpenMath2, which incorporates many of the experiences of the last years, particularly with embedding OpenMath into the OMDoc format.

We will only review the XML encoding of OpenMath objects here, since it is most relevant to the OMDoc format. All elements of the XML encoding live in the namespace http://www.openmath.org/OpenMath, for which we traditionally use the namespace prefix om:.

Element	Attributes		Content
	Required	Optional	
OMOBJ		id, cdbase, class, style	$\langle\!\langle OMel \rangle\!\rangle$?
OMS	cd, name	id, cdbase, class, style	EMPTY
OMV	name	id, class, style	EMPTY
OMA		id, cdbase, class, style	$\langle\!\langle OMel \rangle\!\rangle$*
OMBIND		id, cdbase, class, style	$\langle\!\langle OMel \rangle\!\rangle$,OMBVAR,$\langle\!\langle OMel \rangle\!\rangle$
OMBVAR		id, class, style	(OMV | OMATTR)+
OMFOREIGN		id, cdbase, class, style	ANY
OMATTR		id, cdbase, class, style	$\langle\!\langle OMel \rangle\!\rangle$
OMATP		id, cdbase, class, style	(OMS, ($\langle\!\langle OMel \rangle\!\rangle$|OMFOREIGN))+
OMI		id, class, style	[0-9]*
OMB		id, class, style	#PCDATA
OMF		id, class, style, dec, hex	#PCDATA
OME		id, class, style	$\langle\!\langle OMel \rangle\!\rangle$?
OMR	href		$\langle\!\langle OMel \rangle\!\rangle$?
where $\langle\!\langle OMel \rangle\!\rangle$ is (OMS|OMV|OMI|OMB|OMSTR|OMF|OMA|OMBIND|OME|OMATTR)			

Fig. 13.2. OpenMath objects in OMDoc

13.1.1 The Representational Core of OpenMath

The central construct of the OpenMath is that of an OpenMath **object** (represented by the om:OMOBJ element in the XML encoding), which has a tree-like representation made up of applications (om:OMA), binding structures (om:OMBIND using om:OMBVAR to tag bound variables), variables (om:OMV), and symbols (om:OMS).

The om:OMA element contains representations of the function and its argument in "prefix-" or "Polish notation", i.e. the first child is the representation of the function and all the subsequent ones are representations of the arguments in order.

Objects and concepts that carry meaning independent of the local context (they are called **symbol**s in OpenMath) are represented as om:OMS elements, where the value of the **name** attribute gives the name of the symbol. The **cd** attribute specifies the relevant content dictionary, a document that defines the meaning of a collection of symbols including the one referenced by the om:OMS. This document can either be an original OpenMath content dictionary or an OMDoc document that serves as one (see Subsection 15.6.2 for a discussion). The optional **cdbase** on an om:OMS element contains a URI that can be used to disambiguate the content dictionary. Alternatively, the **cdbase** attribute can be given on an OpenMath element that is a parent to the om:OMS in question: The om:OMS inherits the **cdbase** of the nearest ancestor (inducing the usual XML scoping rules for declarations).

The OPENMATH2 standard proposes the following mechanism for determining a canonical identifying URI for the symbol declaration referenced by an OPENMATH symbol of the form `<OMS cd="foo" name="bar"/>` with the `cdbase`-value e.g. `http://www.openmath.org/cd`: it is the URI reference `http://www.openmath.org/cd/foo#bar`, which by convention identifies an `omcd:CDDefinition` element with a child `omcd:Name` whose value is `bar` in a content dictionary resource `http://www.openmath.org/cd/foo.ocd` (see Subsection 2.1.2 for a very brief introduction to OPENMATH content dictionaries).

Variables are represented as `om:OMV` element. As variables do not carry a meaning independent of their local content, `om:OMV` only carries a `name` attribute (see Section 13.4 for further discussion).

For instance, the formula $\sin(x)$ would be modeled as an application of the sin function (which in turn is represented as an OPENMATH symbol) to a variable:

```
<OMOBJ xmlns="http://www.openmath.org/OpenMath">
 <OMA cdbase="http://www.openmath.org/cd">
  <OMS cd="transc1" name="sin"/>
  <OMV name="x"/>
 </OMA>
</OMOBJ>
```

In our case, the function sin is represented as an `om:OMS` element with name `sin` from the content dictionary `transc1`. The `om:OMS` inherits the `cdbase`-value `http://www.openmath.org/cd`, which shows that it comes from the OPENMATH standard collection of content dictionaries from the `om:OMA` element above. The variable x is represented in an `om:OMV` element with `name`-value `x`.

For the om:OMBIND element consider the following representation of the formula $\forall x.\sin(x) \leq \pi$.

```
<OMOBJ cdbase="http://www.openmath.org/cd">
 <OMBIND>
  <OMS cd="quant1" name="forall"/>
  <OMBVAR><OMV name="x"/></OMBVAR>
  <OMA>
   <OMS cd="arith1" name="leq"/>
   <OMA><OMS cd="transc1" name="sin"/><OMV name="x"/></OMA>
   <OMS cd="nums1" name="pi"/>
  </OMA>
 </OMBIND>
</OMOBJ>
```

The om:OMBIND element has exactly three children, the first one is a "binding operator"[1] — in this case the universal quantifier, the second one is a list of bound variables that must be encapsulated in an om:OMBVAR element, and the third is the body of the binding object, in which the bound variables can be

---

[1] The binding operator must be a symbol which either has the role `binder` assigned by the OPENMATH content dictionary (see [BCC+04] for details) or the symbol declaration in the OMDOC content dictionary must have the value `binder` for the attribute `role` (see Subsection 15.2.1).

used. OPENMATH uses the om:OMBIND element to unambiguously specify the scope of bound variables in expressions: the bound variables in the om:OMBVAR element can be used only inside the mother om:OMBIND element, moreover they can be systematically renamed without changing the meaning of the binding expression. As a consequence, bound variables in the scope of an om:OMBIND are distinct as OPENMATH objects from any variables outside it, even if they share a name.

OPENMATH offers an element for annotating (parts of) formulae with external information (e.g. MATHML or LATEX presentation): the om:OMATTR element that pairs an OPENMATH object with an attribute-value list. To annotate an OPENMATH object, it is embedded as the second child in an om:OMATTR element. The attribute-value list is specified by children of the preceding om:OMATP (Attribute value Pair) element, which has an even number of children: children at odd positions must be om:OMS (specifying the attribute, they are called **keys** or **features**)[2], and children at even positions are the **values** of the keys specified by their immediately preceding siblings. In the OPENMATH fragment in Listing 13.3 the expression $x + \pi$ is annotated with an alternative representation and a color. Listing 13.11 has a more complex one involving types.

Listing 13.3. Associating alternate representations with an OPENMATH object

```
<OMATTR>
  <OMATP>
    <OMS cd="alt−rep" name="ascii"/>
    <OMSTR>(x+1)</OMSTR>
    <OMS cd="alt−rep" name="svg"/>
    <OMFOREIGN encoding="application/svg+xml">
      <svg xmlns='http://www.w3.org/2000/svg'>...</svg>
    </OMFOREIGN>
    <OMS cd="pres" name="color"/>
    <OMS cd="pres" name="red"/>
  </OMATP>
  <OMA>
    <OMS cd="arith1" name="plus"/>
    <OMV name="x"/>
    <OMS cd="nums1" name="pi"/>
  </OMA>
</OMATTR>
```

A special application of the om:OMATTR element is associating non-OPEN-MATH objects with OPENMATH objects. For this, OPENMATH2 allows to use an om:OMFOREIGN element in the even positions of an om:OMATP. This element can be used to hold arbitrary XML content (in our example above SVG: Scalable Vector Graphics [DJ02]), its required **encoding** attribute specifies the format of the content. We recommend a MIME type [FB96] (see Section 19.4 for an application).

---

[2] There are two kinds of keys in OPENMATH distinguished according to the **role** value on their **symbol** declaration in the contentdictionary: **attribution** specifies that this attribute value pair may be ignored by an application, so it should be used for information which does not change the meaning of the attributed OPENMATH object. The **role** is used for keys that modify the meaning of the attributed OPENMATH object and thus cannot be ignored by an application.

13.1.2 Programming Extensions of OPENMATH Objects

For representing objects in computer algebra systems OPENMATH also provides other basic data types: om:OMI for integers, om:OMB for byte arrays, om:OMSTR for strings, and om:OMF for floating point numbers. These do not play a large role in the context of OMDoc, so we refer the reader to the OPENMATH standard [BCC+04] for details.

The om:OME element is used for in-place error markup in OPENMATH objects, it can be used almost everywhere in OPENMATH elements. It has two children; the first one is an error operator[3], i.e. an OPENMATH symbol that specifies the kind of error, and the second one is the faulty OPENMATH object fragment. Note that since the whole object must be a valid OPENMATH object, the second child must be a well-formed OPENMATH object fragment. As a consequence, the om:OME element can only be used for "semantic errors" like non-existing content dictionaries, out-of-bounds errors, etc. XML-well-formedness and DTD-validity errors will have to be handled by the XML tools involved. In the following example, we have marked up two errors in a faulty representation of $\sin(\pi)$. The outer error flags an arity violation (the function sin only allows one argument), and the inner one flags the typo in the representation of the constant π (we used the name po instead of pi).

```
<OME>
  <OMS cd="type-error" name="arity-violation"/>
  <OMA>
    <OMS cd="transc1" name="sin"/>
    <OME>
      <OMS cd="error" name="unexpected_symbol"/>
      <OMS cd="nums1" name="po"/>
    </OME>
    <OMV name="x"/>
  </OMA>
</OME>
```

As we can see in this example, errors can be nested to encode multiple faults found by an OPENMATH application.

13.1.3 Structure Sharing in OPENMATH

As we have seen above, OPENMATH objects are essentially trees, where the leaves are symbols or variables. In many applications mathematical objects can grow to be very large, so that more space-efficient representations are needed. Therefore, OPENMATH2 supports structure sharing[4] in OPENMATH objects. In Figure 13.4 we have contrasted the tree representation of the object $1 + 1 + 1 + 1 + 1 + 1 + 1 + 1$ with the structure-shared one, which

---

[3] An error operator is like a binding operator in footnote 1, only the symbol has role error.

[4] Structure sharing is a well-known technique in computer science that tries to gain space efficiency in algorithms by re-using data structures that have already been created by pointing to them rather than copying.

represents the formula as a directed acyclic graph (DAG). As any DAG can be exploded into a tree by recursively copying all sub-graphs that have more than one incoming graph edge, DAGs can conserve space by structure sharing. In fact the tree on the left in Figure 13.4 is exponentially larger than the corresponding DAG on the right.

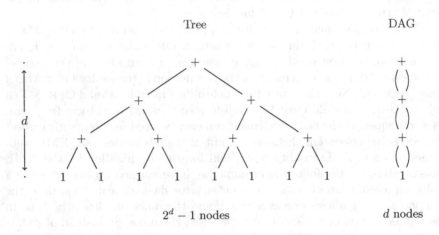

Fig. 13.4. Structure sharing by directed acyclic graphs

To support DAG structures, OPENMATH2 provides the (optional) attribute id on all OPENMATH objects and an element <u>om:OMR</u>[5] for the purpose of cross-referencing. The om:OMR element is empty and has the required attribute href; The OPENMATH element represented by this om:OMR element is a copy of the OPENMATH element pointed to in the href attribute. Note that the representation of the om:OMR element is *structurally equal*, but not identical to the element it points to.

Using the om:OMR element, we can represent the OPENMATH objects in Figure 13.4 as the XML representations in Figure 13.5.

To ensure that the XML representations actually correspond to directed acyclic graphs, the occurrences of the om:OMR must obey the global acyclicity constraint below, where we say that an OPENMATH element **dominates** all its children and all elements they dominate; The om:OMR also dominates its **target**[6], i.e. the element that carries the id attribute pointed to by the href attribute. For instance, in the representation in Figure 13.5 the om:OMA

---

[5] OPENMATH1 and OMDOC 1.0 did now know structure sharing, OMDOC 1.1 added xref attributes to the OPENMATH elements om:OMOBJ, om:OMA, om:OMBIND and om:OMATTR instead of om:OMR elements. This usage is deprecated in OMDOC 1.2, in favor of the om:OMR-based solution from the OPENMATH2 standard. Obviously, both representations are equivalent, and a transformation from xref-based mechanism to the om:OMR-based one is immediate.

[6] The target of an OPENMATH element with an id attribute is defined analogously

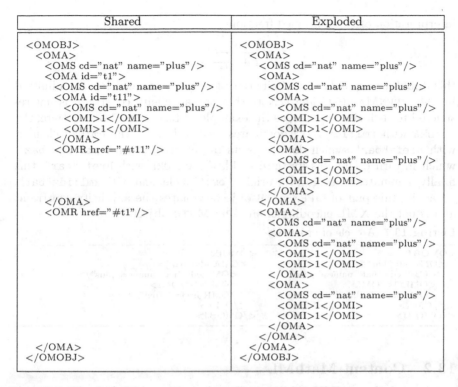

Shared	Exploded

`<OMOBJ>`
 `<OMA>`
  `<OMS cd="nat" name="plus"/>`
  `<OMA id="t1">`
   `<OMS cd="nat" name="plus"/>`
   `<OMA id="t11">`
    `<OMS cd="nat" name="plus"/>`
    `<OMI>1</OMI>`
    `<OMI>1</OMI>`
   `</OMA>`
   `<OMR href="#t11"/>`
  `</OMA>`
  `<OMR href="#t1"/>`
 `</OMA>`
`</OMOBJ>`
	`<OMOBJ>`
	 `<OMA>`
	  `<OMS cd="nat" name="plus"/>`
	  `<OMA>`
	   `<OMS cd="nat" name="plus"/>`
	   `<OMA>`
	    `<OMS cd="nat" name="plus"/>`
	    `<OMI>1</OMI>`
	    `<OMI>1</OMI>`
	   `</OMA>`
	   `<OMA>`
	    `<OMS cd="nat" name="plus"/>`
	    `<OMI>1</OMI>`
	    `<OMI>1</OMI>`
	   `</OMA>`
	  `</OMA>`
	  `<OMA>`
	   `<OMS cd="nat" name="plus"/>`
	   `<OMA>`
	    `<OMS cd="nat" name="plus"/>`
	    `<OMI>1</OMI>`
	    `<OMI>1</OMI>`
	   `</OMA>`
	   `<OMA>`
	    `<OMS cd="nat" name="plus"/>`
	    `<OMI>1</OMI>`
	    `<OMI>1</OMI>`
	   `</OMA>`
	  `</OMA>`
	 `</OMA>`
	`</OMOBJ>`

Fig. 13.5. The OPENMATH objects from Figure 13.4 in XML encoding

element with `xml:id="t1"` and also the second `om:OMA` element dominate the `om:OMA` element with `xml:id="t11"`.

   OPENMATH **Acyclicity Constraint**:
   An OpenMath element may not dominate itself.

Listing 13.6. A simple cycle

```
<OMOBJ>
  <OMA id="foo">
  <OMS cd="nat" name="divide"/>
  <OMI>1</OMI>
  <OMA><OMS cd="nat" name="plus"/>
    <OMI>1</OMI>
    <OMR href="#foo"/>
  </OMA>
  </OMA>
</OMOBJ>
```

In Listing 13.6 the `om:OMA` element with `xml:id="foo"` dominates its third child, which dominates the `om:OMR` with `href="foo"`, which dominates its target: the `om:OMA` element with `xml:id="foo"`. So by transitivity, this element dominates itself, and by the acyclicity constraint, it is not the XML representation of an OPENMATH object. Even though it could be given the interpretation of the continued fraction

$$\cfrac{1}{1 + \cfrac{1}{1 + \cdots}}$$

this would correspond to an infinite tree of applications, which is not admitted by the OPENMATH standard. Note that the acyclicity constraint is not restricted to such simple cases, as the example in Listing 13.7 shows. Here, the `om:OMA` with `xml:id="bar"` dominates its third child, the `om:OMR` element with `href="baz"`, which dominates its target `om:OMA` with `xml:id="baz"`, which in turn dominates its third child, the `om:OMR` with `href="bar"`, this finally dominates its target, the original `om:OMA` element with `xml:id="bar"`. So again, this pair of OPENMATH objects violates the acyclicity constraint and is not the XML encoding of an OPENMATH object.

Listing 13.7. A cycle of order two

```
<OMOBJ>                                 <OMOBJ>
  <OMA id="bar">                          <OMA id="baz">
    <OMS cd="nat" name="plus"/>             <OMS cd="nat" name="plus"/>
    <OMI>1</OMI>                             <OMI>1</OMI>
    <OMR href="#baz"/>                      <OMR href="#bar"/>
  </OMA>                                   </OMA>
</OMOBJ>                                 </OMOBJ>
```

13.2   Content MathML

Content-MATHML is a content markup format that represents the abstract structure of formulae in trees of logical sub-expressions much like OPEN-MATH. However, in contrast to that, Content-MATHML provides a lot of primitive tokens and constructor elements for the K-14 fragment of mathematics (Kindergarten to 14th grade (i.e. undergraduate college level)).

The current released version of the MATHML recommendation is the second edition of MATHML 2.0 [ABC+03], a maintenance release for the MATHML 2.0 recommendation [CIMP01] that cleans up many semantic issues in the content MATHML part. We will now review those parts of MATHML 2.0 that are relevant to OMDoc; for the full story see [ABC+03].

Even though OMDoc allows full Content-MATHML, we will advocate the use of the Content-MATHML fragment described in this section, which is largely isomorphic to OPENMATH (see Subsection 13.2.2 for a discussion).

13.2.1   The Representational Core of Content-MATHML

The top-level element of MATHML is the <u>m:math</u>[7] element, see Figure 13.10 for an example. Like OPENMATH, Content-MATHML organizes the math-

---

[7] For DTD validation OMDoc uses the namespace prefix "m:" for MATHML elements, since the OMDoc DTD needs to include the MATHML DTD with an explicit namespace prefix, as both MATHML and OMDoc have a `selector` element that would clash otherwise (DTDs are not namespace-aware).

Element	Attributes		Content
	Required	Optional	
m:math		id, xlink:href	《CMel》+
m:apply		id, xlink:href	m:bvar?, 《CMel》*
m:csymbol	definitionURL	id, xlink:href	m:EMPTY
m:ci		id, xlink:href	#PCDATA
m:cn		id, xlink:href	([0-9]|,|.)(*|e([0-9]|,|.)*)?
m:bvar		id, xlink:href	m:ci|m:semantics
m:semantics		id, xlink:href, definitionURL	《CMel》,(m:annotation | m:annotation-xml)*
m:annotation		definitionURL, encoding	#PCDATA
m:annotation-xml		definitionURL, encoding	ANY
where 《CMel》 is m:apply|m:csymbol|m:ci|m:cn|m:semantics			

Fig. 13.8. Content-MATHML in OMDOC

ematical objects into a functional tree. The basic objects (MATHML calls them token elements) are

identifiers (element m:ci) corresponding to variables. The content of the m:ci element is arbitrary Presentation-MATHML, used as the name of the identifier.

numbers (element m:cn) for number expressions. The attribute type can be used to specify the mathematical type of the number, e.g. complex, real, or integer. The content of the m:cn element is interpreted as the value of the number expression.

symbols (element m:csymbol) for arbitrary symbols. Their meaning is determined by a definitionURL attribute that is a URI reference that points to a symbol declaration in a defining document. The content of the m:csymbol element is a Presentation-MATHML representation that used to depict the symbol.

Apart from these generic elements, Content-MATHML provides a set of about 80 empty content elements that stand for objects, functions, relations, and constructors from various basic mathematic fields.

The m:apply element does double duty in Content-MATHML: it is not only used to mark up applications, but also represents binding structures if it has an m:bvar child; see Figure 13.10 below for a use case in a universal quantifier.

The m:semantics element provides a way to annotate Content-MATHML elements with arbitrary information. The first child of the m:semantics element is annotated with the information in the m:annotation-xml (for XML-based information) and m:annotation (for other information) elements that follow it. These elements carry definitionURL attributes that point to a "definition" of the kind of information provided by them. The optional encoding is a string that describes the format of the content.

13.2.2   OpenMath vs. Content MathML

OPENMATH and MATHML are well-integrated; there are semantics-preserving converters between the two formats. MATHML supports the `m:semantics` element, that can be used to annotate MATHML presentations of mathematical objects with their OPENMATH encoding. Analogously, OPENMATH supports the `presentation` symbol in the `om:OMATTR` element, that can be used for annotating with MATHML presentation. OPENMATH is the designated extension mechanism for MATHML beyond K-14 mathematics: Any symbol outside can be encoded as a `m:csymbol` element, whose `definitionURL` attribute points to the OPENMATH CD that defines the meaning of the symbol. Moreover all of the MATHML content elements have counterparts in the OPENMATH core content dictionaries [OMC]. For the purposes of OMDOC, we will consider the various representations following four representations of a content symbol in Figure 13.9 as equivalent. Note that the URI in the `definitionURL` attribute does not point to a specific file, but rather uses its base name for the reference. This allows a MATHML (or OMDOC) application to select the format most suitable for it.

`<m:plus/>`
Content-MATHML token element
`<m:plus definitionURL="http://www.openmath.org/cd/arith1#plus"/>`
Content-MATHML token element with explicit pointer
`<m:csymbol definitionURL="http://www.openmath.org/cd/arith1#plus"/>`
empty Content-MATHML `m:csymbol`

`<m:csymbol definitionURL="http://www.openmath.org/cd/arith1#plus">`
   `<m:mo>+</m:mo>`
`</m:csymbol>`

Content-MATHML `m:csymbol` with presentation
`<OMS cdbase="http://www.openmath.org/cd" cd="arith1" name="plus"/>`
OPENMATH symbol

Fig. 13.9. Four equivalent representations of a content symbol

In Figure 13.10 we have put the OPENMATH and content MATHML encoding of the law of commutativity for the real numbers side by side to show the similarities and differences. There is an obvious line-by-line similarity for the tree constructors and token elements. The main difference is the treatment of types and variables.

13.3   Representing Types in Content-MATHML and OPENMATH

Types are representations of certain simple sets that are treated specially in (human or mechanical) reasoning processes. In typed representations vari-

OPENMATH	MATHML

```
<OMOBJ>
 <OMBIND>
  <OMS cd="quant1" name="forall"/>
  <OMBVAR>
   <OMATTR>
    <OMATP>
     <OMS cd="sts" name="type"/>
     <OMS cd="setname1" name="R"/>
    </OMATP>
    <OMV name="a"/>

   </OMATTR>
   <OMATTR>
    <OMATP>
     <OMS cd="sts" name="type"/>
     <OMS cd="setname1" name="R"/>
    </OMATP>

    <OMV name="b"/>
   </OMATTR>
  </OMBVAR>
  <OMA>
   <OMS cd="relation" name="eq"/>
   <OMA>
    <OMS cd="arith1" name="plus"/>
    <OMV name="a"/>
    <OMV name="b"/>
   </OMA>
   <OMA>
    <OMS cd="arith1" name="plus"/>
    <OMV name="b"/>
    <OMV name="a"/>
   </OMA>
  </OMA>
 </OMBIND>
</OMOBJ>
```

```
<m:math>
 <m:apply>
  <m:forall/>
  <m:bvar>




   <m:ci type="real">a</m:ci>
  </m:bvar>



  <m:bvar>
   <m:ci type="real">b</m:ci>

  </m:bvar>
  <m:apply>
   <m:eq/>
   <m:apply>
    <m:plus/>
    <m:ci type="real">a</m:ci>
    <m:ci type="real">b</m:ci>
   </m:apply>
   <m:apply>
    <m:plus/>
    <m:ci type="real">b</m:ci>
    <m:ci type="real">a</m:ci>
   </m:apply>
  </m:apply>
 </m:apply>
</m:math>
```

Fig. 13.10. OPENMATH vs. C-MATHML for commutativity

ables and constants are usually associated with types to support more guided reasoning processes. Types are structurally like mathematical objects (i.e. arbitrary complex trees). Since types are ubiquitous in representations of mathematics, we will briefly review the best practices for representing them in OMDOC.

MATHML supplies the `type` attribute to specify types that can be taken from an open-ended list of type names. OPENMATH uses the `om:OMATTR` element to associate a type (in this case the set of real numbers as specified in the `setname1` content dictionary) with the variable, using the feature symbol `type` from the `sts` content dictionary. This mechanism is much more heavy-weight in our special case, but also more expressive: it allows to use arbitrary content expressions for types, which is necessary if we were to assign e.g. the type $(\mathbb{R} \to \mathbb{R}) \to (\mathbb{R} \to \mathbb{R})$ for functionals on the real numbers. In such cases, the second edition of the MATHML2 Recommendation advises a construction using the `m:semantics` element (see [KD03b] for details). List-

ings 13.11 and 13.12 show the realizations of a quantification over a variable
of functional type in both formats.

Listing 13.11. A complex type in OPENMATH

```
   <OMOBJ>
    <OMBIND>
     <OMS cd="quant1" name="forall"/>
     <OMBVAR>
 5    <OMATTR>
       <OMATP>
        <OMS cd="sts" name="type"/>
        <OMA><OMS cd="sts" name="mapsto"/>
         <OMA><OMS cd="sts" name="mapsto"/>
10         <OMS cd="setname1" name="R"/>
          <OMS cd="setname1" name="R"/>
         </OMA>
         <OMA><OMS cd="sts" name="mapsto"/>
          <OMS cd="setname1" name="R"/>
15         <OMS cd="setname1" name="R"/>
         </OMA>
        </OMA>
       </OMATP>
       <OMV name="F"/>
20     </OMATTR>
     </OMBVAR>
     ...
    </OMBIND>
   </OMOBJ>
```

Note that we have essentially used the same URI (to the `sts` content dictionary) to identify the fact that the annotation to the variable is a type (in a
particular type system).

Listing 13.12. A complex type in Content-MATHML

```
   <m:math>
    <m:apply>
     <m:forall/>
     <m:bvar>
 5    <m:semantics>
       <m:ci>F</m:ci>
       <m:annotation-xml definitionURL="http://www.openmath.org/cd/sts#type">
        <m:apply>
         <m:csymbol definitionURL="http://www.openmath.org/cd/sts#mapsto"/>
10        <m:apply>
          <m:csymbol definitionURL="http://www.openmath.org/cd/sts#mapsto"/>
          <m:csymbol definitionURL="http://www.openmath.org/cd/setname1#real"/>
          <m:csymbol definitionURL="http://www.openmath.org/cd/setname1#real"/>
         </m:apply>
15        <m:apply>
          <m:csymbol definitionURL="http://www.openmath.org/cd/sts#mapsto"/>
          <m:csymbol definitionURL="http://www.openmath.org/cd/setname1#real"/>
          <m:csymbol definitionURL="http://www.openmath.org/cd/setname1#real"/>
         </m:apply>
20       </m:apply>
       </m:annotation-xml>
      </m:semantics>
     </m:bvar>
     ...
25   </m:apply>
   </m:math>
```

13.4   The Semantics of Variables in OPENMATH and Content-MATHML

A more subtle, but nonetheless crucial difference between OPENMATH and MATHML is the handling of variables, symbols, their names, and equality conditions. OPENMATH uses the **name** attribute to identify a variable or symbol, and delegates the presentation of its name to other methods such as style sheets. As a consequence, the elements om:OMS and om:OMV are empty, and we have to understand the value of the **name** attribute as a pointer to a defining occurrence. In case of symbols, this is the symbol declaration in the content dictionary identified in the **cd** attribute. A symbol <OMS cd="$\langle\!\langle cd_1 \rangle\!\rangle$" name="$\langle\!\langle name_1 \rangle\!\rangle$"/> is equal to <OMS cd="$\langle\!\langle cd_2 \rangle\!\rangle$" name="$\langle\!\langle name_2 \rangle\!\rangle$"/>, iff $\langle\!\langle cd_1 \rangle\!\rangle = \langle\!\langle cd_2 \rangle\!\rangle$ and $\langle\!\langle name_1 \rangle\!\rangle = \langle\!\langle name_2 \rangle\!\rangle$ as XML simple names. In case of variables this is more difficult: if the variable is bound by an om:OMBIND element[8], then we interpret all the variables <OMV name="x"/> in the om:OMBIND element as equal and different from any variables <OMV name="x"/> outside. In fact the OPENMATH standard states that bound variables can be renamed without changing the object (α-**conversion**). If <OMV name="x"/> is not bound, then the scope of the variable cannot be reliably defined; so equality with other occurrences of the variable <OMV name="x"/> becomes an ill-defined problem. We therefore discourage the use of unbound variables in OMDOC; they are very simple to avoid by using symbols instead, introducing suitable theories if necessary (see Section 15.6).

MATHML goes a different route: the m:csymbol and m:ci elements have content that is Presentation-MATHML, which is used for the presentation of the variable or symbol name.[9] While this gives us a much better handle on presentation of objects with variables than OPENMATH (where we are basically forced to make due with the ASCII[10] representation of the variable name), the question of scope and equality becomes much more difficult: Are two variables (semantically) the same, even if they have different colors, sizes, or font families? Again, for symbols the situation is simpler, since the definitionURL attribute on the m:csymbol element establishes a global identity criterion (two symbols are equal, iff they have the same definitionURL value (as URI strings; see [BLFM98]).) The second edition of the MATHML standard adopts the same solution for bound variables: it recommends to

---

[8] We say that an om:OMBIND element **binds** an OPENMATH variable <OMV name="x"/>, iff this om:OMBIND element is the nearest one, such that <OMV name="x"/> occurs in (second child of the om:OMATTR element in) the om:OMBVAR child (this is the **defining occurrence** of <OMV name="x"/> here).

[9] Note that surprisingly, the empty Content-MATHML elements are treated more in the OPENMATH spirit.

[10] In the current OPENMATH standard, variable names are restricted to alphanumeric characters starting with a letter. Note that unlike with symbols, we cannot associate presentation information with variables via style sheets, since these are not globally unique (see Section 19.4 for a discussion of the OMDOC solution to this problem).

annotate the m:bvar elements that declare the bound variable with an id attribute and use the definitionURL attribute on the bound occurrences of the m:ci element to point to those. The following example is taken from [KD03a], which has more details.

```
<m:lambda>
  <m:bvar><m:ci xml:id="the−boundvar">complex presentation</m:ci></m:bvar>
  <m:apply>
    <m:plus/>
    <m:ci definitionURL="#the−boundvar">complex presentation</m:ci>
    <m:ci definitionURL="#the−boundvar">complex presentation</m:ci>
  </m:apply>
</m:lambda>
```

For presentation in MATHML, this gives us the best of both approaches, the m:ci content can be used, and the pointer gives a simple semantic equivalence criterion. For presenting OPENMATH and Content-MATHML in other formats OMDOC makes use of the infrastructure introduced in module PRES; see Section 19.4 for a discussion.

13.5   Legacy Representation for Migration

Sometimes, OMDOC is used as a migration format from legacy texts (see Chapter 4 for an example). In such documents it can be too much effort to convert all mathematical objects and formulae into OPENMATH or Content-MATHML form. For this situation OMDOC provides the legacy element, which can contain arbitrary math markup[11]. The legacy element can occur wherever an om:OMOBJ or m:math can and has an optional xml:id attribute for identification. The content is described by a pair of attributes:

- format (required) specifies the format of the content using URI reference. OMDOC does not restrict the possible values, possible values include TeX, pmml, html, and qmath.
- formalism is optional and describes the formalism (if applicable) the content is expressed in. Again, the value is unrestricted character data to allow a URI reference to a definition of a formalism.

For instance in the following legacy element, the identity function is encoded in the untyped λ-calculus, which is characterized by a reference to the relevant Wikipedia article.

```
<legacy format="TeX" formalism="http://en.wikipedia.org/wiki/Lambda_calculus">
  \lambda{x}{x}
</legacy>
```

---

[11] If the content is an XML-based, format like Scalable Vector Graphics [DJ02], the DTD must be augmented accordingly for validation.

Mathematical Text (Modules MTXT and RT)

The everyday mathematical language used in textbooks, conversations, and written onto blackboards all over the world consists of a rigorous, slightly stylized version of natural language interspersed with mathematical formulae, that is sometimes called **mathematical vernacular**[1].

Element	Attributes		D	Content
	Required	Optional	C	
CMP		xml:lang, id	−	《math vernacular》
FMP		xml:id, logic	−	(assumption*, conclusion*) | OMOBJ |m:math |legacy
assumption		xml:id, inductive, class, style	+	(OMOBJ |m:math |legacy)
conclusion		xml:id, class, style	+	(OMOBJ |m:math |legacy)
phrase		xml:id, class, style, index, verbalizes, type	−	《math vernacular》
term	cd, name	role, xml:id, class, style	−	《math vernacular》
omtext		xml:id, type, for, class, style, verbalizes	+	CMP+, FMP*

Fig. 14.1. The OMDOC elements for specifying mathematical properties

14.1 Multilingual Mathematical Vernacular

OMDOC models mathematical vernacular as parsed text interspersed with content-carrying elements. Most prominently, the om:OMOBJ, m:math, and legacy elements are used for mathematical objects, see Chapter 13. Other elements structure the text, such as the phrase and term elements introduced

---

[1] The term "mathematical vernacular" was first introduced by Nicolaas Govert de Bruijn in the 1970s (see [dB94] for a discussion). It derives from the word "vernacular" used in the Catholic church to distinguish the language used by laymen from the official Latin.

in this chapter, or link it to the document structure as the `ref` or `ignore` elements introduced above. In Figure 14.2 we have given an overview over the ones described in this book. The last two modules in Figure 14.2 are optional (see Section 22.3). Other (external or future) OMDOC modules can introduce further elements; natural extensions come when OMDOC is applied to areas outside mathematics, for instance computer science vernacular needs to talk about code fragments (see Section 20.1 and [Koha]), chemistry vernacular about chemical formulae (e.g. represented in Chemical Markup Language [CML05]).

Module	Elements	Comment	see
MOBJ	`om:OMOBJ, m:math, legacy`	mathematical Objects	p. 107
MTXT	`phrase, term`	phrase-level markup	below
DOC	`ref, ignore`	document structure	p. 89
RT	`p, ol, ul, dl, table, link, note, idx`	rich text structure	p. 128
EXT	`omlet`	for applets, images, ...	p. 205

Fig. 14.2. OMDOC modules contributing to mathematical vernacular

To be able to support multilingual documents, the mathematical vernacular is represented as a groups of `CMP`[2] elements which contain the vernacular and have an optional `xml:lang` attribute that specifies the language they are written in. Conforming with the XML recommendation, we use the ISO 639 two-letter country codes (`de` $\hat{=}$ German, `en` $\hat{=}$ English, `fr` $\hat{=}$ French, `nl` $\hat{=}$ Dutch, ...). If no `xml:lang` is given, then `en` is assumed as the default value. It is forbidden to have two or more sibling `CMP` with the same value of `xml:lang`, moreover, `CMP`s that are siblings must be translations of each other.[3] We speak of a multilingual groups of `CMP` elements if this is the case.

Listing 14.3. A multilingual group of `CMP` elements

```
  <CMP>
    Let <OMOBJ id="set"><OMV name="V"/></OMOBJ> be a set.
    A <term role="definiens">unary operation</term> on
    <OMOBJ><OMR href="#set"/></OMOBJ> is a function
5   <OMOBJ id="fun"><OMV name="F"/></OMOBJ> with
    <OMOBJ id="im">
      <OMA>
        <OMS cd="relations1" name="eq"/>
        <OMA><OMS cd="fns1" name="domain"/><OMV name="F"/></OMA>
10      <OMV name="V"/>
      </OMA>
```

---

[2] The name comes from "Commented Mathematical Property" and was originally taken from OPENMATH content dictionaries for continuity reasons. Note that XML does note confuse the two, since they are in different namespaces.

[3] The translation requirement may be alleviated in the future, when further variant relations are encoded in `CMP` groups (see [KK06] for a discussion in the context of "communities of practice"). Then a generalized uniqueness condition must be observed `CMP` groups, so that systems can choose between the supplied variants.

```
        </OMOBJ> and
        <OMOBJ id="ran">
          <OMA>
15          <OMS cd="relations1" name="eq"/>
            <OMA><OMS cd="fns1" name="range"/><OMV name="F"/></OMA>
            <OMV name="V"/>
          </OMA>
        </OMOBJ>.
20    </CMP>
      <CMP xml:lang="de">
        Sei <OMOBJ><OMR href="#set"/></OMOBJ> eine Menge.
        Eine <term role="definiens">unäre Operation</term>
        ist eine Funktion <OMOBJ><OMR href="#fun"/></OMOBJ>, so dass
25      <OMOBJ><OMR href="#im"/></OMOBJ> und
        <OMOBJ><OMR href="#ran"/></OMOBJ>.
      </CMP>
      <CMP xml:lang="fr">
        Soit <OMOBJ><OMR href="#set"/></OMOBJ> un ensemble.
30      Une <term role="definiens">opération unaire</term> sûr
        <OMOBJ><OMR href="#set"/></OMOBJ> est une fonction
        <OMOBJ><OMR href="#fun"/></OMOBJ> avec
        <OMOBJ><OMR href="#im"/></OMOBJ> et
        <OMOBJ><OMR href="#ran"/></OMOBJ>.
35    </CMP>
```

Listing 14.3 shows an example of such a multilingual group. Here, the OPEN-MATH extension by DAG representation (see Section 13.1) facilitates multi-language support: Only the language-dependent parts of the text have to be rewritten, the (language-independent) formulae can simply be re-used by cross-referencing.

14.2 Formal Mathematical Properties

An FMP[4] element is the general element for representing formal mathematical content in the form of OPENMATH objects. FMPs always appear in groups, which can differ in the value of their logic attribute, which specifies the logical formalism. The value of this attribute specifies the logical system used in formalizing the content. All members of the group have to formalize the same mathematical object or property, i.e. they have to be translations of each other, like siblings CMPs, we speak of a **multi-logic FMP group** in this case. Furthermore, if an FMP group has CMP siblings, all must express the same content.

In Listing 14.4 we see two FMP elements, that state the property of being a unary operation in two logics. The first one (fol for first-order logic) uses an equivalence to convey the restriction, the second one (hol for higher-order logic) has λ-abstraction and can therefore define the binary predicate binop directly.

---

[4] The name comes from "Formal Mathematical Properties" and was originally taken from OPENMATH content dictionaries for continuity reasons.

Listing 14.4. A multi-logic `FMP` group for Listing 14.3.

```
<omtext xml:id="binop−def" type="definition">
  ... the content of Listing 14.3 here ...
  <FMP logic="fol">∀V, F.binop(F, V) ⇔ Im(F) = V ∧ Dom(F) = V</FMP>
  <FMP logic="hol">binop = λV, F.Im(F) = V ∧ Dom(F) = V</FMP>
</omtext>
```

As mathematical statements of properties of objects often come as **sequent**s, i.e. as sets of conclusions drawn from a set of assumptions, OMDOC also allows the content of an `FMP` to be a (possibly empty) set of `assumption` elements followed by a (possibly empty) set of `conclusion` elements. The intended meaning is that the FMP asserts that one of the conclusions is entailed by the assumptions together in the current context. As a consequence

```
<FMP><conclusion>A</conclusion></FMP>
```

is equivalent to `<FMP>A</FMP>`, where A is an OPENMATH, Content-MATHML, or `legacy` representation of a mathematical formula. The `assumption` and `conclusion` elements allow to specify the content by an `om:OMOBJ`, `m:math`, or `legacy` element. The `assumption` and `conclusion` elements carry an optional `xml:id` attribute, which can be used to refer to them by `ref` elements in structure sharing. This is important for specifying sequent-style proofs (see Chapter 17), where the assumptions and conclusions of sequents are largely invariant over a proof and would have to be copied otherwise. The `assumption` element carries an additional optional attribute `inductive` for inductive hypotheses.

In the (somewhat contrived) example in Listing 14.5 we show a sequent for a simple fact about set intersection. Here the knowledge in both assumptions (together) is enough to entail one of the conclusions (the first in this case). For details about the `phrase` element see Section 14.4 below.

Listing 14.5. Representing vernacular as an `FMP` sequent

```
<CMP>If a ∈ U and a ∈ V, then a ∈ U ∩ V or
  <phrase index="moon_cheese">the moon is made of green cheese</phrase>.
</CMP>
<FMP>
  <assumption xml:id="A">a ∈ U</assumption>
  <assumption xml:id="B">a ∈ V</assumption>
  <conclusion xml:id="C">a ∈ U ∩ V</conclusion>
  <conclusion xml:id="moon_cheese">made_of(moon, gc)</conclusion>
</FMP>
```

14.3   Text Fragments and Their Rhetoric/Mathematical Roles

As we have explicated above, all mathematical documents state properties of mathematical objects — informally in mathematical vernacular or formally (as logical formulae), or both. OMDOC uses the `omtext` element to mark up text passages that form conceptual units, e.g. paragraphs, statements, or remarks. `omtext` elements have an optional `xml:id` attribute, so that they can

be cross-referenced, the intended purpose of the text fragment in the larger
document context can be described by the optional attribute `type`. This can
take e.g. the values `abstract`, `introduction`, `conclusion`, `comment`, `thesis`,
`antithesis`, `elaboration`, `motivation`, `evidence`, `transition` with the obvious meanings. In the last five cases `omtext` also has the extra attribute `for`,
and in the last one, also an attribute `from`, since these are in reference to other
OMDOC elements.

The content of an `omtext` element is mathematical vernacular contained
in a multi-lingual `CMP` group, followed by an (optional) multi-logic `FMP` group
that expresses the same content. This `CMP` group can be preceded by a
`metadata` element that can be used to specify authorship, give the passage a
title, etc. (see Section 12.1).

We have used the `type` attribute on `omtext` to classify text fragments by
their rhetoric role. This is adequate for much of the generic text that makes
up the narrative and explanatory text in a mathematical textbook. But many
text fragments in mathematical documents directly state properties of mathematical objects (we will call them mathematical statements; see Chapter 15
for a more elaborated markup infrastructure). These are usually classified
as definitions, axioms, etc. Moreover, they are of a form that can (in principle) be formalized up to the level of logical formula; in fact, mathematical
vernacular is seen by mathematicians as a more convenient form of communication for mathematical statements that can ultimately be translated into
a foundational logical system like axiomatic set theory [Ber91]. For such text
fragments, OMDOC reserves the following values for the `type` attribute:

`axiom` (fixes or restricts the meaning of certain symbols or concepts.) An
     axiom is a piece of mathematical knowledge that cannot be derived from
     anything else we know.
`definition` (introduces new concepts or symbols.) A definition is an ax-
     iom that introduces a new symbol or construct, without restricting the
     meaning of others.
`example` (for or against a mathematical property).
`proof` (a proof), i.e. a rigorous — but maybe informal — argument that a
     mathematical statement holds.
`hypothesis` (a local assumption in a proof that will be discharged later) for
     text fragments that come from (parts of) proofs.
`derive` (a step in a proof), we will specify the exact meanings of this and
     the two above in Chapter 17 and present more structured counterparts.

Finally, OMDOC also reserves the values `theorem`, `proposition`, `lemma`,
`corollary`, `postulate`, `conjecture`, `false-conjecture`, and `formula` for
statements that assert properties of mathematical objects (see Figure 15.10
in Subsection 15.3.1 for explanations). Note that the differences between these
values are largely pragmatic or proof-theoretic (conjectures become theorems
once there is a proof). Mathematical `omtext` elements (such with one of these
types) can have additional `FMP` elements (Formal Mathematical Property)

that formally represents the meaning of the descriptive text in the CMPs (if that is feasible).

Further types of text can be specified by providing a URI that points to a description of the text type (much like the `definitionURL` attribute on the `m:csymbol` elements in Content-MATHML).

Of course, the `type` only allows a rough classification of the mathematical statements at the text level, and does not make the underlying content structure explicit or reveals their contribution and interaction with mathematical context. For that purpose OMDOC supplies a set of specialized elements, which we will discuss in Chapter 15. Thus `omtext` elements will be used to give informal accounts of mathematical statements that are better and more fully annotated by the infrastructure introduced in Chapter 15. However, in narrative documents, we often want to be informal, while maintaining a link to the formal element. For this purpose OMDOC provides the optional `verbalizes` attribute on the `omtext` element. Its value is a whitespace-separated list of URI references to formal representations (see Section 15.5 for further discussion).

14.4   Phrase-Level Markup of Mathematical Vernacular

To make the sentence-internal structure of mathematical vernacular more explicit, OMDOC provides an infrastructure to mark up natural language phrases in sentences. Linguistically, a **phrase** is a group of words that functions as a single unit in the syntax of a sentence. Examples include "noun phrases, verb phrases, or prepositional phrases". In OMDOC we adhere to this intuition and restrict the `phrase` element to phrases in this sense. The `term` element is naturally restricted to phrases by construction. The `phrase` element is a general wrapper for sentence-level phrases that allows to mark their specific properties.

The `phrase` element allows the same content as the `CMP` element, so that it can be transparently nested. It has the optional attribute `xml:id` for referencing the text fragment and the CSS attributes `style` and `class` to associate presentation information with it (see the discussion in Sections 10.2 and 19.1). The `type` attribute can be used to specify the (linguistic or mathematical) type of the phrase, currently OMDOC does not make any restrictions on the values of this attribute. Furthermore, the `phrase` element allows the attribute `index` for parallel multilingual markup: Recall that sibling `CMP` elements form multilingual groups of text fragments. We can use the `phrase` element to make the correspondence relation on text fragments more fine-grained: `phrase` elements in sibling `CMP`s that have the same `index` value are considered to be equivalent. Of course, the value of an `index` has to be unique in the dominating `CMP` element (but not beyond). Thus the `index` attributes simplify manipulation of multilingual texts, see Listing 14.10 for an example at the discourse level.

Finally, the `phrase` element can carry a `verbalizes` attribute whose value is a whitespace-separated list of URI references that act as pointers to other

OMDOC elements. This has two applications: the first is another kind of parallel markup where we can state that a phrase corresponds to (and thus "verbalizes") a part of formula in a sibling FMP element.

Listing 14.6. Parallel markup between formal and informal

```
   <CMP>
     If <phrase verbalizes="#isaG"><G, o> is a group</phrase>, then of course
        <phrase verbalizes="#isaM">it is a monoid</phrase> by construction.
   </CMP>
 5 <FMP>
     <OMOBJ>
       <OMA><OMS cd="logic1" name="implies"/>
        <OMA id="isaG"><OMS cd="algebra" name="group"/>
          <OMA id="GG"><OMS cd="set" name="pair">
10          <OMV name="G"/><OMV name="op"/>
          </OMA>
        </OMA>
        <OMA xml:id="isaM"><OMS cd="algebra" name="monoid"/>
          <OMR href="GG"/>
15      </OMA>
       </OMA>
     </OMOBJ>
   </FMP>
```

Another important application of the **verbalizes** is the case of inline mathematical statements, which we will discuss in Section 15.5.

14.5 Technical Terms

In OMDOC we can give the notion of a **technical term** a very precise meaning: it is a phrase representing a concept for which a declaration exists in a content dictionary (see Subsection 15.2.1). In this respect it is the natural language equivalent for an OPENMATH symbol or a Content-MATHML token[5]. Let us consider an example: We can equivalently say "$0 \in \mathbb{N}$" and "the number zero is a natural number". The first rendering in a formula, we would cast as the following OPENMATH object:

```
<OMOBJ>
  <OMA><OMS cd="set1" name="in"/>
    <OMS cd="nat" name="zero"/>
    <OMS cd="nat" name="Nats"/>
  </OMA>
</OMOBJ>
```

with the effect that the components of the formula are disambiguated by pointing to the respective content dictionaries. Moreover, this information can be used by added-value services e.g. to cross-link the symbol presentations in the formula to their definition (see Chapter 25), or to detect logical dependencies. To allow this for mathematical vernacular as well, we provide the **term** element: in our example we might use the following markup.

---

[5] and is subject to the same visibility and scoping conditions as those; see Section 15.6 for details

---

...`<term cd="nat" name="zero">`the number zero`</term>` is an
`<term cd="nat" name="Nats">`natural number`</term>`...

The **term** element has two required attributes: `cd` and `name`, which together determine the meaning of the phrase just like they do for `om:OMS` elements (see the discussion in Section 13.1 and Subsection 15.6.2). The **term** element also allows the attribute `xml:id` for identification of the phrase occurrence, the CSS attributes for styling and the optional `role` attribute that allows to specify the role the respective phrase plays. We reserve the value `definiens` for the defining occurrence of a phrase in a definition. This will in general mark the exact point to point to when presenting other occurrences of the same[6] phrase. Other attribute values for the `role` are possible, OMDOC does not fix them at the current time. Consider for instance the following text fragment from Figure 4.1 in Chapter 4.

> DEFINITION 1. *Let E be a set. A mapping of $E \times E$ is called a* **law of composition** *on E. The value $f(x, y)$ of f for an ordered pair $(x, y) \in E \times E$ is called the* **composition of** x *and* y *under this law. A set with a law of composition is called a magma.*

Here the first boldface term is the definiens for a "law of composition", the second one for the result of applying this to two arguments. It seems that this is not a totally different concept that is defined here, but is derived systematically from the concept of a "law of composition" defined before. Pending a thorough linguistic investigation we will mark up such occurrences with `definiens-applied`, for instance in

Listing 14.7. Marking up the technical terms

Let E be a set. A mapping of $E \times E$ is called a `<term cd="magmas"`
`name="law_of_comp" role="definiens">`law of composition`</term>` on E. The value
$f(x, y)$ of f for an ordered pair $(x, y) \in E \times E$ is called the `<term`
`cd="magmas"name="law_of_comp" role="definiens-applied">`composition of`</term>` x and
5   y under this law.

There are probably more such systematic correlations; we leave their categorization and modeling in OMDOC to the future.

14.6 Rich Text Structure (Module RT)

The infrastructure for mathematical vernacular introduced above assumed the `CMP` elements as atomic fragments of mathematical vernacular allowing for very little discourse-level structure below the level of `CMP`. This would be sufficient, if the `CMP` were only used for text, but as we have seen above, the `CMP` element is also used for mathematical text fragments that correspond to mathematical statements like definitions or theorems, which might have internal text structure and therefore required corresponding structural elements in OMDOC.

---

[6] We understand this to mean with the same `cd` and `name` attributes.

Element	Optional Attributes	DC	Content
p	xml:id, style, class, index, verbalizes	+	⟨⟨math vernacular⟩⟩
ol	xml:id, style, class, index, verbalizes	+	li*
ul	xml:id, style, class, index, verbalizes	+	li*
li	xml:id, style, class, index, verbalizes	+	⟨⟨math vernacular⟩⟩
dl	xml:id, style, class, index, verbalizes	+	di*
di	xml:id, style, class, index, verbalizes	+	dt*,dd*
dt	xml:id, style, class, index, verbalizes	+	⟨⟨math vernacular⟩⟩
dd	xml:id, style, class, index, verbalizes	+	⟨⟨math vernacular⟩⟩
idx	(xml:id|xref)	−	idt?, ide+
ide	index, sort-by, see, seealso, links	−	idp*
idt	style, class	−	⟨⟨math vernacular⟩⟩
idp	sort-by, see, seealso, links	−	⟨⟨math vernacular⟩⟩
table	xml:id, style, class, index, verbalizes	+	tr*
tr	xml:id, style, class, index, verbalizes	+	td*
td	xml:id, style, class, index, verbalizes	+	⟨⟨math vernacular⟩⟩
th	xml:id, style, class, index, verbalizes	+	⟨⟨math vernacular⟩⟩
link	xml:id, style, class, index, verbalizes	−	⟨⟨math vernacular⟩⟩
note	type, xml:id, style, class, index, verbalizes	+	⟨⟨math vernacular⟩⟩

Fig. 14.8. Rich text format OMDOC

In this section we will discuss the OMDOC rich text structure module RT, which introduces text structuring elements for mathematical text below the level of mathematical statements. The elements in this module are loosely patterned after elements from the XHTML specification [Gro00], and can occur as part of mathematical vernacular. Where we do not explicitly discuss the content, it is mathematical vernacular as well. The module RT provides five classes of elements, which we will show in context in Listing 14.9.

Listing 14.9. An example of rich text structure

```
   <CMP>
     <p style="color:red" xml:id="p1">All <idx><idt>animals are dangerous</idt>
       <idp>dangerous</idp><idp seealso="creature">animal</idp></idx>!
       (which is a highly <phrase class="emphasis">unfounded</phrase>
 5     statement). In reality only some animals are, for instance:</p>
     <ul xml:id="l1">
       <li>sharks (they bite) and </li>
       <li>bees (they sting).</li>
     </ul>
10   <p>If we measure danger by the number of deaths, we obtain</p>
     <table>
       <tr>         <th>Culprits</th> <th>Deaths</th> <th>Action</th></tr>
       <tr>         <td>sharks</td> <td>312</td> <td>bite</td></tr>
       <tr xml:id="bn"> <td>bees</td> <td>23</td> <td>sting</td></tr>
15   <tr>         <td>cars</td>  <td>7500</td> <td>crash</td></tr>
     </table>
     <p>So, if we do the numbers <note xml:id="n1" type="ednote">check the
     numbers again</note> we see that animals are dangerous, but they are
     less so than cars but much more photogenic as we can see
20   <link href="http://www.yellowpress.com/killerbee.jpg">here</link>.</p>

     <note type="footnote">From the International Journal of Bee−keeping; numbers only
     available for 2002.</note>
   </CMP>
```

Paragraphs <u>p</u> elements can be used as children in a CMP to divide the text into paragraphs.

Ordered Lists The `ol` element is a constructor for ordered lists, which has `li` elements as children that represent the items. These contain mathematical vernacular as content and are presented as consecutively numbered.

Unordered Lists `ul` is the constructor for unordered or bulleted lists, the in the presentation, list items are indicated by some sort of bullet.

Description Lists Finally, `dl` is a constructor for description lists, which have `di` elements as children. The `di` elements contain an optional `dt` element (description title) followed by a (possibly empty) list of `dd` elements that contain the descriptions.

Tables To mark up simple tables we use the `table` element. Just as in XHTML, it has an arbitrary number of `tr` (table row) elements that contain `td` (table data) and `th` (table header) elements, which contain mathematical vernacular. Note that OMDoc does not support advanced formatting attributes of XHTML, but as tables are mathematical text in the module RT it does support nested tables.

Hyperlinks The `link` element is equivalent to the XHTML `a` element, and carries a required `href`[7] attribute that points to an arbitrary resource in form of a URI reference.

Index Markup The `idx` element is used for index markup in OMDoc. It contains an optional `idt` element that contains the index text, i.e. the phrase that is indexed. The remaining content of the index element specifies what is entered into various indexes. For every index this phrase is registered to there is one `ide` element (index entry); the respective entry is specified by name in its `index` attribute. The `ide` element contains a sequence of index phrases given in `idp` elements. The content of an `idp` element is regular mathematical text. Since index entries are usually sorted, (and mathematical text is difficult to sort), they carry an attribute `sort-by` whose value (a sequence of Unicode characters) can be sorted lexically [DW05]. Moreover, each `idp` and `ide` element carries the attributes `see`, `seealso`, and `links`, that allow to specify extra information on these. The values of the first ones are references to `idx` elements, while the value of the `links` attribute is a whitespace-separated list of (external) URI references. The formatting of the index text is governed by the attributes `style` and `class` on the `idt` element. The `idx` element can carry either an `xml:id` attribute (if this is the defining occurrence of the index text) or an `xref` attribute. In the latter case, all the `ide` elements from the defining `idx` (the one that has the `xml:id` attribute) are imported into the referring `idx` element (the one that has the `xref` attribute).

Notes The `note` element is the closest approximation to a footnote or endnote, where the kind of note is determined by the `type` attribute. OM-

---

[7] It is anticipated that future versions of OMDoc may use simple links from xlink [DMOT01] for such cross-referencing tasks, but at the moment we keep in style to the rest of the specification.

DOC supplies `footnote` as a default value, but does not restrict the range of values. Its `for` attribute allows it to be attached to other OMDOC elements externally where it is not allowed by the OMDOC document type. In our example, we have attached a footnote by reference to a table row, which does not allow `note` children.

All elements in the RT module carry an optional `xml:id` attribute for identification and an `index` attribute for parallel multilingual markup (e.g. Section 14.4 for an explanation and Listing 14.10 for a translation example).

Listing 14.10. Multilingual parallel markup

```
   <omtext xml:id="animals.overview">
     <CMP>
       <p index="intro">Consider the following animals:</p>
       <ul index="animals">
 5       <li index="first">a tiger,</li>
         <li index="second">a dog.</li>
       </ul>
     </CMP>
     <CMP xml:lang="de">
10     <p index="intro">Betrachte die folgenden Tiere:</p>
       <ul index="animals">
         <li index="first">Ein Tiger</li>
         <li index="second">Ein Hund</li>
       </ul>
15   </CMP>
   </omtext>
```

Mathematical Statements (Module ST)

In this chapter we will look at the OMDOC infrastructure to mark up the *functional structure* of mathematical statements and their interaction with a broader mathematical context.

15.1 Types of Statements in Mathematics

In the last chapter we introduced mathematical statements as special text fragments that state properties of the mathematical objects under discussion and categorized them as definitions, theorems, proofs,.... A set of statements about a related set of objects make up the context that is needed to understand other statements. For instance, to understand a particular theorem about finite groups, we need to understand the definition of a group, its properties, and some basic facts about finite groups first. Thus statements interact with context in two ways: the context is built up from (clusters of) statements, and statements only make sense with reference to a context. Of course this dual interaction of statements with *context*[1] applies to any text and to communication in general. In mathematics, where the problem is aggravated by the load of notation and the need for precision for the communicated concepts and objects, contexts are often discussed under the label of **mathematical theories**. We will distinguish two classes of statements with respect to their interaction with theories: We view axioms and definitions as *constitutive* for a given theory, since changing this information will yield a different theory (with different mathematical properties, see the discussion in Section 2.2). Other mathematical statements like theorems or the proofs that support them are not constitutive, since they only illustrate the mathematical objects in the theory by explicitly stating the properties that are implicitly determined by the constitutive statements.

---

[1] In linguistics and the philosophy of language this phenomenon is studied under the heading of "discourse theories", see e.g. [KR93] for a start and references.

To support this notion of context OMDoc supports an infrastructure for theories using special **theory** elements, which we will introduce in Section 15.6 and extend in Chapter 18. Theory-constitutive elements must be contained as children in a **theory** element; we will discuss them in Section 15.2, non-constitutive statements will be defined in Section 15.3. They are allowed to occur outside a **theory** element in OMDoc documents (e.g. as top-level elements), however, if they do they must reference a theory, which we will call their **home theory** in a special **theory** attribute. This situates them into the context provided by this theory and gives them access to all its knowledge. The home theory of theory-constitutive statements is given by the theory they are contained in.

The division of statements into constitutive and non-constitutive ones and the encapsulation of constitutive elements in **theory** elements add a certain measure of safety to the knowledge management aspect of OMDoc. Since XML elements cannot straddle document borders, all constitutive parts of a theory must be contained in a single document; no constitutive elements can be added later (by other authors), since this would change the meaning of the theory on which other documents may depend on.

Before we introduce the OMDoc elements for theory-constitutive statements, let us fortify our intuition by considering some mathematical examples. *Axioms* are assertions about (sets of) mathematical objects and concepts that are assumed to be true. There are many forms of axiomatic restrictions of meaning in mathematics. Maybe the best-known are the five Peano Axioms for natural numbers.

1. 0 is a natural number.
2. The successor $s(n)$ of a natural number n is a natural number.
3. 0 is not a successor of any natural number.
4. The successor function is one-one (i.e. injective).
5. The set $\mathbb{N}$ of natural numbers contains only elements that can be constructed by axioms 1. and 2.

Fig. 15.1. The Peano axioms

The Peano axioms in Figure 15.1 (implicitly) introduce three symbols: the number 0, the successor function s, and the set $\mathbb{N}$ of natural numbers. The five axioms in Figure 15.1 jointly constrain their meaning such that conforming structures exist (the natural numbers we all know and love) any two structures that interpret 0, s, and $\mathbb{N}$ and satisfy these axioms must be isomorphic. This is an ideal situation — the axioms are neither too lax (they allow too many mathematical structures) or too strict (there are no mathematical structures) — which is difficult to obtain. The latter condition (**inconsistent theories**) is especially unsatisfactory, since any statement is a theorem in such theories. As consistency can easily be lost by adding axioms, mathematicians try to keep axiom systems minimal and only add axioms that are safe.

Sometimes, we can determine that an axiom does not destroy consistency of a theory $\mathcal{T}$ by just looking at its form: for instance, axioms of the form $s = \mathbf{A}$, where s is a symbol that does not occur in $\mathcal{T}$ and $\mathbf{A}$ is a formula containing only symbols from $\mathcal{T}$ will introduce no constraints on the meaning of $\mathcal{T}$-symbols. The axiom $s = \mathbf{A}$ only constrains the meaning of the **new symbol** to be a unique object: the one denoted by $\mathbf{A}$. We speak of a **conservative extension** in this case. So, if $\mathcal{T}$ was a consistent theory, the extension of $\mathcal{T}$ with the symbol s and the axiom $s = \mathbf{A}$ must be one too. Thus axioms that result in conservative extensions can be added safely — i.e. without endangering consistency — to theories.

Generally an axiom $\mathcal{A}$ that results in a conservative extension is called a **definition** and any new symbol it introduces a **definiendum** (usually marked e.g. in boldface font in mathematical texts), and we call **definiens** the material in the definition that determines the meaning of the definiendum. We say that a definiendum is **well-defined**, iff the corresponding definiens uniquely determines it; adding such definitions to a theory always results in a conservative extension.

Definiendum	Definiens	Type
The number 1	$1 := s(0)$ (1 is the successor of 0)	simple
The exponential function $e^{\cdot}$	The **exponential function** $e^{\cdot}$ is the solution to the differential equation $\partial f = f$ [where $f(0) = 1$].	implicit
The addition function $+$	**Addition** on the natural numbers is defined by the equations $x + 0 = x$ and $x + s(y) = s(x + y)$.	recursive

Fig. 15.2. Some common definitions

Definitions can have many forms, they can be

- equations where the left hand side is the defined symbol and the right hand side is a term that does not contain it, as in our discussion above or the first case in Figure 15.2. We call such definitions **simple**.
- general statements that uniquely determine the meaning of the objects or concepts in question, as in the second definition in Figure 15.2. We call such definitions **implicit**; the Peano axioms are another example of this category.

  Note that this kind of definitions requires a proof of unique existence to ensure well-definedness. Incidentally, if we leave out the part in square brackets in the second definition in Figure 15.2, the differential equation only characterizes the exponential function up to additive real constants. In this case, the "definition" only restricts the meaning of the exponential

function to a set of possible values. We call such a set of axioms a **loose** definition.
– given as a set of equations, as in the third case of Figure 15.2, even though this is strictly a special case of an implicit definition: it is a sub-case, where well-definedness can be shown by giving an argument why the systematic applications of these equations terminates, e.g. by exhibiting an ordering that makes the left hand sides strictly smaller than the right-hand sides. We call such a definition **inductive**.

15.2   Theory-Constitutive Statements in OMDoc

The OMDoc format provides an infrastructure for four kinds of theory-constitutive statements: symbol declarations, type declarations, (proper) axioms, and definitions. We will take a look at all of them now.

Element	Attributes		D	Content
	Required	Optional	C	
symbol	name	xml:id, role, scope, style, class	+	type*
type		xml:id, system, style, class	−	CMP*, ⟪mobj⟫
axiom		xml:id, for, type, style, class	+	CMP*,FMP*
definition	for	xml:id, type, style, class, uniqueness, existence, consistency, exhaustivity	+	CMP*, (FMP* | requation+ | ⟪mobj⟫)?, measure?, ordering?
requation		xml:id, style, class	−	⟪mobj⟫, ⟪mobj⟫
measure		xml:id, style, class	−	⟪mobj⟫
ordering		xml:id, style, class	−	⟪mobj⟫
where ⟪mobj⟫ is (OMOBJ |m:math |legacy)				

Fig. 15.3. Theory-constitutive elements in OMDoc

15.2.1   Symbol Declarations

The `symbol` element declares a symbol for a mathematical concept, such as 1 for the natural number "one", + for addition, = for equality, or `group` for the property of being a group. Note that we not only use the `symbol` element for mathematical objects that are usually written with mathematical symbols, but also for any concept or object that has a definition or is restricted in its meaning by axioms.

We will refer to the mathematical object declared by a `symbol` element as a "symbol", iff it is usually communicated by specialized notation in mathematical practice, and as a "concept" otherwise. The name "symbol" of the `symbol` element in OMDoc is in accordance with usage in the philosophical literature (see e.g. [NS81]): A **symbol** is a *mental or physical* representation of a concept. In particular, a symbol may, but need not be representable by a (set of) glyphs (symbolic notation). The definiendum objects in Figure 15.2 would be considered as "symbols" while the concept of a "group" in mathematics would be called a "concept".

The symbol element has a required attribute name whose value uniquely identifies it in a theory. Since the value of this attribute will be used as an OPENMATH symbol name, it must be an XML name[2] as defined in XML 1.1 [BPSM+04]. The optional attribute scope takes the values global and local, and allows a simple specification of visibility conditions: if the scope attribute of a symbol has value local, then it is not exported outside the theory. Finally, the optional attribute role that can take the values[3]

binder The symbol may appear as a binding symbol of an binding object, i.e. as the first child of an om:OMBIND object, or as the first child of an m:apply element that has an m:bvar as a second child.

attribution The symbol may be used as key in an OPENMATH om:OMATTR element, i.e. as the first element of a key-value pair, or in an equivalent context (for example to refer to the value of an attribution). This form of attribution may be ignored by an application, so should be used for information which does not change the meaning of the attributed OPENMATH object.

semantic-attribution This is the same as attribution except that it modifies the meaning of the attributed OPENMATH object and thus cannot be ignored by an application.

error The symbol can only appear as the first child of an OPENMATH error object.

application The symbol may appear as the first child of an application object.

constant The symbol cannot be used to construct a compound object.

type The symbol denotes a sets that is used in a type systems to annotate mathematical objects.

sort The symbol is used for a set that are inductively built up from constructor symbols; see Chapter 16.

If the role is not present, the value object is assumed.

The children of the symbol element consist of a multi-system group of type elements (see Subsection 15.2.3 for a discussion). For this group the order does not matter. In Listing 15.4 we have a symbol declaration for the concept of a monoid. Keywords or simple phrases that describes the symbol

---

[2] This limits the characters allowed in a name to a subset of the characters in Unicode 2.0; e.g. the colon : is not allowed. Note that this is not a problem, since the name is just used for identification, and does not necessarily specify how a symbol is presented to the human reader. For that, OMDoc provides the notation definition infrastructure presented in Chapter 19.

[3] The first six values come from the OPENMATH2 standard. They are specified in content dictionaries; therefore OMDoc also supplies them.

in mathematical vernacular can be added in the `metadata` child of `symbol` as `dc:subject` and `dc:descriptions`; the latter have the same content model as the CMP elements, see the discussion in Section 14.1). If the document containing their parent `symbol` element were stored in a data base system, it could be looked up via these metadata. As a consequence the symbol `name` need only be used for identification. In particular, it need not be mnemonic, though it can be, and it need not be language-dependent, since this can be done by suitable `dc:subject` elements.

Listing 15.4. An OMDoc `symbol` declaration

```
<symbol name="monoid">
  <metadata>
    <dc:subject xml:lang="en">monoid</dc:subject>
    <dc:subject xml:lang="de">Monoid</dc:subject>
5   <dc:subject xml:lang="it">monoide</dc:subject>
  </metadata>
  <type system="simply−typed">set[any] → (any → any → any) → any → bool</type>
  <type system="props">
    <OMOBJ><OMS cd="arities" name="ternary−relation"/></OMOBJ>
10  </type>
</symbol>
```

15.2.2 Axioms

The relation between the components of a monoid would typically be specified by a set of axioms (e.g. stating that the base set is closed under the operation). For this purpose OMDoc uses the <u>axiom</u> element, which allows as children a multilingual group of CMPs, which express the mathematical content of the axiom and a multi-logic FMP group that expresses this as a logical formula. `axiom` elements may have a `generated-from` attribute, which points to another OMDoc element (e.g. an `adt`, see Chapter 16) which subsumes it, since it is a more succinct representation of the same mathematical content. Finally the `axiom` element has an optional `for` attribute to specify salient semantic objects it uses as a whitespace-separated list of URI references to symbols declared in the same theory, see Listing 15.5 for an example. Finally, the `axiom` element can have an `type` attribute, whose values we leave unspecified for the moment.

Listing 15.5. An OMDoc `axiom`

```
<axiom xml:id="mon.ax" for="#monoid">
  <CMP>If (M, ∗) is a semigroup with unit e, then (M, ∗, e) is a monoid.</CMP>
</axiom>
```

15.2.3 Type Declarations

Types (also called sorts in some contexts) are representations of certain simple sets that are treated specially in (human or mechanical) reasoning processes. A **type declaration** $e\!:\!t$ makes the information that a symbol or expression e is in a set represented by a type t available to a specified mathematical process. For instance, we might know that 7 is a natural number, or that expressions of the form $\sum_{i=1}^{n} a_i x^i$ are polynomials, if the a_i are real numbers, and exploit this information in mathematical processes like proving, pattern matching, or while choosing intuitive notations. If a type is declared for an expression that is not a symbol, we will speak of a **term declaration**.

OMDOC uses the `type` element for type declarations. The optional attribute `system` contains a URI reference that identifies the type system which interprets the content. There may be various sources of the set membership information conveyed by a type declaration, to justify it this source may be specified in the optional `just-by` attribute. The value of this attribute is a URI reference that points to an `assertion` or `axiom` element that asserts $\forall x_1, \ldots, x_n . e \in t$ for a type declaration $e : t$ with variables $x_1, \ldots, x_n$. If the `just-by` attribute is not present, then the type declaration is considered to be generated by an implicit axiom, which is considered theory-constitutive[4].

The `type` element contains one or two mathematical objects. In the first case, it represents a type declaration for a symbol (we call this a **symbol declaration**), which can be specified in the optional `for` attribute or by embedding the `type` element into the respective `symbol` element. A `type` element with two mathematical objects represents a term declaration $e : t$, where the first object represents the expression e and the second one the type t (see Listing 15.14 for an example). There the type declaration of `monoid` characterizes a monoid as a three-place predicate (taking as arguments the base set, the operation, and a neutral element).

As reasoning processes vary, information pertaining to multiple type systems may be associated with a single symbol and there can be more than one `type` declaration per expression and type system, this just means that the object has more than one type in the respective type system (not all type systems admit principal types).

15.2.4 Definitions

Definitions are a special class axioms that completely fix the meaning of symbols. Therefore `definition` elements that represent definitions carry the required `for` attribute, which contain a whitespace-separated list of names of symbols in the same theory. We call these symbols **defined** and **primitive** otherwise. `definition` contain a multilingual CMP group to describe the meaning of the defined symbols.

In Figure 15.2 we have seen that there are many ways to fix the meaning of a symbol, therefore OMDOC `definition` elements are more complex than `axioms`. In particular, the `definition` element supports several kinds of definition mechanisms with specialized content models specified in the `type` attribute (cf. the discussion at the end of Section 15.1):

---

[4] It is considered good practice to make the axiom explicit in formal contexts, as this allows an extended automation of the knowledge management process.

simple In this case the **definition** contains a mathematical object that can be substituted for the symbol specified in the **for** attribute of the definition. Listing 15.6 gives an example of a simple definition of the number one from the successor function and zero. OMDOC treats the **type** attribute as an optional attribute. If it is not given explicitly, it defaults to **simple**.

Listing 15.6. A simple OMDOC **definition**.

```
  <symbol name="one"/>
  <definition xml:id="one.def" for="#one" type="simple">
   <CMP><OMOBJ><OMS cd="nat" name="one"/></OMOBJ> is the successor of
      <OMOBJ><OMS cd="nat" name="zero"/></OMOBJ>.</CMP>
5  <OMOBJ>
    <OMA>
     <OMS cd="int" name="suc"/>
     <OMS cd="nat" name="zero"/>
    </OMA>
10   </OMOBJ>
  </definition>
```

implicit This kind of definition is often (more accurately) called *"definition by description"*, since the definiendum is described so accurately, that there is exactly one object satisfying the description. The "description" of the defined symbol is given as a multi-system **FMP** group whose content uniquely determines the value of the symbols that are specified in the **for** attribute of the definition. The necessary statement of unique existence can be specified in the **existence** and **uniqueness** attribute, whose values are URI references to to assertional statements (see Subsection 15.3.4) that represent the respective properties. We give an example of an implicit definition in Listing 15.7.

Listing 15.7. An implicit definition of the exponential function

```
  <definition xml:id="exp−def" for="#exp" type="implicit"
          uniqueness="#exp−unique" existence="#exp−exists">
   <FMP>exp' = exp ∧ exp(0) = 1</FMP>
  </definition>
5  <assertion xml:id="exp−unique">
    <CMP>
     There is at most one differentiable function that solves the
     differential equation in definition <ref type="cite" xref="#exp−def"/>.
    </CMP>
10   </assertion>
  <assertion xml:id="exp−exists">
    <CMP>
     The differential equation in <ref type="cite" xref="#exp−def"/> is solvable.
    </CMP>
15   </assertion>
```

inductive This is a variant of the **implicit** case above. It defines a **recursive function** by a set of recursive equations (in **requation** elements) whose left and right hand sides are specified by the two children. The first one is called the **pattern**, and the second one the **value**. The intended meaning of the defined symbol is, that the value (with the variables suitably substituted) can be substituted for a formula that matches the pattern element. In this case, the `definition` element can carry the optional attributes `exhaustivity` and `consistency`, which point to `assertions` stating that the cases spanned by the patterns are exhaustive (i.e. all cases are considered), or that the values are consistent (where the cases overlap, the values are equal).

Listing 15.8 gives an example of a a recursive definition of the addition on the natural numbers.

Listing 15.8. A recursive definition of addition

```
  <definition xml:id="plus.def" for="#plus" type="inductive"
            consistency="#s−not−0" exhaustivity="#s−or−0">
    <metadata><dc:subject>addition</dc:subject></metadata>
    <CMP>Addition is defined by recursion on the second argument.</CMP>
5   <requation>x + 0 ⤳ x</requation>
    <requation>x + s(y) ⤳ s(x + y)</requation>
  </definition>
```

To guarantee termination of the recursive instantiation (necessary to ensure well-definedness), it is possible to specify a measure function and well-founded ordering by the optional <u>measure</u> and <u>ordering</u> elements which contain mathematical objects. The elements contain mathematical objects. The content of the `measure` element specifies a measure function, i.e. a function from argument tuples for the function defined in the parent `definition` element to a space with an ordering relation which is specified in the `ordering` element. This element also carries an optional attribute `terminating` that points to an `assertion` element that states that this ordering relation is a terminating partial ordering.

pattern This is a special degenerate case of the recursive definition. A function is defined by a set of `requation` elements, but the defined function does not occur in the second children. This form of definition is often used instead of `simple` in logical languages that do not have a function constructor. It allows to define a function by its behavior on patterns of arguments. Since termination is trivial in this case, no `measure` and `ordering` elements appear in the body.

15.3   The Unassuming Rest

The bulk of mathematical knowledge is in form of statements that are not theory-constitutive: statements of properties of mathematical objects that are entailed by the theory-constitutive ones. As such, these statements are logically redundant, they do not add new information about the mathematical objects, but they do make their properties explicit. In practice, the entailment is confirmed e.g. by exhibiting a proof of the assertion; we will introduce the infrastructure for proofs in Chapter 17.

Element	Attributes		D	Content
	Required	Optional	C	
assertion		xml:id, for, type, theory, class, style, status, just-by	+	CMP*, FMP*
type	system	xml:id, for, just-by, theory, class, style	−	CMP*, $\langle\!\langle mobj\rangle\!\rangle$, $\langle\!\langle mobj\rangle\!\rangle$
example	for	xml:id, type, assertion, theory, class, style	+	CMP* | $\langle\!\langle mobj\rangle\!\rangle$*
alternative	for, theory, entailed-by, entails, entailed-by-thm, entails-thm	xml:id, type, theory, class, style	+	CMP*, (FMP* | requation* | $\langle\!\langle mobj\rangle\!\rangle$)
where $\langle\!\langle mobj\rangle\!\rangle$ is (OMOBJ |m:math |legacy)				

Fig. 15.9. Assertions, examples, and alternatives in OMDOC

15.3.1   Assertions

OMDOC uses the <u>assertion</u> element for all statements (proven or not) about mathematical objects (see Listing 15.11) that are not axiomatic (i.e. constitutive for the meaning of the concepts or symbols involved). Traditional mathematical documents discern various kinds of these: theorems, lemmata, corollaries, conjectures, problems, etc.

These all have the same structure (formally, a closed logical formula). Their differences are largely pragmatic (e.g. theorems are normally more important in some theory than lemmata) or proof-theoretic (conjectures become theorems once there is a proof). Therefore, we represent them in the general **assertion** element and leave the type distinction to a **type** attribute, which can have the values in Figure 15.10. The **assertion** element also takes an optional **xml:id** element that allows to reference it in a document, an optional **theory** attribute to specify the theory that provides the context for this assertion, and an optional attribute **generated-from**, that points to a higher syntactic construct that generates these assertions, e.g. an abstract data type declaration given by an **adt** element (see Chapter 16).

Listing 15.11. An OMDOC assertion about semigroups

```
<assertion xml:id="ida.c6s1p4.l1" type="lemma">
  <CMP> A semigroup has at most one unit.</CMP>
  <FMP>∀S.sgrp(S) → ∀x,y.unit(x,S) ∧ unit(y,S) → x = y</FMP>
</assertion>
```

To specify its proof-theoretic status of an assertion **assertion** carries the two optional attributes **status** and **just-by**. The first contains a keyword for the status and the second a whitespace-separated list of URI references to OMDOC elements that justify this status of the assertion. For the specification of the status we adapt an ontology for deductive states of assertion from [SZS04] (see Figure 15.12). Note that the states in Figure 15.12 are not mutually exclusive, but have the inclusions depicted in Figure 15.13.

Value	Explanation
theorem, proposition	an important assertion with a proof
Note that the meaning of the **type** (in this case the existence of a proof) is not enforced by OMDOC applications. It can be appropriate to give an assertion the **type theorem**, if the author knows of a proof (e.g. in the literature), but has not formalized it in OMDOC yet.	
lemma	a less important assertion with a proof
The difference of importance specified in this **type** is even softer than the other ones, since e.g. reusing a mathematical paper as a chapter in a larger monograph, may make it necessary to downgrade a theorem (e.g. the main theorem of the paper) and give it the status of a lemma in the overall work.	
corollary	a simple consequence
An assertion is sometimes marked as a corollary to some other statement, if the proof is considered simple. This is often the case for important theorems that are simple to get from technical lemmata.	
postulate, conjecture	an assertion without proof or counter-example
Conjectures are assertions, whose semantic value is not yet decided, but which the author considers likely to be true. In particular, there is no proof or counter-example (see Section 15.4).	
false-conjecture	an assertion with a counter-example
A conjecture that has proven to be false, i.e. it has a counter-example. Such assertions are often kept for illustration and historical purposes.	
obligation, assumption	an assertion on which the proof of another depends
These kinds of assertions are convenient during the exploration of a mathematical theory. They can be used and proven later (or assumed as an axiom).	
formula	if everything else fails
This type is the catch-all if none of the others applies.	

Fig. 15.10. Types of mathematical assertions

15.3.2 Type Assertions

In the last section, we have discussed the **type** elements in **symbol** declarations. These were axiomatic (and thus theory-constitutive) in character, declaring a symbol to be of a certain type, which makes this information available to type checkers that can check well-typedness (and thus plausibility) of the represented mathematical objects.

However, not all type information is axiomatic, it can also be deduced from other sources knowledge. We use the same **type** element we have discussed in Subsection 15.2.3 for such **type assertions**, i.e. non-constitutive statements that inform a type-checker. In this case, the **type** element can occur at top level, and even outside a **theory** element (in which case they have to specify their home theory in the **theory** attribute).

Listing 15.14 contains a type assertion $x + x$: $evens$, which makes the information that doubling an integer number results in an even number available to the reasoning process.

status	just-by points to
tautology *All T-interpretations satisfy A and some C_i*	Proof of $\mathcal{F}$
tautologous-conclusion *All T-interpretations satisfy some C_j*	Proof of $\mathcal{F}_c$.
equivalent *A and C have the same T-models (and there are some)*	Proofs of $\mathcal{F}$ and $\mathcal{F}^{-1}$
theorem *All T-models of A (and there are some) satisfy some C_i*	Proof of $\mathcal{F}$
satisfiable *Some T-models of A (and there are some) satisfy some C_i*	Model of A and some C_i
contradictory-axioms *There are no T-models of A*	Refutation of A
no-consequence *Some T-models of A (and there are some) satisfy some C_i, some satisfy $\overline{C}$*	T-model of A and some C_i, T-model of $A \cup \overline{C}$.
counter-satisfiable *Some T-models of A (and there are some) satisfy $\overline{C}$*	Model of $A \cup \overline{C}$
counter-theorem *All T-models of A (and there are some) satisfy $\overline{C}$*	Proof of $\overline{C}$ from A
counter-equivalent *A and $\overline{C}$ have the same T-models (and there are some)*	Proof of $\overline{C}$ from A and proof of A from $\overline{C}$
unsatisfiable-conclusion *All T-interpretations satisfy $\overline{C}$*	Proof of $\overline{C}$
unsatisfiable *All T-interpretations satisfy A and $\overline{C}$*	Proof of $\neg\mathcal{F}$

Where $\mathcal{F}$ is an assertion whose **FMP** has **assumption** elements $A_1, \ldots, A_n$ and **conclusion** elements $C_1, \ldots, C_m$. Furthermore, let $A := \{A_1, \ldots, A_n\}$ and $C := \{C_1, \ldots, C_m\}$, and $\mathcal{F}^{-1}$ be the sequent that has the C_i as assumptions and the A_i as conclusions. Finally, let $\overline{C} := \{\overline{C_1}, \ldots, \overline{C_m}\}$, where $\overline{C_i}$ is a negation of C_i.

Fig. 15.12. Proof status for assertions in a theory T

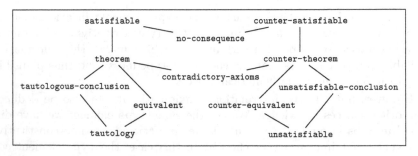

Fig. 15.13. Relations of assertion states

Listing 15.14. A term declaration in OMDOC

```
<type xml:id="double−even.td" system="#POST"
      theory="adv.int" for="#plus" just−by="#double−even">
   <m:math>
      <m:apply><m:plus/>
         <m:ci type="integer">X</m:ci>
         <m:ci type="integer">X</m:ci>
      </m:apply>
   </m:math>
   <m:math>
      <m:csymbol definitionURL="http://www.mathweb.org/cd/integers/evens"/>
   </m:math>
</type>

<assertion xml:id="double−even" type="lemma" theory="adv.int">
   <FMP>
      <m:math>
         <m:apply><m:forall/>
            <m:bvar><m:ci xml:id="x13" type="integer">X</m:ci></m:bvar>
            <m:apply><m:in/>
               <m:apply><m:plus/>
                  <m:ci definitionURL="x13" type="integer">X</m:ci>
                  <m:ci definitionURL="x13" type="integer">X</m:ci>
               </m:apply>
               <m:csymbol definitionURL="http://www.mathweb.org/cd/nat/evens" />
            </m:apply>
         </m:apply>
      </m:math>
   </FMP>
</assertion>
```

The body of a type assertion contains two mathematical objects, first the type of the object and the second one is the object that is asserted to have this type.

15.3.3 Alternative Definitions

In contrast to what we have said about conservative extensions at the end of Subsection 15.2.4, mathematical documents often contain multiple definitions for a concept or mathematical object. However, if they do, they also contain a careful analysis of equivalence among them. OMDOC allows us to model this by providing the **alternative** element. Conceptually, an alternative definition or axiom is just a group of assertions that specify the equivalence of logical formulae. Of course, alternatives can only be added in a consistent way to a body of mathematical knowledge, if it is guaranteed that it is equivalent to the existing ones. The **for** on the **alternative** points to the primary definition or assertion. Therefore, **alternative** has the attributes **entails** and **entailed-by**, that specify **assertions** that state the necessary entailments. It is an integrity condition of OMDOC that any **alternative** element references at least one **definition** or **alternative** element that entails it and one that it is entailed by (more can be given for convenience). The **entails-thm**, and **entailed-by-thm** attributes specify the corresponding assertions. This way we can always reconstruct equivalence of all definitions for a given symbol. As alternative definitions are not theory-constitutive, they can appear outside a `theory` element as long as they have a `theory` attribute.

15.3.4 Assertional Statements

There is another distinction for statements that we will need in the following. Some kinds of mathematical statements add information about the mathematical objects in question, whereas other statements do not. For instance, a symbol declaration only declares an unambiguous name for an object. We will call the following OMDOC elements **assertional**: `axiom` (it asserts central properties about an object), `type` (it asserts type properties about an object), `definition` (this asserts properties of a new object), and of course `assertion`.

The following elements are considered non-assertional: `symbol` (only a name is declared for an object), `alternative` (here the assertional content is carried by the `assertion` elements referenced in the structure-carrying attributes of `alternative`). For the elements introduced below we will discuss whether they are assertional or not in their context. In a nutshell, only statements introduced by the module ADT (see Chapter 16) will be assertional.

15.4 Mathematical Examples in OMDoc

In mathematical practice examples play a great role, e.g. in concept formation as witnesses for definitions or as either supporting evidence, or as counter-examples for conjectures. Therefore examples are given status as primary objects in OMDOC. Conceptually, we model an example $\mathcal{E}$ as a pair $(\mathcal{W}, \mathbf{A})$, where $\mathcal{W} = (\mathcal{W}_1, \ldots, \mathcal{W}_n)$ is an n-tuple of mathematical objects and $\mathbf{A}$ is an assertion. If $\mathcal{E}$ is an example for a mathematical concept given as an OMDOC symbol $\mathbf{S}$, then $\mathbf{A}$ must be of the form $\mathbf{S}(\mathcal{W}_1, \ldots, \mathcal{W}_n)$.

If $\mathcal{E}$ is an example for a conjecture $\mathbf{C}$, then we have to consider the situation more carefully. We assume that $\mathbf{C}$ is of the form $\mathcal{Q}\mathbf{D}$ for some formula $\mathbf{D}$, where $\mathcal{Q}$ is a sequence $\mathcal{Q}_1 W_1, \ldots, \mathcal{Q}_m W_m$ of $m \geq n = \#\mathcal{W}$ quantifications of using quantifiers $\mathcal{Q}_i$ like $\forall$ or $\exists$. Now let $\mathcal{Q}'$ be a sub-sequence of $m - n$ quantifiers of $\mathcal{Q}$ and $\mathbf{D}'$ be $\mathbf{D}$ only that all the W_{i_j} such that the $\mathcal{Q}_{i_j}$ are absent from $\mathcal{Q}'$ have been replaced by $\mathcal{W}_j$ for $1 \leq j \leq n$. If $\mathcal{E} = (\mathcal{W}, \mathbf{A})$ supports $\mathbf{C}$, then $\mathbf{A} = \mathcal{Q}'\mathbf{D}'$ and if $\mathcal{E}$ is a counter-example for $\mathbf{C}$, then $\mathbf{A} = \neg \mathcal{Q}'\mathbf{D}'$.

OMDOC specifies this intuition in an <u>example</u> element that contains a multilingual `CMP` group for the description and n mathematical objects (the witnesses). It has the attributes

for specifying for which concepts or assertions it is an example. This is a reference to a whitespace-separated list of URI references to `symbol`, `definition`, or `assertion` elements.

type specifying the aspect, the value is one of `for` or `against`

assertion a reference to the assertion **A** mentioned above that formally states that the witnesses really form an example for the concept of assertion. In many cases even the statement of this is non-trivial and may require a proof.

example elements are considered non-assertional in OMDoc, since the assertional part is carried by the **assertion** element referenced in the **assertion** attribute.

Note that the list of mathematical objects in an **example** element does not represent multiple examples, but corresponds to the argument list of the symbol, they exemplify. In the example below, the symbol for monoid is a three-place relation (see the type declaration in Listing 15.4), so we have three witnesses.

Listing 15.15. An OMDoc ¿representation of a mathematical example

```
   <symbol name="strings−over"/>
   <definition xml:id="strings.def" for ="#strings−over">... A* ...</definition>
   <symbol name="concat"/>
   <definition xml:id="concat.def" for="#concat">... :: ...</definition>
5  <symbol name="empty−string"/>
   <definition xml:id="empty−string.def" for="#empty−string">... ε ...</definition>
   ...
   <assertion xml:id="string.struct.monoid" type="lemma">
     <CMP>(A*, ::, ε) is a monoid.</CMP>
10   <FMP>mon(A*, ::, ε)</FMP>
   </assertion>
   ...
   <example xml:id="mon.ex1" for="#monoid" type="for"
        assertion="string.struct.monoid">
15   <CMP>The set of strings with concatenation is a monoid.</CMP>
     <OMOBJ>
       <OMA id="nat−strings">
         <OMS cd="strings" name="strings"/>
         <OMS cd="setname1" name="N" />
20     </OMA>
     </OMOBJ>
     <OMOBJ><OMS cd="strings" name="concat"/></OMOBJ>
     <OMOBJ><OMS cd="strings" name="empty−string"/></OMOBJ>
   </example>
25
   <assertion xml:id="monoid.are.groups" type="false−conjecture">
   <CMP>Monoids are groups.</CMP>
   <FMP>∀S, o, e.mon(S, o, e) → ∃i.group(S, o, e, i)</FMP>
   </assertion>
30
   <example xml:id="mon.ex2" for="#monoids.are.groups" type="against"
        assertion="strings.isnt .group">
   <CMP>The set of strings with concatenation is not a group.</CMP>
     <OMOBJ><OMR href="#nat−strings"/></OMOBJ>
35   <OMOBJ><OMS cd="strings" name="strings"/></OMOBJ>
     <OMOBJ><OMS cd="strings" name="concat"/></OMOBJ>
     <OMOBJ><OMS cd="strings" name="empty−string"/></OMOBJ>
   </example>
40 <assertion xml:id="strings.isnt .group" type="theorem">
     <CMP>(A*, ::, ε) is a monoid, but there is no inverse function for it.</CMP>
   </assertion>
```

In Listing 15.15 we show an example of the usage of an **example** element in OMDoc: We declare constructor symbols **strings-over**, that takes an alphabet A as an argument and returns the set A^* of stringss over A, **concat** for strings concatenation (which we will denote by ::), and **empty-string** for the empty string ϵ. Then we state that $W = (A^*, ::, \epsilon)$ is a monoid in an **assertion** with **xml:id="string.struct.monoid"**. The **example** element with **xml:id="mon.ex1"** in Listing 15.15 is an example for the concept of a monoid, since it encodes the pair $(W, \mathbf{A})$ where $\mathbf{A}$ is given by reference to the assertion **string.struct.monoid** in the **assertion** attribute. Example **mon.ex2** uses the pair $(W, \mathbf{A}')$ as a counter-example to the false conjecture **monoids.are.groups** using the assertion **strings.isnt.group** for $\mathbf{A}'$.

15.5   Inline Statements

Note that the infrastructure for statements introduced so far does its best to mark up the interplay of formal and informal elements in mathematical documents, and make explicit the influence of the context and their contribution to it. However, not all statements in mathematical documents can be adequately captured directly. Consider for instance the following situation, which we might find in a typical mathematical textbook.

> **Theorem 3.12**: *In a monoid M the left unit and the right unit coincide, we call it the* **unit** *of M.*

The overt role of this text fragment is that of a mathematical theorem — as indicated by the cue word "**Theorem**", therefore we would be tempted represent it as an **omtext** element with the value **theorem** for the **type** attribute. But the relative clause is clearly a definition (the definiens is even marked in boldface). What we have here is an aggregated verbalization of two mathematical statements. In a simple case like this one, we could represent this as follows:

Listing 15.16. A simple-minded representation of **Theorem 3.12**

```
<assertion type="theorem" style="display=flow">
  <CMP>In a monoid M, the left unit and the right unit coincide,</CMP>
</assertion>
<definition for="#unit" style="display:flow">
  <CMP>we call it the <term role="definiens" name="unit">unit</term> of M</CMP>
</definition>
```

But this representation remains unsatisfactory: the definition is not part of the theorem, which would really make a difference if the theorem continued after the inline definition. The real problem is that the inline definition is linguistically a phrase-level construct, while the **omtext** element is a discourse-level construct. However, as a phrase-level construct, the inline definition cannot really be taken as stand-alone, but only makes sense in the context it is presented in (which is the beauty of it; the re-use of context). With the **phrase** element and its **verbalizes**, we can do the following:

Listing 15.17. an inline definition

```
<assertion xml:id='unit−unique' type="theorem" >
  <CMP>In a monoid M, the left unit and the right unit coincide,
    <phrase verbalizes="#unit−def">we call it the unit of M</phrase>.</CMP>
</assertion>
<symbol name="unit"/>
<definition xml:id="unit−def" for="#unit" just−by='#unit−unique'>
  <CMP>We call the (unique) element of a monoid M that acts as a left
    and right unit the <term role="definiens" name="unit">unit</term> of M.</CMP>
</definition>
```

thus we would have the phrase-level markup in the proper place, and we would have an explicit version of the definition which is standalone[5], and we would have the explicit relation that states that the inline definition is an "abbreviation" of the standalone definition.

15.6   Theories as Structured Contexts

OMDOC provides an infrastructure for mathematical theories as first-class objects that can be used to structure larger bodies of mathematics by functional aspects, to serve as a framework for semantically referencing mathematical objects, and to make parts of mathematical developments reusable in multiple contexts. The module ST presented in this chapter introduces a part of this infrastructure, which can already address the first two concerns. For the latter, we need the machinery for complex theories introduced in Chapter 18.

Theories are specified by the **theory** element in OMDOC, which has a required **xml:id** attribute for referencing the theory. Furthermore, the **theory** element can have the **cdbase** attribute that allows to specify the **cdbase** this theory uses for disambiguation on **om:OMS** elements (see Section 13.1 for a discussion). Additional information about the theory like a title or a short description can be given in the **metadata** element. After this, any top-level OMDOC element can occur, including the theory-constitutive elements introduced in Sections 15.1 and 15.2, even **theory** elements themselves. Note that theory-constitutive elements may *only* occur in **theory** elements.

Theories can be structured like documents e.g. into sections and the like (see Section 11.4 for a discussion) via the **tgroup** element, which behaves exactly like the **omgroup** element introduced in Section 11.4 except that it also allows theory-constitutive elements, but does not allow a **theory** attribute, since this information is already given by the dominating **theory** element.[6]

---

[5] Purists could use the CSS attribute **style** on the **definition** element with value **display:none** to hides it from the document; it might also be placed into another document altogether

[6] This element has been introduced to keep OMDOC validation manageable: We cannot directly use the **omgroup** element,since there is no simple, context-free way to determine whether an **omgroup** is dominated by a **theory** element.

Element	Attributes		D	Content
	Req.	Optional	C	
theory		xml:id, class, style, cdbase, cdversion, cdrevision, cdstatus, cdurl, cdreviewdate	+	($\langle\!\langle top{+}thc\rangle\!\rangle$ | imports)*
imports	from	id, type, class, style	+	
tgroup		xml:id, modules, type, class, style	+	($\langle\!\langle top{+}thc\rangle\!\rangle$)*
where $\langle\!\langle top{+}thc\rangle\!\rangle$ stands for top-level and theory-constitutive elements				

Fig. 15.18. Theories in OMDOC

15.6.1 Simple Inheritance

theory elements can contain imports elements (mixed in with the top-level ones) to specify inheritance: The main idea behind structured theories and specification is that not all theory-constitutive elements need to be explicitly stated in a theory; they can be inherited from other theories. Formally, the set of theory-constitutive elements in a theory is the union of those that are explicitly specified and those that are imported from other theories. This has consequences later on, for instance, these are available for use in proofs. See Section 17.2 for details on availability of assertional statements in proofs and justifications.

The meaning of the imports element is determined by two attributes:

from The value of this attribute is a URI reference that specifies the **source theory**, i.e. the theory we import from. The current theory (the one specified in the parent of the imports element, we will call it the **target theory**) inherits the constitutive elements from the source theory.

type This optional attribute can have the values global and local (the former is assumed, if the attribute is absent): We call constitutive elements **local** to the current theory, if they are explicitly defined as children, and else **inherited**. A **local import** (an imports element with type="local") only imports the local elements of the source theory, a global import also the inherited ones.

The meaning of nested theory elements is given in terms of an implicit imports relation: The inner theory imports from the outer one. Thus

```
 <theory xml:id="a.thy">
   <symbol name="aa"/>
   <theory xml:id="b.thy">
     <symbol name="cc"/>
5    <definition xml:id="cc.def" for="#cc" type="simple">
       <OMOBJ><OMS cd="a.thy" name="af"/></OMOBJ>
     </definition>
   </theory>
 </theory>
```

is equivalent to

```
<theory xml:id="a.thy"><symbol name="aa"/></theory>
<theory xml:id="b.thy">
  <imports from="#a.thy" type="global"/>
  <symbol name="cc"/>
  <definition xml:id="cc.def" for="#cc" type="simple">
    <OMOBJ><OMS cd="a.thy" name="af"/></OMOBJ>
  </definition>
</theory>
```

In particular, the symbol `cc` is visible only in theory `b.thy`, not in the rest of theory `a.thy` in the first representation.

Note that the inherited elements of the current theory can themselves be inherited in the source theory. For instance, in the Listing 15.20 the `left-inv` is the only local axiom of the theory `group`, which has the inherited axioms `closed`, `assoc`, `left-unit`.

In order for this import mechanism to work properly, the inheritance relation, i.e. the relation on theories induced by the `imports` elements, must be acyclic. There is another, more subtle constraint on the inheritance relation concerning multiple inheritance. Consider the situation in Listing 15.19: here theories A and B import theories with `xml:id="mythy"`, but from different URIs. Thus we have no guarantee that the theories are identical, and semantic integrity of the theory C is at risk. Note that this situation might in fact be totally unproblematic, e.g. if both URIs point to the same document, or if the referenced documents are identical or equivalent. But we cannot guarantee this by content markup alone, we have to forbid it to be safe.

Listing 15.19. Problematic multiple inheritance

```
<theory xml:id="A">
  <imports from="http://red.com/theories.omdoc#mythy"/>
</theory>
<theory xml:id="B">
  <imports from="http://blue.org/cd/all.omdoc#mythy"/>
</theory>
<theory xml:id="C"><imports from="#A"/><imports from="#B"/></theory>
```

Let us now formulate the constraint carefully, the **base URI** of an XML document is the URI that has been used to retrieve it. We adapt this to OMDOC theory elements: the base URI of an imported theory is the URI declared in the `cdbase` attribute of the `theory` element (if present) or the base URI of the document which contains it[7]. For theories that are imported along a chain of global imports, which include relative URIs, we need to employ URI normalization to compute the effective URI. Now the constraint is that any two imported theories that have the same value of the `xml:id` attribute must have the same base URI. Note that this does not imply a global unicity constraint for `xml:id` values of `theory` elements, it only means that the mapping of theory identifiers to URIs is unambiguous in the dependency cone of a theory.

---

[7] Note that the base URI of the document is sufficient, since a valid OMDoc document cannot contain more than one `theory` element for a given `xml:id`

In Listing 15.20 we have specified three algebraic theories that gradually build up a theory of groups importing theory-constitutive statements (symbols, axioms, and definitions) from earlier theories and adding their own content. The theory **semigroup** provides symbols for an operation **op** on a base set **set** and has the axioms for closure and associativity of **op**. The theory of monoids imports these without modification and uses them to state the **left-unit** axiom. The theory **monoid** then proceeds to add a symbol **neut** and an axiom that states that it acts as a left unit with respect to **set** and **op**. The theory **group** continues this process by adding a symbol **inv** for the function that gives inverses and an axiom that states its meaning.

Listing 15.20. A structured development of algebraic theories in OMDoc

```
   <theory xml:id="semigroup">
     <symbol name="set"/><symbol name="op"/>
     <axiom xml:id="closed"> ... </axiom><axiom xml:id="assoc"> ... </axiom>
   </theory>
5
   <theory xml:id="monoid">
     <imports from="#semigroup"/>
     <symbol name="neut"/><symbol name="setstar"/>
     <axiom xml:id="left−unit">
10      <CMP>neut is a left unit for op.</CMP><FMP>∀x ∈ set.op(x, neut) = x</FMP>
     </axiom>
     <definition xml:id="setstar.def" for="#setstar" type="implicit">
       <CMP>·* subtracts the unit from a set </CMP><FMP>∀S.S* = S\{unit}</FMP>
     </definition>
15  </theory>

   <theory xml:id="group">
     <imports from="#monoid"/>
     <symbol name="inv"/>
20   <axiom xml:id="left−inv">
       <CMP>For every X ∈ set there is an inverse inv(X) wrt. op.</CMP>
     </axiom>
   </theory>
```

The example in Listing 15.20 shows that with the notion of theory inheritance it is possible to re-use parts of theories and add structure to specifications. For instance it would be very simple to define a theory of Abelian semigroups by adding a commutativity axiom.

The set of symbols, axioms, and definitions available for use in proofs in the importing theory consists of the ones directly specified as **symbol**, **axiom**, and **definition** elements in the target theory itself (we speak of **local** axioms and definitions in this case) and the ones that are inherited from the source theories via **imports** elements. Note that these symbols, axioms, and definitions (we call them **inherited**) can consist of the local ones in the source theories and the ones that are inherited there.

The local and inherited symbols, definitions, and axioms are the only ones available to mathematical statements and proofs. If a symbol is not available in the home theory (the one given by the dominating **theory** element or the one specified in the **theory** attribute of the statement), then it cannot be used since its semantics is not defined.

15.6.2   OMDoc Theories as Content Dictionaries

In Chapter 13, we have introduced the OPENMATH and Content-MATHML representations for mathematical objects and formulae. One of the central concepts there was the notion that the representation of a symbol includes a pointer to a document that defines its meaning. In the original OPENMATH standard, these documents are identified as OPENMATH content dictionaries, the MATHML recommendation is not specific. In the examples above, we have seen that OMDOC documents can contain definitions of mathematical concepts and symbols, thus they are also candidates for "defining documents" for symbols. By the OPENMATH2 standard [BCC+04] suitable classes of OMDOC documents can act as OPENMATH content dictionaries (we call them OMDOC **content dictionaries**; see Subsection 22.3.2). The main distinguishing feature of OMDOC content dictionaries is that they include `theory` elements with symbol declarations (see Section 15.2) that act as the targets for the pointers in the symbol representations in OPENMATH and Content-MATHML. The theory name specified in the `xml:id` attribute of the `theory` element takes the place of the `CDname` defined in the OPENMATH content dictionary.

Furthermore, the URI specified in the `cdbase` attribute is the one used for disambiguation on `om:OMS` elements (see Section 13.1 for a discussion).

For instance the symbol declaration in Listing 15.4 can be referenced as

```
<OMS cd="elAlg" name="monoid" cdbase="http://mathweb.org/algebra.omdoc"/>
```

if it occurs in a theory for elementary algebra whose `xml:id` attribute has the value `elAlg` and which occurs in a resource with the URI `http://mathweb.org/algebra.omdoc` or if the `cdbase` attribute of the `theory` element has the value `http://mathweb.org/algebra.omdoc`.

To be able to act as an OPENMATH2 content dictionary format, OMDOC must be able to express content dictionary metadata (see Listing 5.1 for an example). For this, the `theory` element carries some optional attributes that allow to specify the administrative metadata of OPENMATH content dictionaries.

The `cdstatus` attribute specifies the **content dictionary status**, which can take one of the following values: `official` (i e approved by the OPEN-MATH Society), `experimental` (i.e. under development and thus liable to change), `private` (i.e. used by a private group of OPENMATH users) or `obsolete` (i.e. only for archival purposes). The attributes `cdversion` and `cdrevision` jointly specify the **content dictionary version number**, which consists of two parts, a major **version** and a **revision**, both of which are non-negative integers. For details between the relation between content dictionary status and versions consult the OPENMATH standard [BCC+04].

Furthermore, the `theory` element can have the following attributes:

cdbase for the content dictionary base which, when combined with the content dictionary name, forms a unique identifier for the content dictionary. It may or may not refer to an actual location from which it can be retrieved.

cdurl for a valid URL where the source file for the content dictionary encoding can be found.

cdreviewdate for the **review date** of the content dictionary, i.e. the date until which the content dictionary is guaranteed to remain unchanged.

Abstract Data Types (Module ADT)

Most specification languages for mathematical theories support definition mechanisms for sets that are inductively generated by a set of constructors and recursive functions on these under the heading of **abstract data types**. Prominent examples of abstract data types are natural numbers, lists, trees, etc. The module ADT presented in this chapter extends OMDOC by a concise syntax for abstract data types that follows the model used in the CASL (Common Abstract Specification Language [CoF04]) standard.

Conceptually, an abstract data type declares a collection of symbols and axioms that can be used to construct certain mathematical objects and to group them into sets. For instance, the Peano axioms (see Figure 15.1) introduce the symbols 0 (the number zero), s (the successor function), and $\mathbb{N}$ (the set of natural numbers) and fix their meaning by five axioms. These state that the set $\mathbb{N}$ contains exactly those objects that can be constructed from 0 and s alone (these symbols are called **constructor symbols** and the representations **constructor terms**). Optionally, an abstract data type can also declare **selector symbols**, for (partial) inverses of the constructors. In the case of natural numbers the predecessor function is a selector for s: it "selects" the argument n, from which a (non-zero) number $s(n)$ has been constructed.

Following CASL we will call sets of objects that can be represented as constructor terms **sorts**. A sort is called **free**, iff there are no identities between constructor terms, i.e. two objects represented by different constructor terms can never be equal. The sort $\mathbb{N}$ of natural numbers is a free sort. An example of a sort that is not free is the theory of finite sets given by the constructors $\emptyset$ and the set insertion function ι, since the set $\{a\}$ can be obtained by inserting a into the empty set an arbitrary (positive) number of times; so e.g. $\iota(a, \emptyset) = \iota(a, \iota(a, \emptyset))$. This kind of sort is called **generated**, since it only contains elements that are expressible in the constructors. An abstract data type is called **loose**, if it contains elements besides the ones generated by the constructors. We consider free sorts more **strict** than generated ones, which in turn are more strict than loose ones.

Element	Attributes		D	Content
	Req.	Optional	C	
adt		xml:id, class, style, parameters	+	sortdef+
sortdef	name	type, role, scope, class, style	+	(constructor | insert)*, recognizer?
constructor	name	type, scope, class, style	+	argument*
argument			+	type, selector?
insert	for		–	
selector	name	type, scope, role, total, class, style	+	EMPTY
recognizer	name	type, scope, role, class, style	+	

Fig. 16.1. Abstract data types in OMDoc

In OMDoc, we use the <u>adt</u> element to specify abstract data types possibly consisting of multiple sorts. It is a theory-constitutive statement and can only occur as a child of a **theory** element (see Section 15.1 for a discussion). An **adt** element contains one or more **sortdef** elements that define the sorts and specify their members and it can carry a **parameters** attribute that contains a whitespace-separated list of parameter variable names. If these are present, they declare type variables that can be used in the specification of the new sort and constructor symbols see Section 26.20 for an example.

We will use an augmented representation of the abstract data type of natural numbers as a running example for introduction of the functionality added by the ADT module; Listing 16.2 contains the listing of the OMDoc encoding. In this example, we introduce a second sort $\mathbb{P}$ for positive natural numbers to make it more interesting and to pin down the type of the predecessor function.

A <u>sortdef</u> element is a highly condensed piece of syntax that declares a sort symbol together with the constructor symbols and their selector symbols of the corresponding sort. It has a required **name** attribute that specifies the symbol name, an optional **type** attribute that can have the values **free**, **generated**, and **loose** with the meaning discussed above. A **sortdef** element contains a set of <u>constructor</u> and <u>insert</u> elements. The latter are empty elements which refer to a sort declared elsewhere in a **sortdef** with their **for** attribute: An **insert** element with **for="#⟨name⟩"** specifies that all the constructors of the sort ⟨name⟩ are also constructors for the one defined in the parent **sortdef**. Note that the sort ⟨name⟩ must be declared in a **sortdef** in the same **adt** element. Furthermore, the type of a sort given by a **sortdef** element can only be as strict as the types of any sorts included by its **insert** children.

Listing 16.2 introduces the sort symbols **pos-nats** (positive natural numbers) and **nats** (natural numbers) , the symbol names are given by the required **name** attribute. Since a constructor is in general an n-ary function, a **constructor** element contains n **argument** children that specify the argument sorts of this function along with possible selector functions. The argument sort is given as the first child of the **argument** element: a **type** element as described in Subsection 15.2.3. Note that n may be 0 and thus the constructor element may not have **argument** children (see for instance the **constructor** for **zero** in Listing 16.2). The first **sortdef** element there introduces the constructor symbol **succ@Nat** for the successor function. This function has one argument, which is a natural number (i.e. a member of the sort **nats**).

Sometimes it is convenient to specify the inverses of a constructors that are functions. For this OMDOC offers the possibility to add an empty <u>selector</u> element as the second child of an **argument** child of a **constructor**. The required attribute **name** specifies the symbol name, the optional **total** attribute of the **selector** element specifies whether the function represented by this symbol is total (value **yes**) or partial (value **no**). In Listing 16.2 the **selector** element in the first **sortdef** introduces a selector symbol for the successor function **succ**. As **succ** is a function from **nats** to **pos-nats**, **pred** is a total function from **pos-nats** to **nats**.

Finally, a **sortdef** element can contain a <u>recognizer</u> child that specifies a symbol for a predicate that is true, iff its argument is of the respective sort. The name of the predicate symbol is specified in the required **name** attribute. Listing 16.2 introduces such a **recognizer predicate** as the last child of the **sortdef** element for the sort **pos-nats**.

Note that the **sortdef**, **constructor**, **selector**, and **recognizer** elements define symbols of the name specified by their **name** element in the theory that contains the **adt** element. To govern the visibility, they carry the attribute **scope** (with values **global** and **local**) and the attribute **role** (with values **type**, **sort**, **object**).

Listing 16.2. The natural numbers using **adt** in OMDOC

```
   <theory xml:id="Nat">
     <adt xml:id="nat−adt">
       <metadata>
         <dc:title>Natural Numbers as an Abstract Data Type.</dc:title>
5        <dc:description>The Peano axiomatization of natural numbers.</dc:description>
       </metadata>

       <sortdef name="pos−nats" type="free">
         <metadata>
10         <dc:description>The set of positive natural numbers.</dc:description>
         </metadata>
         <constructor name="succ">
           <metadata><dc:description>The successor function.</dc:description></metadata>
           <argument>
15           <type><OMOBJ><OMS cd='Nat' name="nats"/></OMOBJ></type>
             <selector name="pred" total="yes">
               <metadata><dc:description>The predecessor function.</dc:description></metadata>
             </selector>
           </argument>
20         </constructor>
         <recognizer name="positive">
           <metadata>
             <dc:description>
             The recognizer predicate for positive natural numbers.
25           </dc:description>
```

```
      </metadata>
      </recognizer>
     </sortdef>
30   <sortdef name="nats" type="free">
      <metadata><dc:description>The set of natural numbers</dc:description></metadata>
      <constructor name="zero">
        <metadata><dc:description>The number zero.</dc:description></metadata>
      </constructor>
35    <insort for="#pos-nats"/>
     </sortdef>
    </adt>
   </theory>
```

To summarize Listing 16.2: The abstract data type **nat-adt** is free and defines two sorts **pos-nats** and **nats** for the (positive) natural numbers. The positive numbers (**pos-nats**) are generated by the successor function (which is a constructor) on the natural numbers (all positive natural numbers are successors). On **pos-nats**, the inverse **pred** of **succ** is total. The set **nats** of all natural numbers is defined to be the union of **pos-nats** and the constructor **zero**. Note that this definition implies the five well-known Peano Axioms: the first two specify the constructors, the third and fourth exclude identities between constructor terms, while the induction axiom states that **nats** is generated by **zero** and **succ**. The document that contains the **nat-adt** could also contain the symbols and axioms defined implicitly in the **adt** element explicitly as **symbol** and **axiom** elements for reference. These would then carry the **generated-from** attribute with value **nat-adt**.

Representing Proofs (Module PF)

Proofs form an essential part of mathematics and modern sciences. Conceptually, a **proof** is a representation of uncontroversial evidence for the truth of an assertion.

The question of what exactly constitutes a proof has been controversially discussed (see e.g. [BC01a]). The clearest (and most radical) definition is given by theoretical logic, where a proof is a sequence, or tree, or directed acyclic graph (DAG) of applications of inference rules from a formally defined logical calculus, that meets a certain set of well-formedness conditions. There is a whole zoo of logical calculi that are optimized for various applications. They have in common that they are extremely explicit and verbose, and that the proofs even for simple theorems can become very large. The advantage of having formal and fully explicit proofs is that they can be very easily verified, even by simple computer programs. We will come back to this notion of proof in Section 17.4.

In mathematical practice the notion of a proof is more flexible, and more geared for consumption by humans: any line of argumentation is considered a proof, if it convinces its readers that it could in principle be expanded to a formal proof in the sense given above. As the expansion process is extremely tedious, this option is very seldom carried out explicitly. Moreover, as proofs are geared towards communication among humans, they are given at vastly differing levels of abstraction. From a very informal proof idea for the initiated specialist of the field, who can fill in the details herself, down to a very detailed account for skeptics or novices which will normally be still well above the formal level. Furthermore, proofs will usually be tailored to the specific characteristics of the audience, who may be specialists in one part of a proof while unfamiliar to the material in others. Typically such proofs have a sequence/tree/DAG-like structure, where the leaves are natural language sentences interspersed with mathematical formulae (or mathematical vernacular).

Let us consider a proof and its context (Figure 17.1) as it could be found in a typical elementary math. textbook, only that we have numbered the proof steps for referencing convenience. Figure 17.1 will be used as a running example throughout this chapter.

Theorem: *There are infinitely many prime numbers.*
Proof: We need to prove that the set P of all prime numbers is not finite.

1. We proceed by assuming that P is finite and reaching a contradiction.
2. Let P be finite.
3. Then $P = \{p_1, \ldots, p_n\}$ for some p_i.
4. Let $q \stackrel{def}{=} p_1 \cdots p_n + 1$.
5. Since for each $p_i \in P$ we have $q > p_i$, we conclude $q \notin P$.
6. We prove the absurdity by showing that q is prime:
7. For each $p_i \in P$ we have $q = p_i k + 1$ for some natural number k, so p_i can not divide q;
8. q must be prime as P is the set of all prime numbers.
9. Thus we have contradicted our assumption (2)
10. and proven the assertion.   □

Fig. 17.1. A theorem with a proof

Since proofs can be marked up on several levels, we will introduce the OM-Doc markup for proofs in stages: We will first concentrate on proofs as structured texts, marking up the discourse structure in example Figure 17.1. Then we will concentrate on the justifications of proof steps, and finally we will discuss the scoping and hierarchical structure of proofs.

The development of the representational infrastructure in OMDoc has a long history: From the beginning the format strived to allow structural semantic markup for textbook proofs as well as accommodate a wide range of formal proof systems without over-committing to a particular system. However, the proof representation infrastructure from Version 1.1 of OM-Doc turned out not to be expressive enough to represent the proofs in the Helm library [APCS01]. As a consequence, the PF module has been re-designed [AKC03] as part of the MoWGLI project [AK02]. The current version of the PF module is an adaptation of this proposal to be as compatible as possible with earlier versions of OMDoc. It has been validated by interpreting it as an implementation of the $\overline{\lambda}\mu\tilde{\mu}$ calculus [Coe05] proof representation calculus.

17.1 Proof Structure

In this section, we will concentrate on the structure of proofs apparent in the proof text and introduce the OMDoc infrastructure needed for marking up this aspect. Even if the proof in Figure 17.1 is very short and simple, we can observe several characteristics of a typical mathematical proof. The proof starts with the thesis that is followed by nine main "steps" (numbered from 1 to 10). A very direct representation of the content of Figure 17.1 is given in Listing 17.2.

Listing 17.2. An OMDoc representation of Figure 17.1.

```
<assertion xml:id="a1">
  <CMP>There are infinitely many prime numbers.</CMP>
</assertion>
<proof xml:id="p" for="#a1">
  <omtext xml:id="intro">
    <CMP>We need to prove that the set P of all prime numbers is not finite.</CMP>
  </omtext>
  <derive xml:id="d1">
    <CMP>We proceed by assuming that P is finite and reaching a contradiction.</CMP>
    <method>
      <proof xml:id="p1">
        <hypothesis xml:id="h2"><CMP>Let P be finite.</CMP></hypothesis>
        <derive xml:id="d3">
          <CMP>Then P = {p_1,...,p_n} for some p_i.</CMP>
          <method><premise xref="#h2"/></method>
        </derive>
        <symbol name="q"/>
        <definition xml:id="d4" for="#q" type="informal">
          <CMP>Let q \overset{def}{=} p_1 \cdots p_n + 1</CMP>
        </definition>
        <derive xml:id="d5">
          <CMP> Since for each p_i \in P we have q > p_i, we conclude q \notin P.</CMP>
        </derive>
        <omtext xml:id="c6">
          <CMP>We prove the absurdity by showing that q is prime:</CMP>
        </omtext>
        <derive xml:id="d7">
          <CMP>For each p_i \in P we have q = p_i k + 1 for some
            natural number k, so p_i can not divide q;</CMP>
          <method><premise xref="#d4"/></method>
        </derive>
        <derive xml:id="d8">
          <CMP>q must be prime as P is the set of all prime numbers.</CMP>
          <method><premise xref="#d7"/></method>
        </derive>
        <derive xml:id="d9">
          <CMP>Thus we have contradicted our assumption</CMP>
          <method><premise xref="#d5"/><premise xref="#d8"/></method>
        </derive>
      </proof>
    </method>
  </derive>
  <derive xml:id="d10" type="conclusion">
    <CMP>This proves the assertion.</CMP>
  </derive>
</proof>
```

Proofs are specified by **proof** elements in OMDoc that have the optional attributes `xml:id` and **theory** and the required attribute **for**. The **for** attribute points to the assertion that is justified by this proof (this can be an **assertion** element or a **derive** proof step (see below), thereby making it possible to specify expansions of justifications and thus hierarchical proofs). Note that there can be more than one proof for a given assertion.

Element	Attributes		D	Content
	Req.	Optional	C	
proof	for	theory, xml:id, class, style	+	(omtext | derive | hypothesis | symbol | definition)*
proofobject		xml:id, for, class, style, theory	+	CMP*, (OMOBJ |m:math |legacy)
hypothesis		xml:id, class, style, inductive	−	CMP*, FMP*
derive		xml:id, class, style, type	−	CMP*, FMP*, method?
method		xref	−	(OMOBJ |m:math |legacy | premise | proof | proofobject)*
premise	xref		−	EMPTY

Fig. 17.3. The OMDOC proof elements

The content of a proof consists of a sequence of proof steps, whose DAG structure is given by cross-referencing. These proof steps are specified in four kinds of OMDOC elements:

omtext OMDOC allows this element to allow for intermediate text in proofs that does not have to have a logical correspondence to a proof step, but e.g. guides the reader through the proof. Examples for this are remarks by the proof author, e.g. an explanation why some other proof method will not work. We can see another example in Listing 17.2 in lines 5-7, where the comment gives a preview over the course of the proof.

derive elements specify normal proof steps that derive a new claim from already known ones, from assertions or axioms in the current theory, or from the assumptions of the assertion that is under consideration in the proof. See for example lines $12ff$ in Listing 17.2 for examples of **derive** proof steps that only state the local assertion. We will consider the specification of justifications in detail in Section 17.2 below. The **derive** element carries an optional **xml:id** attribute for identification and an optional **type** to single out special cases of proofs steps.

The value **conclusion** is reserved for the concluding step of a proof[1], i.e. the one that derives the assertion made in the corresponding theorem.

The value **gap** is used for proof steps that are not justified (yet): we call them **gap steps**. Note that the presence of gap steps allows OMDOC to specify incomplete proofs as proofs with gap steps.

hypothesis elements allow to specify local assumptions that allow the hypothetical reasoning discipline needed for instance to specify proof by contradiction, by case analysis, or simply to show that A implies B, by assuming A and then deriving B from this local hypothesis. The scope of an hypothesis extends to the end of the **proof** element containing it. In Listing 17.2 the classification of step 2 from Figure 17.1 as the

---

[1] As the argumentative structure of the proof is encoded in the justification structure to be detailed in Section 17.2, the concluding step of a proof need not be the last child of a proof element.

hypothesis element h2 forces us to embed it into a derive element with a proof grandchild, making a structure apparent that was hidden in the original.

An important special case of hypothesis is the case of "inductive hypothesis", this can be flagged by setting the value of the attribute inductive to yes; the default value is no.

symbol/definition These elements allow to introduce new local symbols that are local to the containing proof element. Their meaning is just as described in Section 15.2, only that the role of the axiom element described there is taken by the hypothesis element. In Listing 17.2 step 4 in the proof is represented by a symbol/definition pair. Like in the hypothesis case, the scope of this symbol extends to the end of the proof element containing it.

These elements contain an informal (natural language) representation of the proof step in a multilingual CMP group and possibly an FMP element that gives a formal representation of the claim made by this proof step. A derive element can furthermore contain a method element that specifies how the assertion is derived from already-known facts (see the next section for details). All of the proof step elements have an optional xml:id attribute for identification and the CSS attributes.

As we have seen above, the content of any proof step is essentially a Gentzen-style sequent; see Listing 17.5 for an example. This mixed representation enhances multi-modal proof presentation [Fie97], and the accumulation of proof information in one structure. Informal proofs can be formalized [Bau99]; formal proofs can be transformed to natural language [HF96]. The first is important, since it will be initially infeasible to totally formalize all mathematical proofs needed for the correctness management of the knowledge base.

17.2 Proof Step Justifications

So far we have only concerned ourselves with the linear structure of the proof, we have identified the proof steps and classified them by their function in the proof. A central property of the derive elements is that their content (the local claim) follows from statements that we consider true. These can be earlier steps in the proof or general knowledge. To convince the reader of a proof, the steps are often accompanied with a **justification**. This can be given either by a logical inference rule or higher-level evidence for the truth of the claim. The evidence can consist in a proof method that can be used to prove the assertion, or in a separate subproof, that could be presented if the consumer was unconvinced. Conceptually, both possibilities are equivalent, since the method can be used to compute the subproof (called its **expansion**). Justifications are represented in OMDOC by the `method` children of `derive` elements[2] (see Listing 17.4 for an example):

The method element contains a structural specification of the justification of the claim made in the FMP of a `derive` element. So the FMP together with the `method` element jointly form the counterpart to the natural language content of the CMP group, they are sibling to: The FMP formalizes the local claim, and the `method` stands for the justification. In Listing 17.4 the formula in the CMP element corresponds to the claim, whereas the part "By ..., we have" is the justification. In other words, a `method` element specifies a proof method or inference rule with its arguments that justifies the assertion made in the FMP elements. It has an optional `xref` attribute whose target is an OMDOC definition of an inference rule or proof method.[3] A method may have `om:OMOBJ`, `m:math`, `legacy`, `premise`, `proof`, and `proofobject`[4] children. These act as parameters to the method, e.g. for the repeated universal instantiation method in Listing 17.4 the parameters are the terms to instantiate the bound variables.

The premise elements are used to refer to already established assertions: other proof steps or statements — e.g. ones given as `assertion`, `definition`, or `axiom` elements — the method was applied to to obtain the local claim of the proof step. The `premise` elements are empty and carry the required attribute `xref`, which contains the URI of the assertion. Thus the `premise` elements specify the DAG structure of the proof. Note that even if we do not mark up the method in a justification (e.g. if it is unknown or obvious) it can still make sense to structure the argument in `premise` elements. We have done so in Listing 17.2 to make the dependencies of the argumentation explicit.

If a `derive` step is a logically (or even mathematically) complex step, an expansion into sub-steps can be specified in a `proof` or `proofobject` element embedded into the justifying `method` element. An embedded proof allows us to specify generic markup for the hierarchic structure of proofs. Expansions of nodes justified by method applications are computed, but the information about the method itself is not discarded in the process as in

---

[2] The structural and formal justification elements discussed in this section are derived from hierarchical data structures developed for semi-automated theorem proving (satisfying the logical side). They allow natural language representations at every level (allowing for natural representation of mathematical vernacular at multiple levels of abstraction). This proof representation (see [BCF+97] for a discussion and pointers) is a DAG of nodes which represent the proof steps.

[3] At the moment OMDOC does not provide markup for such objects, so that they should best be represented by `symbols` with `definition` where the inference rule is explained in the CMP (see the lower part of Listing 17.4), and the FMP holds a content representation for the inference rule, e.g. using the content dictionary [Koh05d]. A good enhancement is to encapsulate system-specific encodings of the inference rules in `private` or `code` elements and have the `xref` attribute point to these.

[4] This object is an alternative representation of certain proofs, see Section 17.4.

tactical theorem provers like ISABELLE [Pau94] or NUPRL [CAB+86]. Thus, proof nodes may have justifications at multiple levels of abstraction in an hierarchical proof data structure. Thus the **method** elements allow to augment the linear structure of the proof by a tree/DAG-like secondary structure given by the **premise** links. Due to the complex hierarchical structure of proofs, we cannot directly utilize the tree-like structure provided by XML, but use cross-referencing. The **derive** step in Listing 17.4 represents an inner node of the proof tree/DAG with three children (the elements with identifiers A2, A4, and A5).

Listing 17.4. A **derive** proof step

```
  <proof xml:id="proof.2.1.2.proof.D2.1" for="#assertion.2.1.2">
    ...
    <derive xml:id="D2.1">
      <CMP>By <ref type="cite" xref="#A2"/>, <ref type="cite" xref="#A4"/>, and
5         <ref type="cite" xref="#A5"/> we have z + (a + (−a)) = (z + a) + (−a).</CMP>
      <FMP>z + (a + (−a)) = (z + a) + (−a)</FMP>
      <method xref="nk−sorts.omdoc#NK−Sorts.forallistar">
        <OMOBJ><OMV name="z"/></OMOBJ>
        <OMOBJ><OMV name="a"/></OMOBJ>
10        <OMOBJ>−a</OMOBJ>
        <premise xref="#A2"/><premise xref="#A4"/><premise xref="#A5"/>
      </method>
    </derive>
    ...
15 </proof>
    ...
  <theory xml:id="NK−Sorts">
    <metadata>
      <dc:title>Natural Deduction for Sorted Logic</dc:title>
20    </metadata>

    <symbol name="forallistar">
      <metadata>
        <dc:description>Repeated Universal Instantiation</dc:description>
25      </metadata>
    </symbol>
    <definition xml:id="forallistar.def" for="#forallistar" type="informal">
      <CMP>Given n parameters, the inference rule ∀I* instantiates
        the first n universal quantifications in the antecedent with them.</CMP>
30    </definition>
    ...
  </theory>
```

In OMDOC the **premise** elements must reference proof steps in the current proof or statements (**assertion** or **axiom** elements) in the scope of the current theory: A statement is **in scope of** the current theory, if its home theory is the current theory or imported (directly or indirectly) by the current theory.

Furthermore note that a proof containing a **premise** element is not self-contained evidence for the validity of the **assertion** it proves. Of course it is only evidence for the validity at all (we call such a proof grounded), if all the statements that are targets of **premise** references have grounded proofs themselves[5] and the reference relation does not contain cycles. A grounded

---

[5] For **assertion** targets this requirement is obvious. Obviously, **axioms** do not need proofs, but certain forms of definitions need well-definedness proofs (see Subsection 15.2.4). These are included in the definition of a grounded proof.

proof can be made self-contained by inserting the target statements as `derive` elements before the referencing `premise` and embedding at least one `proof` into the `derive` as a justification.

Let us now consider another proof example (Listing 17.5) to fortify our intuition.

Listing 17.5. An OMDoc representation of a proof by cases

```
     <assertion xml:id="t1" theory="sets">
       <CMP>If $a \in U$ or $a \in V$, then $a \in U \cup V$.</CMP>
       <FMP>
         <assumption xml:id="t1_a">$a \in U \vee a \in V$</assumption>
5        <conclusion xml:id="t1_c">$a \in U \cup V$</conclusion>
       </FMP>
     </assertion>
     <proof xml:id="t1_p1" for="#t1" theory="sets">
       <omtext xml:id="t1_p1_m1">
10         <CMP> We prove the assertion by a case analysis.</CMP>
       </omtext>
       <derive xml:id="t1_p1_l1">
         <CMP>If $a \in U$, then $a \in U \cup V$.</CMP>
         <FMP>
15           <assumption xml:id="t1_p1_l1_a">$a \in U$</assumption>
             <conclusion xml:id="t1_p1_l1_c">$a \in U \cup V$</conclusion>
         </FMP>
         <method xref="sk.omdoc#SK.by_definition">∪</method>
       </derive>
20       <derive xml:id="t1_p1_l2">
         <CMP>If $a \in V$, then $a \in U \cup V$.</CMP>
         <FMP>
           <assumption xml:id="t1_p1_l2_a">$a \in V$</assumption>
           <conclusion xml:id="t1_p1_l2_c">$a \in U \cup V$</conclusion>
25         </FMP>
         <method xref="sk.omdoc#SK.by_definition">∪</method>
       </derive>
       <derive xml:id="t1_p1_c">
         <CMP> We have considered both cases, so we have $a \in U \cup V$.</CMP>
30       </derive>
     </proof>
```

This proof is in **sequent style**: The statement of all local claims is in self-contained FMPs that mark up the statement in `assumption`/`conclusion` form, which makes the logical dependencies explicit. In this example we use inference rules from the calculus "SK",Gentzen's sequent calculus for classical first-order logic [Gen35], which we assume to be formalized in a theory SK. Note that local assumptions from the FMP should not be referenced outside the `derive` step they were made in. In effect, the `derive` element serves as a grouping device for local assumptions.

Note that the same effect as embedding a `proof` element into a `derive` step can be obtained by specifying the `proof` at top-level and using the optional `for` attribute to refer to the identity of the enclosing proof step (given by its optional `xml:id` attribute), we have done this in the proof in Listing 17.6, which expands the `derive` step with identifier t1_p1_l1 in Listing 17.5.

Listing 17.6. An external expansion of step t_1_p1_l1 in Listing 17.5

```
   <definition xml:id="union.def" for="#union">
   <OMOBJ>∀P, Q, x.x ∈ P ∪ Q ⇔ x ∈ P ∨ x ∈ Q</OMOBJ>
   </definition>

5  <proof xml:id="t1_p1_l1.exp" for="#t1_p1_l1">
     <derive xml:id="t1_p1_l1.d1">
       <FMP>
         <assumption xml:id="t1_p1_l1.d1.a">a ∈ U</assumption>
         <conclusion xml:id="t1_p1_l1.d1.c">a ∈ U</conclusion>
10     </FMP>
       <method xref="sk.omdoc#SK.axiom"/>
     </derive>
     <derive xml:id="t1_p1_l1.l1.d2">
       <FMP>
15       <assumption xml:id="t1_p1_l1.d2.a">a ∈ U</assumption>
         <conclusion xml:id="t1_p1_l1.d2.c">a ∈ U ∨ a ∈ V</conclusion>
       </FMP>
       <method xref="sk.omdoc#SK.orR"><premise xref="#t1_p1_l1.d1"/></method>
     </derive>
20   <derive xml:id="t1_p1_l1.d3">
       <FMP>
         <assumption xml:id="t1_p1_l1.d3.a">a ∈ U ∨ a ∈ V</assumption>
         <conclusion xml:id="t1_p1_l1.d3.c">a ∈ U ∪ V</conclusion>
       </FMP>
25     <method xref="sk.omdoc#SK.definition−rl">U, V, a
         <premise xref="#unif.def"/>
       </method>
     </derive>
     <derive xml:id="t1_p1_l1.d4">
30     <FMP>
         <assumption xml:id="t1_p1_l1.d3.a">a ∈ U</assumption>
         <conclusion xml:id="t1_p1_l1.d3.c">a ∈ U ∪ V</conclusion>
       </FMP>
       <method xref="sk.omdoc#SK.cut">
35       <premise xref="#t1_p1_l1.d2"/>
         <premise xref="#t1_p1_l1.d3"/>
       </method>
     </derive>
   </proof>
```

17.3   Scoping and Context in a Proof

Unlike the sequent style proofs we discussed in the last section, many informal proofs use the **natural deduction style** [Gen35], which allows to reason from local assumptions. We have already seen such hypotheses as hypothesis elements in Listing 17.2. The main new feature is that hypotheses can be introduced at some point in the proof, and are discharged later. As a consequence, they can only be used in certain parts of the proof. The hypothesis is inaccessible for inference outside the nearest ancestor proof element of the hypothesis.

Let us now reconsider the proof in Figure 17.1. Some of the steps (2, 3, 4, 5, 7) leave the thesis unmodified; these are called **forward reasoning** or **bottom-up proof steps**, since they are used to derive new knowledge from the available one with the aim of reaching the conclusion. Some other steps (1,

6) are used to conclude the (current) thesis by opening new subproofs, each one characterized with a new local thesis. These steps are called **backward reasoning** or **top-down proof step**s steps, since they are used to reduce a complex problem (proving the thesis) to several simpler problems (the subproofs). In our example, both backward reasoning steps open just one new subproof: Step 1 reduces the goal to proving that the finiteness of P implies a contradiction; step 5 reduces the goal to proving that q is prime.

Step 2 is used to introduce a new hypothesis, whose scope extends from the point where it is introduced to the end of the current subproof, covering also all the steps inbetween and in particular all subproofs that are introduced in these. In our example the scope of the hypothesis that P is finite (step 2 in Figure 17.1) are steps 3 – 8. In an inductive proof, for instance, the scope of the inductive hypothesis covers only the proof of the inductive step and not the proof of the base case (independently from the order adopted to present them to the user).

Step 4 is similar, it introduces a new symbol q, which is a local declaration that has scope over lines 4 – 9. The difference between a hypothesis and a local declaration is that the latter is used to introduce a variable as a new element in a given set or type, whereas the former, is used to locally state some property of the variables in scope. For example, *"let n be a natural number"* is a declaration, while *"suppose n to be a multiple of 2"* is a hypothesis. The introduction of a new hypothesis or local declaration should always be justified by a proof step that discharges it. In our example the declaration P is discharged in step 10. Note that in contrast to the representation in Listing 17.2 we have chosen to view step 6 in Figure 17.1 as a top-down proof step rather than a proof comment.

To sum up, every proof step is characterized by a current thesis and a *context*, which is the set of all the local declarations, hypotheses, and local definitions in scope. Furthermore, a step can either introduce a new hypothesis, definition, or declaration or can just be a forward or backward reasoning step. It is a forward reasoning `derive` step if it leaves the current thesis as it is. It is a backward reasoning `derive` step if it opens new subproofs, each one characterized by a new thesis and possibly a new context.

Listing 17.7. A top-down representation of the proof in Figure 17.1

```
   <assertion xml:id="a1">
     <CMP>There are infinitely many prime numbers.</CMP>
   </assertion>
   <proof for="#a1">
5    <omtext xml:id="c0">
       <CMP>We need to prove that the set P of all prime numbers is not finite.</CMP>
     </omtext>
     <derive xml:id="d1">
       <CMP> We proceed by assuming that P is finite and reaching a contradiction.</CMP>
10     <method xref="nk.omdoc#NK.by−contradiction">
         <proof>
           <hypothesis xml:id="h2"><CMP>Let P be finite.</CMP></hypothesis>
           <derive xml:id="d3"><CMP>Then P = {p_1,...,p_n} for some n</CMP></derive>
           <symbol name="q"/>
15         <definition xml:id="d4" for="#q" type="informal">
```

```
                  <CMP>Let q =def p1 ··· pn + 1</CMP>
                </definition>
                <derive xml:id="d5a">
                  <CMP>For each pi ∈ P we have q > pi</CMP>
 20               <method xref="#Trivial"><premise xref="#d4"/></method>
                </derive>
                <derive xml:id="d5b">
                  <CMP>q ∉ P</CMP>
                  <method xref="#Trivial"><premise xref="#d5"/></method>
 25             </derive>
                <derive xml:id="d6">
                  <CMP>We show absurdity by showing that q is prime</CMP>
                  <FMP>⊥</FMP>
                  <method xref="#Contradiction">
 30                 <premise xref="#d5b"/>
                    <proof>
                      <derive xml:id="d7a">
                        <CMP>
                          For each pi ∈ P we have q = pi k + 1 for a given natural number k.
 35                     </CMP>
                        <method xref="#By_Definition"><premise xref="#d1"/></method>
                      </derive>
                      <derive xml:id="d7b">
                        <CMP>Each pi ∈ P does not divide q</CMP>
 40                   </derive>
                      <derive xml:id="d8">
                        <CMP>q is prime</CMP>
                        <method xref="#Trivial">
                          <premise xref="#h2"/>
 45                       <premise xref="#p4"/>
                        </method>
                      </derive>
                    </proof>
                  </method>
 50             </derive>
              </proof>
            </method>
          </derive>
        </proof>
```

proof elements are considered to be non-assertional in OMDoc, since they
do not make assertions about mathematical objects themselves, but only justify such assertions. The assertional elements inside the proofs are governed
by the scoping mechanisms discussed there, so that using them in a context
where assertional elements are needed, can be forbidden.

17.4 Formal Proofs as Mathematical Objects

In OMDoc, the notion of fully formal proofs is accommodated by the
proofobject element. In logic, the term **proof object** is used for term representations for formal proofs via the Curry/Howard/DeBruijn Isomorphism
(see e.g. [Tho91] for an introduction and Figure 17.8 for an example). λ-terms
are among the most succinct representations of calculus-level proofs as they
only document the inference rules. Since they are fully formal, they are very
difficult to read and need specialized proof presentation systems for human
consumption. In proof objects inference rules are represented as mathematical symbols, in our example in Figure 17.8 we have assumed a theory PLOND

for the calculus of natural deduction in propositional logic which provides the necessary symbols (see Listing 17.9).

The `proofobject` element contains an optional multilingual group of CMP elements which describes the formal proof as well as a proof object which can be an `om:OMOBJ`, `m:math`, or `legacy` element.

```
                                          <proofobject xml:id="ac.p" for="#and−comm">
                                           <metadata>
                                            <dc:description>
                                            Assuming A ∧ B we have B and A
                                            from which we can derive B ∧ A.
                                            </dc:description>
                                           </metadata>
                                           <OMOBJ>
                                            <OMBIND id="andcom.pf">
                                            <OMS cd="PL0ND" name="impliesI"/>
                                            <OMBVAR>
                                             <OMATTR>
                                              <OMATP>
                                               <OMS cd="PL0ND" name="type"/>
                                               A ∧ B
                                              </OMATP>
                                              <OMV name="X"/>
                                             </OMATTR>
                                            </OMBVAR>
                                            <OMA>
                                             <OMS cd="PL0ND" name="andI"/>
                                             <OMA>
                                              <OMS cd="PL0ND" name="andEr"/>
                                              <OMV name="X"/>
                                             </OMA>
                                             <OMA>
                                              <OMS cd="PL0ND" name="andEl"/>
                                              <OMV name="X"/>
                                             </OMA>
                                            </OMA>
                                            </OMBIND>
                                           </OMOBJ>
                                          </proofobject>
```

$$\dfrac{\dfrac{[A \wedge B]}{B}\wedge E_r \quad \dfrac{[A \wedge B]}{A}\wedge E_l}{\dfrac{B \wedge A}{A \wedge B \Rightarrow B \wedge A}\Rightarrow I}\wedge I$$

The schema on the left shows the proof as a natural deduction proof tree, the OMDoc representation gives the proof object as a λ term. This term would be written as the following term in traditional (mathematical) notation: $\Rightarrow I(\lambda X : A \wedge B. \wedge I(\wedge E_r(X), \wedge E_l(X)))$

Fig. 17.8. A proof object for the commutativity of conjunction

Note that using OMDoc symbols for inference rules and mathematical objects for proofs reifies them to the object level and allows us to treat them at par with any other mathematical objects. We might have the following theory for natural deduction in propositional logic as a reference target for the second inference rule in Figure 17.8.

Listing 17.9. A theory for propositional natural deduction

```
<theory xml:id="PL0ND">
  <metadata>
    <dc:description>The Natural Deduction Calculus for Propositional Logic</dc:description>
  </metadata>
  ...
  <symbol name="andI">
    <metadata><dc:subject>Conjunction Introduction</dc:subject></metadata>
    <type system="prop−as−types">A → B → (A ∧ B)</type>
  </symbol>

  <definition xml:id="andI.def" for="#andi">
    <CMP>Conjunction introduction, if we can derive A and B,
      then we can conclude A ∧ B.</CMP>
  </definition>
  ...
</theory>
```

In particular, it is possible to use a `definition` element to define a derived inference rule by simply specifying the proof term as a definiens:

```
<symbol name="andcom">
  <metadata><dc:description>Commutativity for ∧</dc:description></metadata>
  <type system="prop−as−types">(A ∧ B) → (B ∧ A)</type>
</symbol>
<definition xml:id="andcom.def" for="#andcom" type="simple">
  <OMOBJ><OMR href="#andcom.pf"/></OMOBJ>
</definition>
```

Like `proofs`, `proofobjects` elements are considered to be non-assertional in OMDoc, since they do not make assertions about mathematical objects themselves, but only justify such assertions.

Complex Theories (Modules CTH and DG)

In Section 15.6 we have presented a notion of theory and inheritance that is sufficient for simple applications like content dictionaries that informally (though presumably rigorously) define the static meaning of symbols. Experience in e.g. program verification has shown that this infrastructure is insufficient for large-scale developments of formal specifications, where reusability of formal components is the key to managing complexity. For instance, for a theory of rings we cannot simply inherit the same theory of monoids as both the additive and multiplicative structure.

In this chapter, we will generalize the inheritance relation from Section 15.6 to that of "theory inclusions", also called "theory morphisms" or "theory interpretations" elsewhere [Far93]. This infrastructure allows to structure a collection of theories into a complex theory graph that particularly supports modularization and reuse of parts of specifications and theories. This gives rise to the name "complex theories" of the OMDOC module.

Element	Attributes		D	Content
	Required	Optional	C	
theory		xml:id, class, style	+	(《top-level》 | imports | inclusion)*
imports	from	xml:id, type, class, style, conservativity, conservativity-just	+	morphism?
morphism		xml:id, base, class, style, type, hiding, consistency, exhaustivity	−	requation*, measure?, ordering?
inclusion	via	xml:id, conservativity, conservativity-just	−	EMPTY
theory-inclusion	from, to	xml:id, class, style, conservativity, conservativity-just	+	(CMP*,FMP*, morphism, obligation*)
axiom-inclusion	from, to	xml:id, class, style, conservativity, conservativity-just	+	morphism?, obligation*

Fig. 18.1. Complex theories in OMDOC

18.1   Inheritance via Translations

Literal inheritance of symbols is often insufficient to re-use mathematical structures and theories efficiently. Consider for instance the situation in the elementary algebraic hierarchy: for a theory of rings, we should be able to inherit the additive group structure from the theory **group** of groups and the structure of a multiplicative monoid from the theory **monoid**: A ring is a set R together with two operations $+$ and $*$, such that $(R, +)$ is a group with unit 0 and inverse operation $-$ and $(R^*, *)$ is a monoid with unit 1 and base set $R^* := \{r \in R \mid r \neq 0\}$. Using the literal inheritance regime introduced so far, would lead us into a duplication of efforts as we have to define theories for semigroups and monoids for the operations $+$ and $*$ (see Figure 18.2).

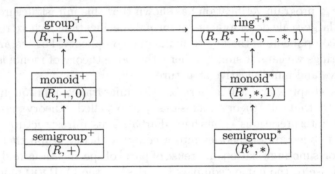

Fig. 18.2. A theory of rings via simple inheritance

This problem[1] can be alleviated by allowing theory inheritance via translations. Instead of literally inheriting the symbols and axioms from the source theory, we involve a symbol mapping function (we call this a **morphism**) in the process. This function maps source formulae (i.e. built up exclusively from symbols visible in the source theory) into formulae in the target theory by translating the source symbols.

Figure 18.3 shows a theory graph that defines a theory of rings by importing the monoid axioms via the morphism σ. With this translation, we do not have to duplicate the **monoid** and **semigroup** theories and can even move the definition of $\cdot^*$ operator into the theory of monoids, where it intuitively belongs[2].

Formally, we extend the notion of inheritance given in Section 15.6 by allowing a target theory to import another a source theory **via a morphism**:

---

[1] which seems negligible in this simple example, but in real life, each instance of multiple inheritance leads to a *multiplication* of all dependent theories, which becomes an exponentially redundant management nightmare.

[2] On any monoid $M = (S, \circ, e)$, we have the $\cdot^*$ operator, which converts a set $S \subseteq M$ in to $S^* := \{r \in S \mid r \neq e\}$

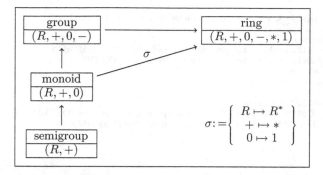

Fig. 18.3. A theory of rings via morphisms

Let $\mathcal{S}$ be a theory with theory-constitutive elements[3] $t_1, \ldots, t_n$ and $\sigma \colon \mathcal{S} \to \mathcal{T}$ a morphism, if we declare that $\mathcal{T}$ imports $\mathcal{S}$ via σ, then $\mathcal{T}$ **inherits** the theory-constitutive statements $\sigma(t_i)$ from $\mathcal{S}$. For instance, the theory of rings inherits the axiom $\forall x . x + 0 = x$ from the theory of monoids as $\sigma(\forall x . x + 0 = x) = \forall x . x * 1 = x$.

To specify the formula mapping function, module CTH extends the `imports` element by allowing it to have a child element **morphism**, which specifies a formula mapping by a set of recursive equations using the `requation` element described in Section 15.2. The optional attribute **type** allows to specify whether the function is really recursive (value **recursive**) or pattern-defined (value **pattern**). As in the case of the **definition** element, termination of the defined function can be specified using the optional child elements **measure** and **ordering**, or the optional attributes **uniqueness** and **existence**, which point to uniqueness and existence assertions. Consistency and exhaustivity of the recursive equations are specified by the the optional attributes **consistency** and **exhaustivity**.

Listing 18.4 gives the OMDOC representation of the theory graph in Figure 18.3, assuming the theories in Listing 15.20.

Listing 18.4. A theory of rings by inheritance via renaming

```
   <theory xml:id="ring">
     <symbol name="times"/><symbol name="one"/>
     <imports xml:id="add.import" from="#group" type="global"/>
     <imports xml:id="mult.import" from="#monoid" type="global">
5      <morphism>
         <requation>
           <OMOBJ><OMS cd="monoid" name="set"/></OMOBJ>
           <OMOBJ>
             <OMA><OMS cd="monoid" name="setstar"/>
10             <OMS cd="semigroup" name="set"/>
             </OMA>
           </OMOBJ>
         </requation>
         <requation>
15         <OMOBJ><OMS cd="monoid" name="op"/></OMOBJ>
```

---

[3] which may in turn be inherited from other theories

```
        <OMOBJ><OMS cd="ring" name="times"/></OMOBJ>
      </requation>
      <requation>
        <OMOBJ><OMS cd="monoid" name="neut"/></OMOBJ>
20      <OMOBJ><OMS cd="ring" name="one"/></OMOBJ>
      </requation>
    </morphism>
  </imports>
  <axiom xml:id="ring.distribution">
25    <CMP><OMOBJ><OMS cd="semigroup" name="op"/></OMOBJ> distributes over
        <OMOBJ><OMS cd="ring" name="times"/></OMOBJ>
    </CMP>
  </axiom>
</theory>
```

To conserve space and avoid redundancy, OMDOC morphisms need only specify the values of symbols that are translated; all other symbols are inherited literally. Thus the set of symbols inherited by an **imports** element consists of the symbols of the source theory that are not in the domain of the morphism. In our example, the symbols R, $+$, 0, $-$, $*$, 1 are visible in the theory of rings (and any other symbols the theory of semigroups may have inherited). Note that we do not have a name clash from multiple inheritance.

Finally, it is possible to hide symbols from the source theory by specifying them in the **hiding** attribute. The intended meaning is that the underlying signature mapping is defined (total) on all symbols in the source theory except on the hidden ones. This allows to define symbols that are local to a given theory, which helps achieve data protection. Unfortunately, there is no simple interpretation of hiding in the general case in terms of formula translations, see [CoF04, MAH06] for details. If we restrict ourselves to hiding defined symbols, then the situation becomes simpler to understand: A morphism that hides a (defined) symbol s will translate the theory-constitutive elements of the source theory by expanding definitions. Thus s will not be present in the target theory, but all the contributions of the theory-constitutive elements of the source theory will have been inherited. Say, we want to define the concept of a sorting function, i.e. a function that — given a list L as input — returns a returns a permutation L' of L that is ordered. In the situation depicted in Figure 18.6, we would the concept of an ordering function (a function that returns a permutation of the input list that is ordered) with the help of predicates **perm** and **ordered**. Since these are only of interest in the context of the definition of the latter, they would typically be hidden in order to refrain from polluting the name space.

As morphisms often contain common prefixes, the **morphism** element has an optional **base** attribute, which points to a chain of morphisms, whose composition is taken to be the base of this morphism. The intended meaning is that the new morphism coincides as a function with the base morphism, wherever the specified pattern do not match, otherwise their corresponding values take precedence over those in the base morphism. Concretely, the **base** contains a whitespace-separated list of URI references to **theory-inclusion**, **axiom-inclusion**, and **imports** elements. Note that the order of the references matters: they are ordered in order of the path in the local chain, i.e if we have globals="...#⟪*ref1*⟫ #⟪*ref2*⟫ ..." there must be theory inclusions σ_i with xml:id="⟪*refi*⟫", such that the target theory of σ_{i-1} is the source theory of σ_i.

Finally, the CTH module adds two the optional attributes conservativity and conservativity-just to the imports element for stating and justifying conservativity (see the discussion below).

18.2 Postulated Theory Inclusions

We have seen that inheritance via morphisms provides a powerful mechanism for structuring and re-using theories and contexts. It turns out that the distinguishing feature of theory morphisms is that all theory-constitutive elements of the source theory are valid in the target theory (possibly after translation). This can be generalized to obtain even more structuring relations and thus possibilities for reuse among theories. Before we go into the OMDOC infrastructure, we will briefly introduce the mathematical model (see e.g. [Hut00] for details).

A **theory inclusion** from a **source theory** S to a **target theory** T is a mapping σ from S objects[4] to those of T, such that for every theory-constitutive statement **S** of S, $\sigma(\mathbf{S})$ is provable in T (we say that $\sigma(\mathbf{S})$ is a T-theorem).

In OMDOC, we weaken this logical property to a structural one: We say that a theory-constitutive statement **S** in theory S is **structurally included** in theory T via σ, if there is an assertional statement **T** in T, such that the content of **T** is $\sigma(\mathbf{S})$. Note that strictly speaking, σ is only defined on formulae, so that if a statement **S** is only given by a CMP, $\sigma(\mathbf{S})$ is not defined. In such cases, we assume $\sigma(\mathbf{S})$ to contain a CMP element containing suitably translated mathematical vernacular. In this view, a **structural theory inclusion** from S to T is a morphism $\sigma: S \rightarrow T$, such that every theory-constitutive element is structurally included in T.

Note that an imports element in a theory T with source theory S as discussed in Section 18.1 induces a theory inclusion from S into T[5] (the theory-constitutive statements of S are accessible in T after translation and are therefore structurally included trivially). We call this kind of theory inclusion **definitional**, since it is a theory inclusion by virtue of the definition of the target theory. For all other theory inclusions (we call them **postulated theory inclusions**), we have to establish the theory inclusion property by

---

[4] Mathematical objects that can be represented using the only symbols of the source theory S.

[5] Note that in contrast to the inheritance relation induced by the imports elements the relation induced by general theory inclusions may be cyclic. A cycle just means that the theories participating in it are semantically equivalent.

proving the translations of the theory-constitutive statements of the source theory (we call these translated formulae **proof obligation**).

The benefit of a theory inclusion is that all theorems, proofs, and proof methods of the source theory can be used (after translation) in the target theory (see Section 18.4). Obviously, the transfer approach only depends on the theorem inclusion property, and we can extend its utility by augmenting the theory graph by more theory morphisms than just the definitional ones (see [FGT93] for a description of the IMPS theorem proving system that makes heavy use of this idea). We use the infrastructure presented in this chapter to structure a collection of theories as a graph — the **theory graph** — where the nodes are theories and the links are theory inclusions (definitional and postulated ones).

We call a theory inclusion $\sigma: S \to T$ **conservative**, iff **A** is already a S-theorem for for all T-theorems of the from $\sigma(\mathbf{A})$. If the morphism σ is the identity, then this means the local axioms in T only affect the local symbols of T, and do not the part inherited from S. In particular, conservative extensions of consistent theories cannot be inconsistent. For instance, if all the local theory-constitutive elements in T are symbol declarations with definitions, then conservativity is guaranteed by the special form of the definitions. We can specify conservativity of a theory inclusion via the `conservativity`. The values `conservative` and `conservative` are used for the two cases discussed above. There is a third value: `conservative`, which we will not explain here, but refer the reader to [MAH06].

OMDOC implements the concept of postulated theory inclusions in the top-level `theory-inclusion` element. It has the required attributes `from` and `to`, which point to the source- and target theories and contains a `morphism` child element as described above to define the translation function. A subsequent (possibly empty) set of `obligation` elements can be used to mark up proof obligations for the theory-constitutive elements of the source theory.

An `obligation` is an empty element whose `assertion` attribute points to an `assertion` element that states that the theory-constitutive statement specified by the `induced-by` (translated by the morphism in the parent `theory-inclusion`) is provable in the target theory. Note that a `theory-inclusion` element must contain `obligation` elements for all theory-constitutive elements (inherited or local) of the source theory to be correct.

Listing 18.5 shows a theory inclusion from the theory `group` defined in Listing 15.20 to itself. The morphism just maps each element of the base set to its inverse. A good application for this kind of theory morphism is to import claims for symmetric (e.g. with respect to the function `inv`, which serves as an involution in `group`) cases via this theory morphism to avoid explicitly having to prove them (see Section 18.4).

Listing 18.5. A theory inclusion for groups

```
<assertion xml:id="conv.assoc">∀x, y, z ∈ M.z ∘ (y ∘ x) = (z ∘ y) ∘ x</assertion>
<assertion xml:id="conv.closed" theory="semigroup">∀x, y ∈ M.y ∘ x ∈ M</assertion>
<assertion xml:id="left.unit" theory="monoid">∀x ∈ M.e ∘ x = x</assertion>
```

```
   <assertion xml:id="conv.inv" theory="group">∀x, y ∈ M.x ∘ x⁻¹ = e</assertion>
5  <theory−inclusion xml:id="grp−conv−grp" from="#group" to="#group">
     <morphism><requation>X ∘ Y ⇝ Y ∘ X</requation></morphism>
     <obligation assertion="#conv.closed" induced−by="#closed.ax"/>
     <obligation assertion="#conv.assoc" induced−by="#assoc.ax"/>
     <obligation assertion="#left.unit" induced−by="#unit.ax"/>
10   <obligation assertion="#conv.inv" induced−by="#inv.ax"/>
   </theory−inclusion>
```

18.3 Local- and Required Theory Inclusions

In some situations, we need to pose well-definedness conditions on theories, e.g. that a specification of a program follows a certain security model, or that a parameter theory used for actualization satisfies the assumptions made in the formal parameter theory; (see Chapter 6 for a discussion). If these conditions are not met, the theory intuitively does not make sense. So rather than simply stating (or importing) these assumptions as theory-constitutive statements — which would make the theory inconsistent, when they are not met — they can be stated as well-definedness conditions. Usually, these conditions can be posited as theory inclusions, so checking these conditions is a purely structural matter, and comes into the realm of OMDOC's structural methods.

OMDOC provides the empty **inclusion** element for this purpose. It can occur anywhere as a child of a **theory** element and its **via** attribute points to a theory inclusion, which is required to hold in order for the parent theory to be well-defined.

If we consider for instance the situation in Figure 18.6[6]. There we have a theory OrdList of lists that is generic in the elements (which is assumed to be a totally ordered set, since we want to talk about ordered lists). We want to to instantiate OrdList by applying it to the theory NatOrd of natural numbers and obtain a theory NatOrdList of lists of natural numbers by importing the theory OrdList in NatOrdList. This only makes sense, if NatOrd is a totally ordered set, so we add an **inclusion** element in the statement of theory NatOrdList that points to a theory inclusion of TOSet into OrdNat, which forces us to verify the axioms of TOSet in OrdNat.

Furthermore note, that the inclusion of OrdList into NatOrdList should not include the TOSet axioms on orderings, since this would defeat the purpose of making them a precondition to well-definedness of the theory NatOrdList. Therefore OMDOC follows the "development graph model" put forward in [Hut00] and generalizes the notion of theory inclusions even further: A formula mapping between theories S and T is called a **local theory inclusion** or **axiom inclusion**, if the theory inclusion property holds for the local theory-constitutive statements of the source theory. To distinguish this from the notion of a proper theory inclusion — where the theory inclusion property holds for all theory constitutive statements of S (even the inherited

---

[6] This example is covered in detail in Chapter 6.

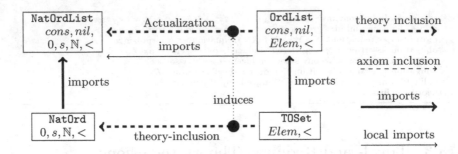

Fig. 18.6. A structured specification of lists (of natural numbers)

ones) — we call the latter one **global**. Of course all global theory inclusions are also local ones, so that the new notion is a true generalization. Note that the structural inclusions of an axiom inclusion are not enough to justify translated source theorems in the target theory.

To allow for a local variant of inheritance, the CTH module adds an attribute `type` to the `imports` element. This can take the values `global` (the default) and `local`. In the latter case, only the theory-constitutive statements that are local to the source theory are imported.

Furthermore, the CTH module introduces the <u>axiom-inclusion</u> element for local theory inclusions. This has the same attributes as `theory-inclusion`: `from` to specify source theory, `to` for the target theory. It also allows `obligation` elements as children.

18.4   Induced Assertions and Expositions

The main motivation of theory inclusions is to be able to transport mathematical statements from the source theory to the target theory. In OMDOC, this operation can be made explicit by the attributes `generated-from` and `generated-via` that the module CTH adds to all mathematical statements. On a statement **T**, the second attribute points to a theory inclusion σ whose target is (imported into the) current theory, the first attribute points to a statement **S** in that theory which is of the same type (i.e. has the same OM-DOC element name) as **T**. The content of **T** must be (equivalent to) the content of **S** translated by the morphism of σ.

In the context of the theory inclusion in Listing 18.5, we might have the following situation:

Listing 18.7. Translating a statement via a theory inclusion

```
<assertion xml:id="foo" type="theorem">...</assertion>
<proof xml:id="foo.pf" for="#foo">...</proof>
<assertion xml:id="target" induced-by="#foo" induced-via="#grp-conv-grp">
 ...
5  </assertion>
```

Here, the second assertion is induced by the first one via the theory inclusion in Listing 18.5, the statement of the theorem is about the inverses. In particular, the proof of the second theorem comes for free, since it can also be induced from the proof of the first one.

In particular we see that in OMDOC documents, not all statements are automatically generated by translation e.g. the proof of the second assertion is not explicitly stated. Mathematical knowledge management systems like knowledge bases might choose to do so, but at the document level we do not mandate this, as it would lead to an explosion of the document sizes. Of course we could cache the transformed proof giving it the same "cache attribute state".

Note that not only statements like assertions and proofs can be translated via theory inclusions, but also whole documents: Say that we have course materials for elementary algebra introducing monoids and groups via left units and left inverses, but want to use examples and exercises from a book that introduces them using right units and right inverses. Assuming that both are formalized in OMDOC, we can just establish a theory morphism much like the one in Listing 18.5. Then we can automatically translate the exercises and examples via this theory inclusion to our own setting by just applying the morphism to all formulae in the text[7] and obtain exercises and examples that mesh well with our introduction. Of course there is also a theory inclusion in the other direction, which is an inverse, so our colleague can reuse our course materials in his right-leaning setting.

Another example is the presence of different normalization factors in physics or branch cuts in elementary complex functions. In both cases there is a plethora of definitions, which all describe essentially the same objects (see e.g. [BCD+02] for an overview over the branch cut situation). Reading materials that are based on the "wrong" definition is a nuisance at best, and can lead to serious errors. Being able to adapt documents by translating them from the author theory to the user theory by a previously established theory morphism can alleviate both.

Mathematics and science are full of such situations, where objects can be viewed from different angles or in different representations. Moreover, no single representation is "better" than the other, since different views reveal or highlight different aspects of the object (see [KK05] for a systematic account). Theory inclusions seem uniquely suited to formalize the structure of different views in mathematics and their interplay, and the structural markup for theories in OMDOC seems an ideal platform for offering added-value services that feed on these structures without committing to a particular formalization or foundation of mathematics.

---

[7] There may be problems, if mathematical statements are verbalized; this can currently not be translated directly, since it would involve language processing tools much beyond the content processing tools described in this book. For the moment, we assume that the materials are written in a controlled subset of mathematical vernacular that avoids these problems.

18.5   Development Graphs (Module DG)

The OMDOC module DG for **development graphs** complements module
CTH with high-level justifications for the theory inclusions. Concretely, the
module provides an infrastructure for dealing efficiently with the proof obligations induced by theory inclusions and forms the basis for a management
of theory change. We anticipate that the elements introduced in this chapter
will largely be hidden from the casual user of mathematical software systems,
but will form the basis for high-level document- and mathematical knowledge
management services.

18.5.1   Introduction

As we have seen in the example in Listing 18.5, the burden of specifying
an `obligation` element for each theory-constitutive element of the source
theory can make the establishment of a theory inclusion quite cumbersome
— theories high up in inheritance hierarchies can have a lot (often hundreds)
of inherited, theory-constitutive statements. Even more problematically, such
obligations are a source of redundancy and non-local dependencies, since
many of the theory-constitutive elements are actually inherited from other
theories.

Consider for instance the situation in Figure 18.8, where we are interested
in the top theory inclusion Γ. On the basis of theories T_1 and T_2, theory C_1
is built up via theories A_1 and B_1. Similarly, theory C_2 is built up via A_2
and B_2 (in the latter, we have a non-trivial non-trivial morphism σ). Let us
assume for the sake of this argument that for $X_i \in \{A, B, C\}$ theories X_1 and
X_2 are so similar that axiom inclusions (they are indicated by thin dashed
arrows in Figure 18.8 and have the formula-mappings α, β, and γ) are easy
to prove[8].

To justify Γ, we must prove that the Γ-translations of all the theory-
constitutive statements of C_1 are provable in C_2. So let statement **B** be theory-
constitutive for C_1, say that it is local in B_1, then we already know that $\beta(\mathbf{B})$ is
provable in B_2 since β is an axiom inclusion. Moreover, we know that $\sigma(\beta(\mathbf{B}))$
is provable in C_2, since σ is a (definitional, global) theory inclusion. So, if we
have $\Gamma = \sigma \circ \beta$, then we are done for **B** and in fact for all local statements
of B_1, since the argument is independent of **B**. Thus, we have established
the existence of an axiom inclusion from B_1 to C_2 simply by finding suitable
inclusions and checking translation compatibility.

We will call a situation, where a theory T can be reached by an axiom
inclusion with a subsequent chain of theory inclusions a **local chain** (with

---

[8] A common source of situations like this is where the X_2 are variants of the X_1
theories. Here we might be interested whether C_2 still proves the same theories
(and often also in the converse theory inclusion Γ^{-1} that would prove that the
variants are equivalent).

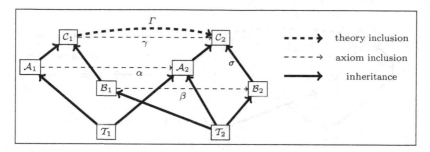

Fig. 18.8. A development graph with theory inclusions

morphism $\tau := \sigma_n \circ \cdots \circ \sigma_1 \circ \sigma)$, if $\mathcal{S} \xrightarrow{\sigma} \mathcal{T}_1$ is an axiom inclusion or (local theory import) and $\mathcal{T}_i \xrightarrow{\sigma_i} \mathcal{T}_{i+1}$ are theory inclusions (or local theory import).

$$\mathcal{S} \xrightarrow[\sigma]{} \mathcal{T}_1 \xdashrightarrow[\sigma_1]{} \mathcal{T}_2 \xdashrightarrow[\sigma_2]{} \cdots \xdashrightarrow[\sigma_{n-1}]{} \mathcal{T}_n \xdashrightarrow[\sigma_n]{} \mathcal{T}$$
$$\tau = \sigma_n \circ \cdots \circ \sigma_1 \circ \sigma$$

Note that by an argument like the one for **B** above, a local chain justifies an axiom inclusion from $\mathcal{S}$ into $\mathcal{T}$: all the τ-translations of the local theory-constitutive statements in $\mathcal{S}$ are provable in $\mathcal{T}$.

In our example in Figure 18.8 — given the obvious compatibility assumptions on the morphisms which we have not marked in the figure, — we can justify four new axiom inclusions from the theories $\mathcal{T}_1$, $\mathcal{T}_2$, $\mathcal{A}_1$, and $\mathcal{B}_1$ into $\mathcal{C}_2$ by the following local chains[9].

$$\mathcal{T}_2 \longrightarrow \mathcal{B}_2 \xrightarrow{\sigma} \mathcal{C}_2 \qquad \mathcal{B}_1 \xdashrightarrow{\beta} \mathcal{B}_2 \xrightarrow{\sigma} \mathcal{C}_2$$

$$\mathcal{T}_1 \longrightarrow \mathcal{A}_2 \longrightarrow \mathcal{C}_2 \qquad \mathcal{A}_1 \xdashrightarrow{\alpha} \mathcal{A}_2 \longrightarrow \mathcal{C}_2$$

Thus, for each theory $\mathcal{X}$ that $\mathcal{C}_1$ inherits from, there is an axiom inclusion into $\mathcal{C}_2$. So for any theory-constitutive statement in $\mathcal{C}_1$ (it must be local in one of the $\mathcal{X}$) we know that it is provable in $\mathcal{C}_2$; in other words Γ is a theory inclusion if it is compatible with the morphisms of these axiom inclusions. We have depicted the situation in Figure 18.9.

We call a situation where we have a formula mapping $\mathcal{S} \xrightarrow{\sigma} \mathcal{T}$, and an axiom inclusion $\mathcal{X} \xrightarrow{\sigma_{\mathcal{X}}} \mathcal{T}$ for every theory $\mathcal{X}$ that $\mathcal{S}$ inherits from a **decomposition** for σ, if the $\sigma_{\mathcal{X}}$ and σ are compatible. As we have seen in the example above, a decomposition for σ can be used to justify that σ a theory inclusion: all theory-constitutive elements in $\mathcal{S}$ are local in itself or one of the theories $\mathcal{X}$ it inherits from. So if we have axiom inclusions from all of these to $\mathcal{T}$, then all obligations induced by them are justified and σ is indeed a theory inclusion.

---

[9] Note for the leftmost two chains use the fact that theory inclusions (in our case definitional ones) are also axiom inclusions by definition.

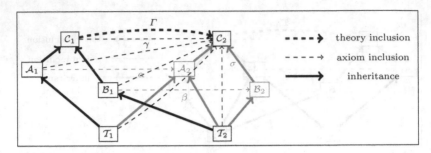

Fig. 18.9. A decomposition for the theory inclusion Γ

18.5.2   An OMDoc Infrastructure for Development Graphs (Module DG)

The DG module provides the decomposition element to model justification by decomposition situations. This empty element can occur at top-level or inside a theory-inclusion element.

The decomposition element can occur as a child to a theory-inclusion element and carries the required attribute links that contains a whitespace-separated list of URI references to the axiom- and theory-inclusion elements that make up the decomposition situation justifying the parent theory-inclusion element. Note that the order of references in links is irrelevant. If the decomposition appears on top-level, then the optional for attribute must be used to point to the theory-inclusion it justifies. In this situation the decomposition element behaves towards a theory-inclusion much like a proof for an assertion.

Element	Attributes		D	Content
	Required	Optional	C	
decomposition	links		−	EMPTY
path-just	local, globals	for	−	EMPTY
theory-inclusion	from, to, by	xml:id, class, style	+	(CMP*,FMP*, morphism, (decomposition* | obligation*))
axiom-inclusion	from, to	xml:id, class, style	+	morphism?, (path-just* | obligation*)

Fig. 18.10. Development graphs in OMDOC

Furthermore module DG provides path-just elements as children to the axiom-inclusion elements to justify that this relation holds, much like a proof element provides a justification for an assertion element for some property of mathematical objects.

A path-just element justifies an axiom-inclusion by reference to other axiom- or theory-inclusions. Local chains are encoded in the empty path-just element via the required attributes local (for the first axiom-inclusion) and the attribute globals attribute, which contains a white-space-separated list of URI references to theory-inclusions. Note that the order of the references in the globals matters: they are ordered in order of the path in the local chain, i.e if we have globals="... #ref1 #ref2 ..." there must be theory inclusions σ_i with xml:id="ref$i$", such that the target theory of σ_1 is the source theory of σ_2.

Like the decomposition element, path-just can appear at top-level, if it specifies the axiom-inclusion it justifies in the (otherwise optional) for attribute.

Let us now fortify our intuition by casting the situation in Listings 18.11 to 18.5.2 in OMDoc syntax. Another — more mathematical — example is carried out in detail in Chapter 7.

Listing 18.11. The OMDoc representation of the theories in Figure 18.8

```
   <theory xml:id="t1">...</theory>        <theory xml:id="t2">...</theory>

   <theory xml:id="a1">                     <theory xml:id="b1">
     <imports xml:id="ima1" from="#t1"/>      <imports xml:id="imb1" from="#t2"/>
5    <axiom xml:id="axa11">...</axiom>        <axiom xml:id="axb11">...</axiom>
     <axiom xml:id="axa12">...</axiom>      </theory>
   </theory>

   <theory xml:id="a2">                     <theory xml:id="b2">
10   <imports xml:id="im1a2" from="#t1"/>     <imports xml:id="imb2" from="#t2"/>
     <imports xml:id="im2a2" from="#t2"/>
     <axiom xml:id="axa21">...</axiom>        <axiom xml:id="axb21">...</axiom>
   </theory>                                 </theory>

15 <theory xml:id="c1">                     <theory xml:id="c2">
     <imports xml:id="im1c1" from="#a1"/>     <imports xml:id="im1c2" from="#a2"/>
     <imports xml:id="im2c1" from="#b1"/>     <imports xml:id="im2c2" from="#b2"/>
     <axiom xml:id="axc11">...</axiom>        <axiom xml:id="axc21">...</axiom>
   </theory>                                 </theory>
```

Here we set up the theory structure with the theory inclusions given by the imports elements (without morphism to simplify the presentation). Note that these have xml:id attributes, since we need them to construct axiom- and theory inclusions later. We have also added axioms to induce proof obligations in the axiom inclusions:

Listing 18.12. The OMDoc Representation of the inclusions in Figure 18.8

```
  <axiom-inclusion xml:id="aia" from="#a1" to="#a2">
    <obligation induced-by="#axa11" assertion="#th-axa11"/>
    <obligation induced-by="#axa12" assertion="#th-axa12"/>
  </axiom-inclusion>

5
  <axiom-inclusion xml:id="bib" from="#b1" to="#b2">
    <obligation induced-by="#axb11" assertion="#th-axb1"/>
  </axiom-inclusion>

10 <axiom-inclusion xml:id="cic" from="#c1" to="#c2">
    <obligation induced-by="#axc11" assertion="#th-axc1"/>
  </axiom-inclusion>
```

We leave out the actual assertions that justify the `obligations` to conserve space. From the axiom inclusions, we can now build four more via path justifications:

Listing 18.13. The induced axiom inclusions in Figure 18.8

```
   <axiom−inclusion xml:id="t1ic" from="#t1" to="#c2">
     <path−just local="#im1a2" globals="#im1c2"/>
   </axiom−inclusion>

5  <axiom−inclusion xml:id="t2ic" from="#t2" to="#c2">
     <path−just local="#imb2" globals="#im2c2"/>
   </axiom−inclusion>

   <axiom−inclusion xml:id="aic" from="#a1" to="#c2">
10   <path−just local="#aia" globals="#im1c2"/>
   </axiom−inclusion>

   <axiom−inclusion xml:id="bic" from="#b1" to="#c2">
     <path−just local="#bib" globals="#im2c2"/>
15 </axiom−inclusion>
```

Note that we could also have justified the axiom inclusion `t2ic` with two local paths: via the theory $\mathcal{A}_2$ and via $\mathcal{B}_2$ (assuming the translations work out). These alternative justifications make the development graph more robust against change; if one fails, the axiom inclusion still remains justified. Finally, we can assemble all of this information into a decomposition that justifies the theory inclusion Γ:

```
   <theory−inclusion xml:id="tcic" from="#c1" to="#c2">
     <decomposition links="#t1ic #t2ic #aic #bic #cic"/>
   </theory−inclusion>
```

Notation and Presentation (Module PRES)

As we have seen, OMDOC is concerned mainly with the content and structure of mathematical documents, and offers a complex infrastructure for dealing with that. However, mathematical texts often carry typographic conventions that cannot be determined by general principles alone. Moreover, non-standard presentations of fragments of mathematical texts sometimes carry meanings that do not correspond to the mathematical content or structure proper. In order to accommodate this, OMDOC provides a limited functionality for embedding style information into the document.

Element	Attributes		Content
	Required	Optional	
omstyle	element	for, xml:id, xref, class, style	(style|xslt)*
presentation	for	xml:id, xref, fixity, role, lbrack, rbrack, separator, bracket-style, class, style, precedence, crossref-symbol	(use | xslt | style)*
xslt	format	xml:lang, requires, xref	XSLT fragment
use	format	xml:lang, requires, fixity, lbrack, rbrack, separator, element, attributes, crossref-symbol	(element | text | recurse | map | value-of)*

Fig. 19.1. The OMDOC elements for notation information

The normal (but of course not the only) way to generate presentation from XML documents is to use XSLT style sheets (see Chapter 25 for other applications). XSLT [Dea99] is a general transformation language for XML. XSLT programs (often called **style sheets**) consist of a set of **templates** (rules for the transformation of certain nodes in the XML tree). These templates are recursively applied to the input tree to produce the desired output.

The general approach to presentation and notation in OMDOC is not to provide general-purpose presentational primitives that can be sprinkled over the document, since that would distract the author from the mathematical content, but to support the specification of general style information for OMDOC elements and mathematical symbols in separate elements.

In the case of a single OMDOC document it is possible to write a specialized style sheet that transforms the content-oriented markup used in the document into mathematical notation. However, if we have to deal with a large collection of OMDOC representations, then we can either write a specialized style sheet for each document (this is clearly infeasible to do by hand), or we can develop a style sheet for the whole collection (such style sheets tend to get large and unmanageable).

The OMDOC format allows to generate specialized style sheets that are tailored to the presentation of (collections of) OMDOC documents. The mechanism will be discussed in Chapter 25, here we only concern ourselves with the OMDOC primitives for representing the necessary data. In the next section, we will address the specification of style information for OMDOC elements by `omstyle` elements, and then the question of the specification of notation for mathematical symbols in `presentation` elements.

19.1   Specifying Style Information for OMDOC Elements

OMDOC provides the `omstyle`[1] elements for specifying style information for OMDOC elements. An `omstyle` element has the attributes

element This required attribute specifies the OMDOC element this style information should be applied to. The value of this attribute must be the full qualified name (i.e. including the namespace) of the element.
for This optional attribute allows to further restrict the OMDOC element to a single instance. The value of this attribute is a URI reference to a single element.
xref This optional attribute can be used to refer to another existing `omstyle` element (in another document via a URI reference), sometimes avoiding double specification: If an `omstyle` element carries an `xref` attribute, its attributes and content is disregarded, and those of the target `omstyle` element is considered instead.
class This optional attribute is an additional parameter that controls the output style. Remember that all OMDOC elements that have `xml:id` attributes also carry a `class` attribute, which allows to specify different notational conventions (see Section 10.2): In the presentation of an OMDOC element only those `omstyle` elements are taken into account that have the same value in the `class` attribute.
Note that the choice of notational style is not a content-carrying feature, and should not be depended on, indeed the value of the `class` need not be respected by output routines, but can be overwritten.

---

[1] This element would perhaps be more aptly be named `omclass`, since its function is more similar to the CSS class concept, but we keep the name `omstyle` for backwards compatibility in OMDOC 1.2.

In the presentation process described in Section 25.3 the information specified in the body of this element is then used to generate XSLT templates that are included then into the generated style sheets. This information is either given directly in XSLT using the <u>xslt</u> element, or in a `style` element using an OMDoc-internal equivalent of a small subset of XSL<u>T</u>. The latter is used if the full power of XSLT is not needed, and has the advantage that it can be transformed into the input of other formatting engines. The `xslt` and `style` elements share the following attributes:

`format` This required attribute specifies the output format. Its value is a set of format specifiers divided by the | character. We use the specifiers `TeX` for TEX and LATEX, `pmml` for Presentation-MATHML, `cmml` for Content-MATHML, `html` for HTML, `mathematica` for MATHEMATICA® notebooks. Other formats can be specified at liberty. Finally, there is the pseudo format-specifier `default`, which will be taken, if no other format is defined. Note that case matters in these specifiers, so `TeX` is not the same as `tex`. Furthermore, `default` is not a regular format specifier, so it cannot appear in the disjunctions.

`xml:lang` This specifies the languages for which this notation is used. Note that it is used differently than e.g. in the CMP element: on `omstyle`, the attribute `xml:lang` contains a whitespace-separated list of language specifiers and it does not have a default value `en`, if the attribute is not present, this means that this element is not language-specific.

`requires` This attribute contains a URI reference that points to a `code` element that contains a code fragment that is needed to be included for the presentation engine. For instance, the body of the `omstyle` element may contain TEX macros that need to be defined. Their definitions would need to be included in the output document by the presentation style sheet before they can be used.

Listing 19.2 shows a very simple example, where a `phrase` element is used to mark a text passage as "important". Its `class` attribute is picked up by the `omstyle` element to prompt special treatment in the output.

Listing 19.2. Specifying style information with the `phrase` element

```
   <CMP>
     I want to mark <phrase xml:id="w1" class="important">this important
     text</phrase> as special.<phrase class="linebreak"/>
   </CMP>

5
   <omstyle element="omdoc:phrase" class="important">
     <style format='html|pmml'><element name="em"><recurse/></element></style>
     <xslt format='TeX' xmlns:xsl="http://www.w3.org/1999/XSL/Transform">
       <xsl:text>\emph{</xsl:text>
10        <xsl:apply-templates/>
       <xsl:text>}</xsl:text>
     </xslt>
   </omstyle>

15 <omstyle element="omdoc:phrase" class="linebreak">
     <style format='html|pmml'><element name="br"/></style>
     <style format='TeX'><text>\par\noindent</text></style>
   </omstyle>
```

19.2   A Restricted Style Language

Let us now have a closer look at the presentation-language used in `style`
elements. In the first `omstyle` element in Listing 19.2 we see that the content
of an `xslt` element is an XSLT fragment. Note that when referring to OM-
Doc elements, the XSLT must use the full qualified name (i.e. including the
namespace) of the elements for the presentation to work.[2]

Element	Attributes		Content
	Required	Optional	
style	format	xml:lang, requires, xref	(element | text | recurse | map | value-of)*
element	name	crid, cr, ns	(attribute | element | text | value-of | recurse | map)*
attribute	name		(value-of | text)*
text			(#PCDATA)
value-of	select		EMPTY
recurse		select	EMPTY
map		select	separator?, (element | text | recurse | map)
separator			(element | text | recurse | map)

Fig. 19.3. The OMDoc elements for styling

Let us analyze the example to see the presentation in action before we define
it. In the first `style` element in the `omstyle` for `linebreak` in Listing 19.2
we see that the `element` element can be used to insert an XML element
into the output; in this case it is the empty HTML element `<br/>`. In the
second `style` child the `text` element (it does not have attributes) allows to
add arbitrary text into the output (in this case some TEX macros). In the
first `omstyle` element, we see that the `element` element may be non-empty, it
contains the element `recurse`, which corresponds to the directive to continue
presentation generation recursively over the children of the element specified
in the dominating `omstyle` element. The effect of this is that the content of
the first `phrase` element is encased in the HTML `em` element.

   Textual material can be added to the output in two ways: by copying
it from the source, or supplying it in the transformation. For the latter,
OMDoc supplies the <u>text</u> element (it does not have attributes), which allows
to add arbitrary text (its body) into the output. For the former, we have
the <u>value-of</u> element, an empty element that carries the required attribute
`select`, whose value is an XPATH expression. It adds the value (a string) to
the XML node specified by the expression to the output.

   The <u>element</u> element allows to generate XML elements. It has a required
attribute **name**, which contains its (local) name, and the optional attribute **ns**
to specify the namespace. Attributes of the resulting element can be specified

---

[2] For DTD validation the XSLT fragments must be encoded using the `xsl:`
namespace prefix, unless the DTD has been adapted to a different prefix by
setting the appropriate parameter entity.

by the **attribute** element: any `attribute` element adds an attribute-value pair of the form ⟨*name*⟩="⟨*value*⟩" to the output element specified by the enclosing `element` element, where the local part ⟨*name*⟩ is the value of the `name` attribute (its namespace URI given by the value the optional `ns` attribute), and ⟨*value*⟩ is either the result of presentation on the content of the `attribute` element or (iff that is empty), the value of the XPATH expression in the optional `select` attribute.

To navigate the OMDOC structure to be transformed, we have two elements: the `recurse` allows to specify a fragment continues presentation on a sub-element, and the `map` element that maps directives over a set of sub-elements. The **recurse** element is empty, and can have the attribute `select`, which contains an XPATH [Cla99b] expression specifying a set of OMDOC elements the presentation should continue with recursively. If this attribute is missing, presentation continues on the children as in Listing 19.2. The **map** element (see Listing 19.5 for an example) has the optional attribute `select` and contains a combination of the transformation directive elements `element`, `text`, `recurse`, `map` after an optional `separator` child. The `map` element directs the presentation engine to map the body directives[3] over the list of elements specified by the XPATH expression in the `select`, between any two elements, the result of styling the body of the `separator` element is inserted between the result node sets. In Listing 19.5 the `map` element recursively styles the children of the `om:OMBVAR` element and separates them by commata. Furthermore, the `map` element can have the attributes `precedence`, `lbrack`, and `rbrack` to specify brackets (with precedence-based elision) around the result. This is useful for generating argument groups.

Note that this OMDOC-internalized subset of XSLT restricts the expressivity of the presentation style by leaving out the computational features of XSLT. Firstly, the infrastructure for iteration, recursion, variable declaration, ... is not present, and secondly, path expressions are restricted to pure XPATH [Cla99b], leaving out all XSLT extensions (e.g. functions calls), again leaving us with a more declarative subset of XSLT.

19.3 Specifying the Notation of Symbols

In this section we discuss the problem of specifying the notation of mathematical symbols in OMDOC. The approach taken is very similar to the one for OMDOC elements presented in the previous section. The mathematical concepts and symbols introduced in an OMDOC document (by `symbol` elements or implicitly by abstract data types) often carry typographic conventions that cannot be determined by general principles alone. Therefore, these need to be specified, so that pleasing presentations can be generated.

---

[3] i.e. those elements after the `separator` element

We have already seen the use of `style` and `xslt` elements for specifying the presentation of general OMDOC elements in the last section. Here we will present yet another way to specify presentation information that is specialized to notations of mathematical symbols. The main idea is to specify the properties of mathematical symbols in relation to the representations of their children and siblings.

19.3.1 Specifying Notation via Templates

Let us build up our intuition by an example: For the notation information for the universal quantifier we would use an XSLT template like the one shown in Listing 19.4.

Listing 19.4. An XSLT template for the universal quantifier

```
<xsl:template match="OMBIND[OMS[position()=1 and @name='forall' and @cd='quant1']]">
  <xsl:text>∀</xsl:text>
  <xsl:for−each select="OMBVAR"/>
   <xsl:apply−templates/>
   <xsl:if  test="position()!=last()">,</xsl:if>
  </xsl:for−each>
  <xsl:text>.</xsl:text>
   <xsl:apply−templates select="*[3]"/>
</xsl:template>
```

The XPATH expression in the `match` attribute (the **template head**) specifies that this template acts as a presentation rule for `om:OMBIND` elements, where the first child is of the form `<OMS cd="quant1" name="forall"/>`. Applied to such a node, the body of the template will be executed: it will print the quantifier ∀, then the bound variables as a comma-separated list (for each of the children of `om:OMBVAR` it recursively applies XSLT templates from the style sheet), print a dot, and then recurse on the third child of the `om:OMBIND` element. Thus this template will print the OPENMATH expression below as $\forall P, Q. P \vee Q \Rightarrow Q \vee P$ assuming appropriate templates for implication and disjunction.

```
<OMBIND>
  <OMS cd="quant1" name="forall"/>
  <OMBVAR><OMV name="P"/><OMV name="Q"/></OMBVAR>
  <OMA>
    <OMS cd="logic1" name="implies"/>
    <OMA><OMS cd="logic1" name="or"/>
      <OMV name="P"/>
      <OMV name="Q"/>
    </OMA>
    <OMA><OMS cd="logic1" name="or"/>
      <OMV name="Q"/>
      <OMV name="P"/>
    </OMA>
  </OMA>
</OMBIND>
```

To annotate a symbol with notation information OMDOC supplies the `presentation` element. It is a top-level element whose `for` attribute points to the symbol in question. It contains a multilingual CMP group that allows to specify the notation[4]. Like the `omstyle` element, it has children that specify the presentation: The `xslt` element can be used to literally include the body of the template, and the `style` can express the presentation directives natively in OMDOC. In Listing 19.5 we have juxtaposed the presentational content from Listing 19.4 in `xslt` and `style` elements. Note that the directives in their body share much of the structure; the directives in the `style` are somewhat more succinct. The main difference to the XSLT template in Listing 19.4 is the specification of the template head: the attributes in the `presentation` element carry all the information necessary to identify the application conditions.

Listing 19.5. A simple `presentation` element for the universal quantifier

```
   <presentation for="#quant1.forall" role="binding">
      <CMP>We write
         <OMOBJ>
            <OMBIND><OMS cd="quant1" name="forall"/>
5            <OMBVAR><OMV name="X"></OMBVAR>
               <OMV name="A"/>
            </OMBIND>
         </OMOBJ>
         for the phrase "A holds for all  X".
10      </CMP>
      <xslt format="default" xmlns:xsl="http://www.w3.org/1999/XSL/Transform">
         <xsl:text>∀</xsl:text>
         <xsl:for-each select="OMBVAR"/>
          <xsl:apply-templates/>
15         <xsl:if test="position()!=last()">,</xsl:if>
         </xsl:for-each>
         <xsl:text>.</xsl:text>
         <xsl:apply-templates select="*[3]"/>
      </xslt>
20      <style format="html">
         <text>&#8704;</text>
         <map select="OMBVAR/*">
            <separator><text>,</text></separator>
            <recurse/>
25      </map>
         <text>.</text>
         <recurse select="*[3]"/>
      </style>
      <style format="pmml">
30         <element crid="." name="mrow" ns="http://www.w3.org/1998/Math/MathML">
            <element crid="*[1]" cr="yes" name="mo"><text>&#8704;</text></element>
            <element name="mrow" crid="*[2]">
               <map select="OMBVAR/*">
                  <separator>
35                     <element name="mo" cr="yes">
                        <attribute name="separator"><text>true</text></attribute>
                        <text>,</text>
                     </element>
                  </separator>
40                  <recurse/>
               </map>
            </element>
            <recurse select="*[3]"/>
         </element>
45      </style>
   </presentation>
```

---

[4] Of course in the content markup in OMDOC, this looks somewhat awkward, since the representation relies on the fact that it will be rendered in the correct way. In the source, the whole markup looks somewhat circular.

The `element` element can have the `crid` attribute which specifies the role of the generated element in parallel markup of mathematical formulae (see Subsection 2.1.1). The value of this element (if present) must be a XPATH fragment (see [Cla99b]) pointing to the element in the source that semantically corresponds to the generated element (see Listing 19.5[5]). Finally, the `element` element can carry the `cr` attribute, which (if its value is `yes`) instructs the presentation system to to set an `xlink:href` attribute on the result element that acts as a cross-reference to the symbol declaration.

19.3.2 Specifying Notation via Syntactic Roles

Note that hand-coding XSLT-templates is a tedious and error-prone process, and that we need a template for each output format (e.g. LATEX, HTML, Presentation-MATHML, ASCII), and even various output languages (for instance the greatest common divisor of two integers is expressed by the symbol *gcd* in English but *ggT* ("größter gemeinsamer Teiler") in German). Obviously, the respective templates for all of these transformations share a great deal of structure (in our example, they only differ in the representation of the glyph for the quantifier itself).

Therefore OMDOC goes another step and supplies a set of abbreviations that are sufficient for most presentation applications via the `use` elements that can occur as children of `presentation` elements. The user only needs to specify the relevant information in the `use` elements and a separate translation process generates the needed XSLT templates from that (see Chapter 25). The `use` elements make use of the same symbolic attributes and specialize (over-define) these attributes according to the respective format and language. The following set of attributes are particular to the `presentation`, since they are independent of the language and the output format.

`for`, `xref`, `class` (see the specification for `omstyle` in the last section)
`role` This attribute specifies to which roles of the symbol the `presentation` element applies. The value of this attribute can be one of
    `applied` for situations, where the symbol occurs as a function symbol that is applied to a list of arguments, i.e. as the first child of an `om:OMA` or an `m:apply` element.
    `binding` for situations, where the symbol occurs as a binding symbol, i.e as the first child of an `om:OMBIND` element or an `m:apply` element that is followed by an `m:bvar` element.
    `key` for situations, where the symbol occurs as a key in an attribution, i.e. as a child of an `om:OMATTR` element at an odd position (Content-MATHML does not have the attribution construct).

---

[5] There the top-level generated `mrow` element corresponds to the application as specified by the path ".", whereas its first child corresponds to the quantifier symbol, and the bound variables correspond to each other.

In the examples in Figure 19.7 we have assumed the head to be an om:OMA element (for functional application). It can also be an om:OMBIND as in the case of a quantifier in Figure 19.8.

fixity This optional attribute can be one of the keywords prefix (the default), infix, postfix, and assoc. The value assoc has two variants: infixl and infixr, which have the same presentation; infixl is used for a binary infix operator that associates to the left like the list constructor in Standard ML, infixr is the right-leaning analogon.

If the fixity attribute is given, then it determines the placement of the symbol specified in the for attribute. For prefix it is placed in front of the arguments, (this is the generic mathematical function notation). For postfix the function is put behind the arguments, e.g. for derivatives: f'. The case infix is reserved for binary operators, where the function is inserted between the two arguments. Finally, assoc is used for associative operators like addition, it puts the function symbol between any two arguments.

Note that infix is almost a special case of assoc, but since it is reserved for binary operators, it disregards any arguments but the first two.

bracket-style The fixity information can be combined with the bracketing style, which can be either lisp (LISP-style brackets) or math (generic mathematical function notation which is the default).

Figure 19.7 shows some combinations of attributes and their results on the function style.

precedence allows us to specify the operator precedence in order to elide unnecessary brackets. The OMDoc presentation system orients itself on the PROLOG standard: lower precedences mean stronger binding, and brackets can be omitted. If we set the default precedence to 1000, and other precedences as specified in Figure 19.6, then the formulae below are presented as $(x + 2)^2$ and $x + y^2$, respectively.

```
<OMA>                                <OMA>
  <OMS cd="arith1" name="power"/>      <OMS cd="arith1" name="plus"/>
  <OMA>                                <OMV name="x"/>
    <OMS cd="arith1" name="plus"/>     <OMA>
    <OMV name="x"/>                      <OMS cd="arith1" name="power"/>
    <OMV name="y"/>                      <OMV name="y"/>
  </OMA>                                 <OMI>2</OMI>
  <OMI>2</OMI>                         </OMA>
</OMA>                                </OMA>
```

The next set of attributes can occur both in presentation and use elements. If they occur in both, then the values of those specified on the use elements take precedence over those specified in the dominating presentation element.

lbrack/rbrack These two attributes handle the brackets to be used in presentation of a complex expression. They will be used unless elided according to the precedence.

Precedence	Operators	Comment
200	+,-	unary
200	^	exponentiation
400	$*, \wedge, \cap$	multiplicative
500	$+, -, \vee, \cup$	additive
600	/	fraction
700	$=, \neq, \leq, <, >, \geq$	relation

Fig. 19.6. Common operator precedences

`separator` This specifies the separator in the argument list of a function. The default for `separator` is the comma. See Figure 19.7 for some combinations.

fixity	bracket-style	separator	yields
prefix	lisp	" "	$(f\ 1\ 2\ 3)$
postfix	lisp	" "	$(1\ 2\ 3\ f)$
prefix	math	","	$f(1, 2, 3)$
postfix	math	","	$(1, 2, 3)f$
assuming lbrack="(" and rbrack=")"			

Fig. 19.7. Attribute-combination and function style

`crossref-symbol` This attribute specifies to which parts of the symbol's presentation cross-references should be attached to: in some formats like HTML, and recently also in LATEX (thanks to the `hyperref.sty` package), it may be useful to attach a hyperlink from the presentation of the symbol to its definition. Some symbols are constructed by using the `lbrack` and `rbrack`, or the `separator` attributes as part of the symbol presentation. For instance, in the notation (a, b) for pairs, the binary function symbol for pairing is really composed of three parts "(", ")", and ",", which should all be cross-referenced. The attribute's values `no`, `yes`, `brackets`, `separator`, `lbrack`, `rbrack all` can be used to specify this behavior. `no` means cross-referencing is forbidden, `yes` – which is the default value – means cross-referencing only on the print-form of the function symbol, `lbrack`, `rbrack`, `brackets`, only on the left/right/both brackets, `separator`, on the separator, and finally `all` on all presentation parts.

In Figure 19.8, the effect of the default `yes` can be seen in the lower part of the figure: the LATEX and the HTML presentations have attached hyperlinks to the representation of the universal quantifier.

The next set of attributes can only appear on the `use` attribute, since they are only meaningful for selected output formats.

`format`, `xml:lang`, `requires` (see the specification for `xslt` and `style` above).

Notation specification	Example

`<presentation for="#forall"`
          `role="binding"`
          `separator=".">`
  `<use format="TeX">\forall</use>`
  `<use format="html">&#8704;</use>`
`</presentation>`
	`<OMBIND>`
	  `<OMS cd="quant1" name="forall"/>`
	  `<OMBVAR>`
	    `<OMV name="X"/>`
	  `</OMBVAR>`
	  `<OMS cd="logic1" name="true"/>`
	`</OMBIND>`

Using XSLT templates induced from the **presentation** element on the OPEN-MATH expression yields $\forall X.\text{true}$, where the glyph $\forall$ carries a hyperlink[6] to it definition, as the **crossref-symbol** on the **presentation** element has the default value **yes**. Internally, the hyperlinks are format-dependent, we have:

LATEX: `\href{../ocd/logic1.ps#true}{\forall}X.`
        `\href{../ocd/logic1.ps#true}{{\sf true}}`
HTML: `<a href="../ocd/logic1.html#forall">&#8704;</a> X.`
        `<a href="../ocd/logic1.html#true"><b>true</b></a>`

Fig. 19.8. Notation for `forall` (cf. Listing 19.4) using `presentation`

element, attributes, bracket-style These attributes simplify the specification of notations in XML-based formats like MATHML. The **element** attribute contains the name and the **attributes** the attribute declarations of an XML element that takes the place of the brackets specified in the attributes **lbrack** and **rbrack**. If the attribute **fixity** is used on a **use** element in conjunction with the **element** and **attributes** attributes, then it specifies the position of the element brackets rather than the brackets specified in the **lbrack** and **rbrack** attributes.

For instance, the binomial coefficient is some presented as $\binom{n}{m}$ (spoken "n choose m") and represented as

```
<mfrac linethickness='0'><mi>n</mi><mi>m</mi></frac>
```

in Presentation-MATHML. The first **presentation** element in Listing 19.9 shows a **presentation** element that has this effect. The second **presentation** element in Listing 19.9 shows a notation declaration, which applied to

```
<OMA><OMS cd="arith" name="power"/>
  <OMI>3</OMI><OMI>5</OMI>
</OMA>
```

would yield `3<sup>5</sup>` for the target **html**.

Listing 19.9. Presentation for binomial coefficients

```
<presentation for="#binomial" role="applied">
  <use format="default" fixity="infix">choose</use>
  <use format="TeX" lbrack="\bigl({" rbrack="}\bigr)">\atop</use>
  <use format="pmml" element="mfrac" attributes="linethickness='0'"/>
</presentation>

<presentation for="#power" role="applied" fixity="infix"
```

```
         crossref −symbol="no" precedence="200" bracket−style="lisp">
         <use format="html" fixity="prefix" bracket−style="math" element="sup"/>
10       <use format="TeX">^</use>
         <use format="pmml" element="msup" fixity="prefix"/>
       </presentation>
```

Conceptually, the attributes of the **presentation** and **use** elements form a meta-language for XSLT style sheets that aims at covering the most common notations succinctly and legibly. In situations, where this language does not suffice, we must fall back to to **style** or even **xslt** elements.

19.4 Presenting Bound Variables

As we have seen in Section 13.4, the presentation approaches for symbols do not work for (bound) variables[7], as there is no independent place to put the **presentation** element. In this section, we will present the OMDOC solution to this problem. The main idea is simply to annotate defining occurrences of variables with notation information. Without this, we are forced to use the ASCII variable name in OPENMATH and a translation of the Presentation-MATHML in the **m:ci** element for other formats in MATHML. This is hardly adequate for modern mathematics, where variables are numbered, decorated with primes or change marks, and cast in other colors or font families for better recognition.

In OMDOC we follow the spirit of the OPENMATH standard [BCC+04] which suggests to annotate (via om:OMATTR parts of) the OPENMATH objects with notation information by **presentation** elements. Unlike OPENMATH, we restrict this practice to defining occurrences of bound variables, since all the other constructs can be handled with the methods introduced above. We use the symbol `<OMS cd="omdoc" name="notation"/>` symbol to identify the following object as a notation declaration and the om:OMFOREIGN element to hold it.

Listing 19.10. Notation for bound variables in OPENMATH

```
   <OMOBJ>
     <OMBIND>
       <OMS cd="quant1" name="forall"/>
       <OMBVAR>
5        <OMATTR>
           <OMATP>
             <OMS cd="omdoc" name="notation"/>
             <OMFOREIGN encoding="application/omdoc+xml">
               <presentation for="#X">
10               <use format="TeX">X_4</use>
```

---

[7] We say that an om:OMBIND element binds a variable `<OMV name="x"/>`, iff this om:OMBIND element is the nearest one, such that `<OMV name="x"/>` occurs in (second child of the om:OMATTR element in) the om:OMBVAR child (this is the **defining occurrence** of `<OMV name="x"/>`). For content MATHML, the definition is analogous, only that an m:apply element with m:bvar child takes the role of the om:OMBIND and om:OMBVAR elements.

```
         <use format="pmml">
            <msub><mi>X</mi><mn>4</mn></msub>
         </use>
         <use format="html">X<sub>4</sub></use>
15       </presentation>
      </OMFOREIGN>
     </OMATP>
     <OMV name="X4"/>
    </OMATTR>
20   </OMBVAR>
    <OMA><OMS cd="relation1" name="eq"/>
    <OMV name="X4"/>
    <OMV name="X4"/>
    </OMA>
25  </OMBIND>
  </OMOBJ>
```

To represent binding objects in Content-MATHML we follow a very similar
strategy, using the `m:semantics` element to associate the defining occurrence
of the bound variable with its notation declaration, which is embedded into
the `m:annotation-xml` child.

Listing 19.11. Notation for bound variables in Content-MATHML

```
<m:math>
  <m:apply>
   <m:forall/>
   <m:bvar>
5    <m:semantics>
       <m:ci><m:msub><m:mi>X</m:mi><m:mn>4</m:mn></m:msub></m:ci>
       <m:annotation-xml encoding="application/xml+OMDoc"
         definitionURL="http://www.mathweb.org/omdoc/omdoc.omdoc#notation">
         <presentation for="#X">
10          <use format="TeX">X_4</use>
           <style format="pmml">
             <element name="msub" ns="http://www.w3.org/1998/Math/MathML">
               <element name="mi" ns="http://www.w3.org/1998/Math/MathML">
                 <text>X</text>
15             </element>
               <element name="mn" ns="http://www.w3.org/1998/Math/MathML">
                 <text>4</text>
               </element>
             </element>
20           </style>
           <style format="html">
             <text>X</text>
             <element name="sub" ns="http://www.w3.org/1999/xhtml">
               <text>4</text>
25           </element>
           </style>
         </presentation>
       </m:annotation-xml>
     </m:semantics>
30   </m:bvar>
   <m:apply><m:eq/><m:cn>4</m:cn><m:cn>4</m:cn></m:apply>
  </m:apply>
</m:math>
```

With these declarations, all the variables in the scope of the universal quantifier would be represented as X_4, yielding $\forall X_4 . X_4 = X_4$ which is exactly
what we wanted. Note that if we want to specify notations for function variables (OMDOC does not prevent the user from doing this), we need to also
specify notations for the non-applied occurrences of the symbol — otherwise

a fallback using the variable name has to be used. For instance, to make the (false) conjecture that all relations are symmetric we could use the following representation:

Listing 19.12. Notation for bound variables in OPENMATH

```
<OMOBJ xmlns="http://www.openmath.org/OpenMath">
  <OMBIND>
    <OMS cd="quant1" name="forall"/>
    <OMBVAR>
      <OMATTR>
        <OMATP>
          <OMS cd="omdoc" name="notation"/>
          <OMFOREIGN encoding="application/omdoc+xml">
            <presentation xmlns="http://www.mathweb.org/omdoc"
                          for="#R" role="applied" precedence="500" fixity="infix">
              <use format="TeX">\prec</use>
              <use format="pmml|html">&#x022DE;</use>
            </presentation>
            <presentation xmlns="http://www.mathweb.org/omdoc" for="#R">
              <use format="TeX">{}\prec{}</use>
              <use format="pmml|html">&#x022DE;</use>
            </presentation>
          </OMFOREIGN>
        </OMATP>
        <OMV name="R"/>
      </OMATTR>
      <OMV name="X"/>
    </OMBVAR>
    <OMA><OMV name="R"/><OMV name="X"/><OMV name="X"/></OMA>
  </OMBIND>
</OMOBJ>
```

This would give us the presentation $\forall \prec, X.X \prec X$. Here, the first occurrence of the variable $\prec$ is handled by the second notation declaration (it does not occur in applied position), the second occurrence of $\prec$ is in applied position, so the second notation declaration governs this and puts it in to infix position. Note that while OMDOC allows to specify this kind of notation declarations, they should be used with great care and discretion. In this particular case, the infix notation of $\prec$ de-emphasizes the variable nature, and might lead to confusion; moreover, the particular choice of the glyph $\prec$ may suggest irreflexivity, which may or may not be intended.

Auxiliary Elements (Module EXT)

Up to now, we have been mainly concerned with providing elements for marking up the inherent structure of mathematical knowledge in mathematical statements and theories. Now, we interface OMDOC documents with the Internet in general and mathematical software systems in particular. We can thereby generate presentations from OMDOC documents where formulae, statements or even theories that are active components that can directly be manipulated by the user or mathematical software systems. We call these documents **active documents**. For this we have to solve two problems: an abstract interface for calls to external (web) services[1] and a way of storing application-specific data in OMDOC documents (e.g. as arguments to the system calls).

The module EXT provides a basic infrastructure for these tasks in OM-DOC. The main purpose of this module is to serve as an initial point of entry. We envision that over time, more sophisticated replacements will be developed driven by applications.

Element	Attributes		D	Content
	Req.	Optional	C	
private		xml:id, for, theory, requires, type, reformulates, class, style	+	CMP*, data+
code		xml:id, for, theory, requires, type, class, style	+	CMP*, input?, output?, effect?, data+
input		xml:id, style, class	+	CMP*, FMP*
output		xml:id, style, class	+	CMP*, FMP*
effect		xml:id, style, class	+	CMP*, FMP*
data		format, href, size, original, pto, pto-version	−	<![CDATA[...]]>

Fig. 20.1. The OMDOC auxiliary elements for non-XML data

---

[1] Compare Chapter 9 in the OMDOC Primer.

20.1   Non-XML Data and Program Code in OMDoc

The representational infrastructure for mathematical knowledge provided by OMDoc is sufficient as an output- and library format for mathematical software systems like computer algebra systems, theorem provers, or theory development systems. In particular, having a standardized output- and library format like OMDoc will enhance system interoperability, and allows to build and deploy general storage and library management systems (see Section 26.4 for an OMDoc example). In fact this was one of the original motivations for developing the format.

However, most mathematical software systems need to store and communicate system-specific data that cannot be standardized in a general knowledge-representation format like OMDoc. Examples of this are pieces of program code, like tactics or proof search heuristics of tactical theorem provers or linguistic data of proof presentation systems. Only if these data can be integrated into OMDoc, it will become a full storage and communication format for mathematical software systems. One characteristic of such system-specific data is that it is often not in XML syntax, or its format is not fixed enough to warrant for a general XML encoding.

For this kind of data, OMDoc provides the `private` and code elements. As the name suggests, the latter is intended for program code[2] and the former for system-specific data that is not program code.

The attributes of these elements are almost identical and contain metadata information identifying system requirements and relations to other OMDoc elements. We will first describe the shared attributes and then describe the elements themselves.

`xml:id` for identification.

`theory` specifies the mathematical theory (see Section 15.6) that the data is associated with.

`for` allows to attach data to some other OMDoc element. Attaching `private` elements to OMDoc elements is the main mechanism for system-specific extension of OMDoc.

`requires` specifies other data this element depends upon as a whitespace-separated list of URI references. This allows to factor private data into smaller parts, allowing more flexible data storage and retrieval which is useful for program code or private data that relies on program code. Such data can be broken up into procedures and the call-hierarchy can be encoded in `requires` attributes. With this information, a storage application based on OMDoc can always communicate a minimal complete code set to the requesting application.

---

[2] There is a more elaborate proposal for treating program code in the OMDoc arena at [Koha], which may be integrated into OMDoc as a separate module in the future, for the moment we stick to the basic approach.

reformulates (`private` only) specifies a set of OMDoc elements whose
knowledge content is reformulated by the `private` element as a whitespace-separated list of URI references. For instance, the knowledge in the
assertion in Listing 20.2 can be used as an algebraic simplification rule in
the Analytica theorem prover [CKOS03] based on the Mathematica
computer algebra system.

The `private` and `code` elements contain an optional `metadata` element and
a set of `data` elements that contain or reference the actual data.

Listing 20.2. Reformulating mathematical knowledge

```
<assertion xml:id="ALGX0">
  <CMP>If a, b, c, d are numbers, then we have a + b(c + d) = a + bc + bd.</CMP>
</assertion>
<private xml:id="alg−expr−1" pto="Analytica" reformulates="ALGX0">
  <data format="mathematica−5.0">
    <![CDATA[SIMPLIFYRULES[a_ + b_*(c_ + d_) :> a + b*c + b*d /; NumberQ[b]]]]>
  </data>
</private>
```

The **data** element contains the data in a `CDATA` section. Its `pto` attribute
contains a whitespace-separated list of URI references which specifies the set
of systems to which the data are related. The intention of this field is that
the data is visible to all systems, but should only manipulated by a system
that is mentioned here. The `pto-version` attribute contains a whitespace-
separated list of version number strings; this only makes sense, if the value
of the corresponding `pto` is a singleton. Specifying this may be necessary, if
the data or even their format change with versions.

If the content of the `data` element is too large to store directly in the
OMDoc or changes often, then the `data` element can be augmented by a
link, specified by a URI reference in the `href` attribute. If the `data` element
is non-empty and there is a `href`[3], then the optional attribute `original`
specifies whether the `data` content (value `local`) or the external resource
(value `external`) is the original. The optional `size` attribute can be used
to specify the content size (if known) or the resource identified in the `href`
attribute. The `data` element has the (optional) attribute `format` to specify
the format the data are in, e.g. `image/jpeg` or `image/gif` for image data,
`text/plain` for text data, `binary` for system-specific binary data, etc. It
is good practice to use the MIME types [FB96] for this purpose whenever
applicable. Note that in a `private` or `code` element, the `data` elements must
differ in their format attribute. Their order carries no meaning.

In Listing 20.3 we use a `private` element to specify data for an image[4]
in various formats, which is useful in a content markup format like OMDoc
as the transformation process can then choose the most suitable one for the
target.

---

[3] e.g. if the **data** content serves as a cache for the data at the URI, or the **data**
content fixes a snapshot of the resource at the URI

[4] actually Figure 4.1 from Chapter 4

Listing 20.3. A `private` element for an image

```
<private xml:id="legacy">
  <metadata>
    <dc:title>A fragment of Bourbaki's Algebra</dc:title>
    <dc:creator role="trl">Michael Kohlhase</dc:creator>
5   <dc:date action="created">2002-01-03T0703</dc:date>
    <dc:description>A fragment of Bourbaki's Algebra</dc:description>
    <dc:source>Nicolas Bourbaki, Algebra, Springer Verlag 1974</dc:source>
    <dc:type>Text</dc:type>
  </metadata>
10  <data format="application/x-latex" href="legacy.tex"/>
    <data format="image/jpg" href="legacy.jpeg"/>
    <data format="application/postscript" href="legacy.ps"/>
    <data format="application/pdf" href="legacy.pdf"/>
</private>
```

The `code` element is used for embedding pieces of program code into an
OMDOC document. It contains the documentation elements `input`, `output`,
and `effect` that specify the behavior of the procedure defined by the code
fragment. The `input` element describes the structure and scope of the input
arguments, `output` the outputs produced by calling this code on these elements, and `effect` any side effects the procedure may have. They contain
a multilingual group of `CMP` elements with an optional `FMP` group for a formal description. The latter may be used for program verification purposes.
If any of these elements are missing it means that we may not make any
assumptions about them, not that there are no inputs, outputs or effects. For
instance, to specify that a procedure has no side-effects we need to specify
something like

```
<effect><CMP>None.</CMP></effect>
```

These documentation elements are followed by a set of `data` elements that
contain or reference the program code itself. Listing 20.7 shows an example
of a `code` element used to store Java code for an applet.

Listing 20.4. The program code for a Java applet

```
<code xml:id="callMint" requires="org.riaca.cas">
  <metadata>
    <dc:description>
      The multiple integrator applet. It puts up a user interface, queries the user for a
5     function, which it then integrates by calling one of several computer algebra systems.
    </dc:description>
  </metadata>
  <data format="application/x-java-applet">
    <![CDATA[... ⟪the callMint code goes here⟫ ...]]>
10  </data>
  <input><CMP>None: the applet handles input itself.</CMP></input>
  <output><CMP>The result of the integration.</CMP></output>
  <effect><CMP>None.</CMP></effect>
</code>
```

20.2 Applets and External Objects in OMDoc

Web-based text markup formats like HTML have the concept of an external object or "applet", i.e. a program that can in some way be executed in the browser or web client during document manipulation. This is one of the primary format-independent ways used to enliven parts of the document. Other ways are to change the document object model via an embedded programming language (e.g. JavaScript). As this method (dynamic HTML) is format-dependent[5], it seems difficult to support in a content markup format like OMDoc.

The challenge here is to come up with a format-independent representation of the applet functionality, so that the OMDoc representation can be transformed into the specific form needed by the respective presentation format. Most user agents for these presentation formats have built-in mechanisms for processing common data types such as text and various image types. In some instances the user agent may pass the processing to an external application ("plug-ins"). These need information about the location of the object data, the MIME type associated with the object data, and additional values required for the appropriate processing of the object data by the object handler at run-time.

Element	Attributes		D	Content
	Req.	Optional	C	
omlet	data,	xml:id, action, show, actuate, class, style	+	((《CMP content》 | param)*,data*
param	name	value, valuetype	-	EMPTY

Fig. 20.5. The OMDoc elements for external objects

In OMDoc, we use the <u>omlet</u> element for applets. It generalizes the HTML applet concept in two ways: The computational engine is not restricted to plug-ins of the browser (we do not know what the result format and presentation engine will be) and the program code can be included in the OMDoc document, making document-centered computation easier to manage.

Like the xhtml:object tag, the omlet element can be used to wrap any text. In the OMDoc context, this means that the children of the omlet element can be any elements or text that can occur in the CMP element together with param elements to specify the arguments. The main presentation intuition is that the applet reserves a rectangular space of a given pre-defined size (specified in the CSS markup in the style attribute; see Listing 20.7) in the result document presentation, and hands off the presentation and interaction with the document in this space to the applet process. The data for the external object is referenced in two possible ways. Either via the data

---

[5] In particular, the JavaScript references the HTML DOM, which in our model is created by a presentation engine on the fly.

attribute, which contains a URI reference that points to an OMDOC code or private element that is accessible (e.g. in the same OMDOC) or by embedding the respective code or private elements as children at the end of the omlet element. This indirection allows us to reuse the machinery for storing code in OMDOCs. For a simple example see Listing 20.7.

The behavior of the external object is specified in the attributes action, show and actuate attributes[6].

The action specified the intended action to be performed with the data. For most objects, this is clear from the MIME type. Images are to be displayed, audio formats will be played, and application-specific formats are passed on to the appropriate plug-in. However, for the latter (and in particular for program code), we might actually be interested to display the data in its raw (or suitably presented) form. The action addresses this need, it has the possible values execute (pass the data to the appropriate plug-in or execute the program code), display (display it to the user in audio- or visual form), and other (the action is left unspecified).

The show attribute is used to communicate the desired presentation of the ending resource on traversal from the starting resource. It has one of the values new (display the object in a new document), replace (replace the current document with the presentation of the external object), embed (replace the omlet element with the presentation of the external object in the current document), and other (the presentation is left unspecified).

The actuate attribute is used to communicate the desired timing of the action specified in the action attribute. Recall that OMDOC documents as content representations are not intended for direct viewing by the user, but appropriate presentation formats are derived from it by a "presentation process" (which may or may not be incorporated into the user agent). Therefore the actuate attribute can take the values onPresent (when the presentation document is generated), onLoad (when the user loads the presentation document), onRequest (when the user requests it, e.g. by clicking in the presentation document), and other (the timing is left unspecified).

The simplest form of an omlet is just the embedding of an external object like an image as in Listing 20.6, where the data attribute points to the private element in Listing 20.3. For presentation, e.g. as XHTML in a modern browser, this would be transformed into an xhtml:object element [Gro00], whose specific attributes are determined by the information in the omlet element here and those data children of the private element specified in the data attribute of the omlet that are chosen for presentation in XHTML. If the action specified in the action attribute is impossible (e.g. if the contents of the data target cannot be presented), then the content of the omlet element is processed as a fallback.

---

[6] These latter two attributes are modeled after the XLINK [DMOT01] attributes show and actuate.

Listing 20.6. An `omlet` for an image

```
<omlet data="#legacy" show="embed">A Fragment of Bourbaki's Algebra</omlet>
```

In Listing 20.7 we present an example of a conventional Java applet in a mathematical text: the `data` attribute points to a `code` element, which will be executed (if the value of the `action` attribute were `display`, the code would be displayed).

Listing 20.7. An `omlet` that Calls the Java Applet from Listing 20.4.

```
<omtext xml:id="monp_1">
  <CMP>
    <p>Let practice integration!</p>
    <p><omlet data="#callMint" action="execute" style="width:320;height:200">
       No plug-in found for callMint!
    </omlet></p>
  </CMP>
</omtext>
```

In this example, the Java applet did not need any parameters (compare the documentation in the `input` element in Listing 20.4).

In the applet in Listing 20.8 we assume a code fragment or plug-in (in a `code` element whose `xml:id` attribute has the value `sendtoTP`, which we have not shown) that processes a set of named arguments (parameter passing with keywords) and calls the theorem prover, e.g. via a web-service as described in Chapter 9.

Listing 20.8. An `omlet` for connecting to a theorem prover

```
<CMP> Let us prove it interactively:
  <omlet data="#sendtoTP" action="display">
    <param name="timeout" value="30" valuetype="data"/>
    <param name="performative" value="prove"/>
    <param name="problem" value="#ALGX0" valuetype="object"/>
    <param name="description" value="http://example.org/prob17.html" valuetype="ref"/>
    <param name="instance">
      <OMOBJ>
        <OMA><OMS name="root" cd="arith1"/>
          <OMI>3</OMI><OMI>3</OMI>
        </OMA>
      </OMOBJ>
    </param>
    Sorry, no theorem prover available!
  </omlet>
</CMP>
```

For parameter passing, we use the `param` elements which specify a set of values that may be required to process the object data by a plug-in at run-time. Any number of `param` elements may appear in the content of an `omlet` element. Their order does not carry any meaning. The `param` element carries the attributes

name This required attribute defines the name of a run-time parameter, assumed to be known by the plug-in. Any two `param` children of an `omlet` element must have different `name` values.

value This attribute specifies the value of a run-time parameter passed to the plug-in for the key name. Property values have no meaning to OMDoc; their meaning is determined by the plug-in in question.

valuetype This attribute specifies the type of the value attribute. The value data (the default) means that the value of the value will be passed to the plug-in as a string. The value ref specifies that the value of the value attribute is to be interpreted as a URI reference that designates a resource where run-time values are stored. Finally, the value object specifies that the value value points to a private or code element that contains a multi-format collection of data elements that carry the data.

If the param element does not have a value attribute, then it may contain a list of mathematical objects encoded as om:OMOBJ, m:mathml, or legacy elements.

Exercises (Module QUIZ)

Exercises and study problems are vital parts of mathematical documents like textbooks or exams, in particular, mathematical exercises contain mathematical vernacular and pose the same requirements on context like mathematical statements. Therefore markup for exercises has to be tightly integrated into the document format, so OMDOC provides a module for them.

Note that the functionality provided in this module is very limited, and largely serves as a place-holder for more pedagogically informed developments in the future (see Section 26.8 and [GMUC03] for an example in the OMDOC framework).

Element	Attributes		D	Content
	Req.	Optional	C	
exercise		xml:id, class, style	+	CMP*,FMP*,hint?,(solution*|mc*)
hint		xml:id, class, style	+	CMP*, FMP*
solution		xml:id, for, class, style	+	⟪top-level element⟫
mc		xml:id, for, class, style		choice, hint?, answer
choice		xml:id, class, style	+	CMP*, FMP*
answer	verdict	xml:id, class, style	+	CMP*, FMP*

Fig. 21.1. The OMDOC auxiliary elements for exercises

The QUIZ module provides the top-level elements **exercise**, hint, and solution. The first one is used for exercises and assessments. The question statement is represented in the multilingual CMP group followed by a multi-logic FMP group. This information can be augmented by hints (using the hint element) and a solution/assessment block (using the solution and mc elements).

The hint and solution elements can occur as children of exercise; or outside, referencing it in their optional for attribute. This allows a flexible positioning of the hints and solutions, e.g. in separate documents that can be distributed separately from the exercise elements. The hint element contains a CMP/FMP group for the hint text. The solution element can contain any number of OMDOC top-level elements to explain and justify the solution.

This is the case, where the question contains an assertion whose proof is not displayed and left to the reader. Here, the `solution` contains a proof.

Listing 21.2. An exercise from the TEXBook

```
   <exercise xml:id="TeXBook−18−22">
     <CMP>
       <p>Sometimes the condition that defines a set is given as a  fairly  long
         English description ; for  example consider '{p|p and p+2 are prime}'. An
5        hbox would do the job:</p>

       <p style="display:block;font−family:fixed">
         $\{\,p\mid\hbox{$p$ and $p+2$ are prime}\,\}$
       </p>
10
       <p>but a long formula like this is  troublesome in a paragraph, since an hbox cannot
         be broken between lines, and since the glue  inside  the
         <phrase style="font−family:fixed">\hbox</phrase> does not vary with the inter−word
         glue  in  the  line  that contains it . Explain how the given formula could be
15       typeset with line breaks.</p>
     </CMP>  <hint>
       <CMP>Go back and forth between math mode and horizontal mode.</CMP>
     </hint>
     <solution>
20     <CMP>
         <phrase style="font−family:fixed">
           $\{\,p\mid p$~and $p+2$ are prime$\,\}$
         </phrase>,
         assuming that <phrase style="font−family:fixed">\mathsurround</phrase> is
25       zero . The more difficult  alternative  '<phrase style="font−family:fixed">
         $\{\,p\mid p\\ {\rm and}\ p+2\rm\ are\ prime\,\}$</phrase>'
         is  not a solution , because line  breaks do not occur at
         <phrase style="font−family:fixed">\_</phrase> (or at glue of any
         kin) within math formulas. Of course it  may be best to display a formula  like
30       this , instead  of  breaking  it  between lines.
       </CMP>
     </solution>
   </exercise>
```

Multiple-choice exercises (see Listing 21.3) are represented by a group of `mc` elements inside an `exercise` element. An `mc` element represents a single choice in a multiple choice element. It contains the elements below (in this order).

choice for the description of the choice (the text the user gets to see and is asked to make a decision on). The <u>choice</u> element carries the `xml:id`, `style`, and `class` attributes and contains a `CMP/FMP` group for the text.

hint (optional) for a hint to the user, see above for a description.

answer for the feedback to the user. This can be the correct answer, or some other feedback (e.g. another hint, without revealing the correct answer). The `verdict` attribute specifies the truth of the answer, it can have the values `true` or `false`. This element is required, inside a `mc`, since the `verdict` is needed. It can be empty if no feedback is available. Furthermore, the <u>answer</u> element carries the `xml:id`, `style`, and `class` attributes and contains a `CMP/FMP` group for the text.

Listing 21.3. A multiple-choice exercise in OMDoc

```
   <exercise for="#ida.c6s1p4.l1" xml:id="ida.c6s1p4.mc1">
     <CMP>
     What is the unit element of the semi−group Q with operation a ∗ b = 3ab?
     </CMP>
5    <mc>
       <choice><FMP><OMOBJ><OMI>1</OMI></OMOBJ></FMP></choice>
       <answer verdict="false"><CMP>No, 1 ∗ 1 = 3 and not 1</CMP></answer>
     </mc>
     <mc>
10     <choice><CMP>1/3</CMP></choice>
       <answer verdict="true"></answer>
     </mc>
     <mc>
       <choice><CMP>It has no unit.</CMP></choice>
15     <answer verdict="false"><CMP>No, try another answer</CMP></answer>
     </mc>
   </exercise>
```

Document Models for OMDoc

In almost all XML applications, there is a tension between the document view
and the object view of data; after all, XML is a document-oriented interoperability framework for exchanging data objects. The question, which view is
the correct one for XML in general is hotly debated among XML theorists.
In OMDoc, actually both views make sense in various ways. Mathematical
documents are the objects we try to formalize, they contain knowledge about
mathematical objects that are encoded as formulae, and we arrive at content markup for mathematical documents by treating knowledge fragments
(statements and theories) as objects in their own right that can be inspected
and reasoned about.

In Chapters 13 to 21, we have defined what OMDoc documents look like
and motivated this by the mathematical objects they encode. But we have
not really defined the properties of these documents as objects themselves
(we will speak of the OMDoc **document object model** (OMDOM)). To
get a feeling for the issues involved, let us take stock of what we mean by the
object view of data. In mathematics, when we define a class of mathematical
objects (e.g. vector spaces), we have to say which objects belong to this class,
and when they are to be considered equal (e.g. vector spaces are equal, iff
they are isomorphic). When defining the intended behavior of operations,
we need to care only about objects of this class, and we can only make
use of properties that are invariant under object equality. In particular, we
cannot use properties of a particular realization of a vector space that are not
preserved under isomorphism. For document models, we do the same, only
that the objects are documents.

22.1 XML Document Models

XML supports the task of defining a particular class of documents (e.g. the
class of OMDoc documents) with formal grammars such as the document
type definition (DTD) or an XML schema, that can be used for mechanical

document validation. Surprisingly, XML leaves the task of specifying document equality to be clarified in the (informal) specifications, such as this OMDOC specification. As a consequence, current practice for XML applications is quite varied. For instance, the OPENMATH standard (see [BCC+04] and Section 13.1) gives a mathematical object model for OPENMATH objects that is specified independently of the XML encoding. Other XML applications like e.g. presentation MATHML [ABC+03] or XHTML [Gro00] specify models in form of the intended screen presentation, while still others like the XSLT [Dea99] give the operational semantics.

For a formal definition let $\mathcal{K}$ be a set of documents. We take a **document model** to be a partial equivalence relation[1] $\mathcal{X}$ on documents, such that $\{d|d\mathcal{X}d\} = \mathcal{K}$. In particular, a relation $\mathcal{X}$ is an equivalence relation on $\mathcal{K}$. For a given document model $\mathcal{X}$, let us say that two documents d and d' are $\mathcal{X}$**-equal**, iff $d\mathcal{X}d'$. We call a property p $\mathcal{X}$**-invariant**, iff for all $d\mathcal{X}d'$, p holds on d whenever p holds on d'.

A possible source of confusion is that documents can admit more than one document model (see [KK05] for an exploration of possible document models for mathematics). Concretely, OMDOC documents admit the OM-DOC document model that we will specify in section Section 22.2 and also the following four XML document models that can be restricted to OMDOC documents (as a relation).[2]

The binary document model interprets files as sequences of bytes. Two documents are equal, iff they are equal as byte sequence. This is the most concrete and fine-grained (and thus weakest) document model imaginable.

The lexical document model interprets binary files as sequences of Unicode characters [Inc03] using an encoding table. Two files may be considered equal by this document model even though they differ as binary files, if they have different encodings that map the byte sequences to the same sequence of UNICODE characters.

The XML syntax document model interprets UNICODE Files as sequences consisting of an XML declaration, a DOCTYPE declaration, tags, entity references, character references, CDATA sections, PCDATA comments, and processing instructions. At this level, for instance, whitespace characters between XML tags are irrelevant, and XML documents may be considered the same, if they are different as UNICODE sequences.

The XML structure document model interprets documents as XML trees of elements, attributes, text nodes, processing instructions, and sometimes comments. In this document model the order of attribute declarations in XML elements is immaterial, double and single quotes can be used interchangeably for strings, and XML comments (`<!--...-->`) are ignored.

---

[1] A partial equivalence relation is a symmetric transitive relation. We will use $[d]_{\mathcal{X}}$ for the **equivalence class** of d, i.e. $[d]_{\mathcal{X}} := \{e|d\mathcal{X}e\}$

[2] Here we follow Eliotte Rusty Harold's classification of layers of XML processing in [Har03], where he distinguishes the binary, lexical, sequence, structure, and semantic layer, the latter being the document model of the XML application

Each of these document models, is suitable for different applications, for instance the lexical document model is the appropriate one for Unicode-aware editors that interpret the encoding string in the XML declaration and present the appropriate glyphs to the user, while the binary document model would be appropriate for a simple ASCII editor. Since the last three document models are refinements of the XML document model, we will recap this in the next section and define the OMDOC document model in Section 22.2.

To get a feeling for the issues involved, let us compare the OMDOC elements in Listings 22.1 to 22.3 below. For instance, the serialization in Listing 22.2 is XML-equal to the one in Listing 22.1, but not to the one in Listing 22.3.

Listing 22.1. An OMDOC definition

```
<definition xml:id="comm−def" for="#comm">
  <CMP xml:lang="en">
    An operation <OMOBJ id="op"><OMV name="op"/></OMOBJ>
    is called commutative, iff
5   <OMOBJ id="comm1">
      <OMA><OMS cd="relation1" name="eq"/>
        <OMA><OMV name="op"/><OMV name="X"/><OMV name="Y"/></OMA>
        <OMA><OMV name="op"/><OMV name="Y"/><OMV name="X"/></OMA>
      </OMA>
10    </OMOBJ> for all <OMOBJ id="x"><OMV name="X"/></OMOBJ>
    and <OMOBJ id="y"><OMV name="Y"/></OMOBJ>.
  </CMP>
  <CMP xml:lang="de">
    Eine Operation <OMOBJ><OMR href="#op"/></OMOBJ> heißt kommutativ, falls
15  <OMOBJ><OMR href="#comm1"/></OMOBJ> für alle
    <OMOBJ><OMR href="#x"/></OMOBJ> und
    <OMOBJ><OMR href="#y"/></OMOBJ>.
  </CMP>
</definition>
```

Listing 22.2. An XML-equal serialization for Listing 22.1

```
<definition for="#comm" xml:id="comm−def" >
  ...
  <CMP xml:lang='de'> <!−− Note the unabbreviated empty element −−>
    Eine Operation <OMOBJ><OMR href="#op"/></OMOBJ> heißt
5   kommutativ, falls <OMOBJ><OMR href='comm1'/></OMOBJ> für alle
    <OMOBJ><OMR href="#x"/></OMOBJ> und
    <OMOBJ><OMR href='y'/></OMOBJ>.
  </CMP>
</definition>
```

22.2 The OMDoc Document Model

The OMDOC document model extends the XML structure document model in various ways. We will specify the equality relation in the table below, and discuss a few general issues here.

The OMDOC document model is guided by the notion of content markup for mathematical documents. Thus, two document fragments will only be considered equal, if they have the same abstract structure. For instance, the order of `CMP` children of an `omtext` element is irrelevant, since they form a multilingual group which form the base for multilingual text assembly. Other facets of the OMDoc document model are motivated by presentation-independence, for instance the distribution of whitespace is irrelevant even in text nodes, to allow formatting and reflow in the source code, which is not considered to change the information content of a text.

Listing 22.3. An OMDoc-equal representation for Listings 22.1 and 22.2

```
   <definition xml:id="comm−def" for="#comm">
     <CMP xml:lang="de">Eine Operation <OMOBJ><OMR href="#op"/></OMOBJ>
       heißt kommutativ, falls
       <OMOBJ id="comm1">
5        <OMA><OMS cd="relation1" name="eq"/>
           <OMA><OMV name="op"/><OMV name="X"/><OMV name="Y"/></OMA>
           <OMA><OMV name="op"/><OMV name="Y"/><OMV name="X"/></OMA>
         </OMA>
       </OMOBJ> für alle <OMOBJ><OMR href="#x"/></OMOBJ> und
10     <OMOBJ><OMR href="#y"/></OMOBJ>.
     </CMP>
     <CMP xml:lang="en">
       An operation <OMOBJ id="op"><OMV name="op"/></OMOBJ>
       is called commutative, iff <OMOBJ><OMR href="#comm1"/></OMOBJ>
15     for all <OMOBJ id="x"><OMV name="X"/></OMOBJ> and
       <OMOBJ id="y"><OMV name="Y"/></OMOBJ>.
     </CMP>
   </definition>
```

Compared to other document models, this is a rather weak (but general) notion of equality. Note in particular, that the OMDoc document model does *not* use mathematical equality here, which would make the formula $X + Y = Y + X$ (the om:OMOBJ with `xml:id="comm1"` in Listing 22.3 instantiated with addition for `op`) mathematically equal to the trivial condition $X+Y = X+Y$, obtained by exchanging the right hand side $Y + X$ of the equality by $X + Y$, which is mathematically equal (but not OMDoc-equal).

Let us now specify (part of) the equality relation by the rules in the table in Figure 22.4. We have discussed a machine-readable form of these equality constraints in the XML schema for OMDoc in [KA03].

#	Rule	comment	elements
1	unordered	The order of children of this element is irrelevant (as far as permitted by the content model). For instance only the order of obligation elements in the axiom-inclusion element is arbitrary, since the others must precede them in the content model.	`adt  axiom-inclusion` `metadata  symbol  code` `private  presentation` `omstyle`
2	multi-group	The order between siblings elements does not matter, as long as the values of the key attributes differ.	`CMP  FMP  requation` `dc:description  sortdef` `data dc:title solution`
3	DAG encoding	Directed acyclic graphs built up using om:OMR elements are equal, iff their tree expansions are equal.	`om:OMR ref`
4	Dataset	If the content of the dc:type element is `Dataset`, then the order of the siblings of the parent `metadata` element is irrelevant.	`dc:type`

Fig. 22.4. The OMDoc document model

The last rule in Figure 22.4 is probably the most interesting, as we have seen in Chapter 11, OMDOC documents have both formal and informal aspects, they can contain *narrative* as well as *narrative-structured* information. The latter kind of document contains a formalization of a mathematical theory, as a reference for automated theorem proving systems. There, logical dependencies play a much greater role than the order of serialization in mathematical objects. We call such documents **content OMDoc** and specify the value `Dataset` in the `dc:type` element of the OMDOC metadata for such documents. On the other extreme we have human-oriented presentations of mathematical knowledge, e.g. for educational purposes, where didactic considerations determine the order of presentation. We call such documents **narrative-structured** and specify this by the value `Text` (also see the discussion in Section 12.1)

22.3   OMDoc Sub-Languages

In the last chapters we have described the OMDOC modules. Together, they make up the OMDOC document format, a very rich format for marking up the content of a wide variety of mathematical documents. (see Part II for some worked examples). Of course not all documents need the full breadth of OMDOC functionality, and on the other hand, not all OMDOC applications (see Part IV for examples) support the whole language.

One of the advantages of a modular language design is that it becomes easy to address this situation by specifying sub-languages that only include part of the functionality. We will discuss plausible OMDOC sub-languages and their applications that can be obtained by dropping optional modules from OMDOC. Figure 22.5 visualizes the sub-languages we will present in this chapter. The full language OMDOC is at the top, at the bottom is a minimal sub-language OMDOC Basic, which only contains the required modules (mathematical documents without them do not really make sense). The arrows signify language inclusion and are marked with the modules acquired in the extension.

The sub-language identifiers can be used as values of the `modules` attribute on the `omgroup` and `omdoc` elements. Used there, they abbreviate the list of modules these sub-languages contain.

22.3.1   Basic OMDOC

Basic OMDOC is sufficient for very simple mathematical documents that do not introduce new symbols or concepts, or for early (and non-specific) stages in the migration process from legacy representations of mathematical material (see Section 4.2). This OMDOC sub-language consists of five modules: we need module MOBJ for mathematical objects and formulae, which are present in almost all mathematical documents. Module DOC provides the

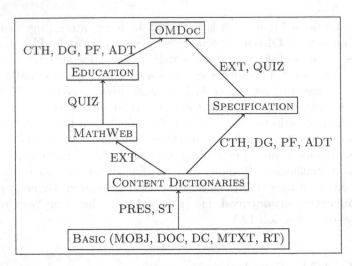

Fig. 22.5. OMDoc sub-languages and modules

document infrastructure, and in particular, the root element `omdoc`. We need DC for titles, descriptions, and administrative metadata, and module MTXT so we can state properties about the mathematical objects in `omtext` element. Finally, module RT allows to structured text below the `omtext` level. This module is not strictly needed for basic OMDoc, but we have included it for convenience.

22.3.2   OMDoc Content Dictionaries

Content Dictionaries are used to define the meaning of symbols in the OPEN-MATH standard [BCC+04], they are the mathematical documents referred to in the `cd` attribute of the `om:OMS` element. To express content dictionaries in OMDoc, we need to add the module ST to Basic OMDoc. It provides the possibility to specify the meaning of basic mathematical objects (symbols) by axioms and definitions together with the infrastructure for inheritance, and grouping, and allows to reference the symbols defined via their home theory (see the discussion in Section 15.6).

With this extension alone, OMDoc content dictionaries add support for multilingual text, simple inheritance for theories, and document structure to the functionality of OPENMATH content dictionaries. Furthermore, OMDoc content dictionaries allow the conceptual separation of mathematical properties into constitutive ones and logically redundant ones. The latter of these are not strictly essential for content dictionaries, but enhance maintainability and readability, they are included in OPENMATH content dictionaries for documentation and explanation.

The sub-language for OMDoc content dictionaries also allows the specification of notations for the introduced symbols (by module PRES). So the resulting documents can be used for referencing (as in OPENMATH) and as a resource for deriving presentation information for the symbols defined here. To get a feeling for this sub-language, see the example in the OMDOC variant of the OPENMATH content dictionary `arith1` in Chapter 5, which shows that the OPENMATH content dictionary format is (isomorphic to) a subset of the OMDOC format. In fact, the OPENMATH2 standard only presents the content dictionary format used here as one of many encodings and specifies abstract conditions on content dictionaries that the OMDOC encoding below also meets. Thus OMDOC is a valid content dictionary encoding.

22.3.3   Specification OMDOC

OMDOC content dictionaries are still a relatively lightweight format for the specification of meaning of mathematical symbols and objects. Large scale formal specification efforts, e.g. for program verification need more structure to be practical. Specification languages like CASL (Common Algebraic Specification Language [CoF04]) offer the necessary infrastructure, but have a syntax that is not integrated with web standards.

The Specification OMDOC sub-language adds the modules ADT and CTH to the language of OMDOC content dictionaries. The resulting language is equivalent to the CASL standard, see [AHMS00, Hut00, MAH06] for the necessary theory.

The structured definition schemata from module ADT allow to specify abstract data types, sets of objects that are inductively defined from constructor symbols. The development graph structure built on the theory morphisms from module CTH allow to make inclusion assertions about theories that structure fragments of mathematical developments and support a management of change.

22.3.4   MathWeb OMDOC

OMDOC can be used as a content-oriented basis for web publishing of mathematics. Documents for the web often contain images, applets, code fragments, and other data, together with mathematical statements and theories.

The OMDOC sub-language MathWeb OMDOC extends the language for OMDOC content dictionaries by the module EXT, which adds infrastructure for images, applets, code fragments, and other data.

22.3.5   Educational OMDOC

OMDOC is currently used as a content-oriented basis for various systems for mathematics education (see e.g. Chapter 8 for an example and discussion). The OMDOC sub-language Educational OMDOC extends MathWeb OMDOC by the module QUIZ, which adds infrastructure for exercises and assessments.

22.3.6 Reusing OMDoc **Modules in Other Formats**

Another application of the modular language design is to share modules with other XML applications. For instance, formats like DocBook [WM99] or XHTML [Gro00] could be extended with the OMDoc statement level. Including modules MOBJ, DC, and (parts of) MTXT, but not RT and DOC would result in content formats that mix the document-level structure of these formats. Another example is the combination of XML-RPC envelopes and OMDoc documents used for interoperability in Chapter 9.

OMDoc Applications, Tools, and Projects

In this part we will address current applications, tools and projects using the OMDoc format. We will first discuss the possibilities and tools of processing documents in the OMDoc format via style sheets with the purpose of generating documents specialized for consumption by other mathematical software systems, and by humans. Then we will present three projects descriptions that use OMDoc at the core.

23

OMDoc Resources

In this chapter we will describe various public resources for working with the OMDOC format.

23.1 The OMDoc Web Site, Wiki, and Mailing List

The main web site for the OMDOC format is `http://www.mathweb.org/omdoc`. It hosts news about developments, applications, collaborators, and events, provides access to an list of "frequently asked questions" (FAQ), and current and old OMDOC specifications and provides pre-generated examples from the OMDOC distribution.

There are two mailing lists for discussion of the OMDOC format:

`omdoc@mathweb.org` is for announcements and discussions of the OMDOC format on the user level. Direct your questions to this list.

`omdoc-dev@mathweb.org` is for developer discussions.

For subscription and archiving details see the OMDOC resources page for mailing lists [Kohd].

Finally, the OMDOC web site hosts a Wiki [OMDb] for user-driven documentation and discussion.

23.2 The OMDoc Distribution

All resources on the OMDOC web site are available from the MATHWEB SUB-VERSION repository [OMDa] for anonymous download. SUBVERSION (SVN) is a collaborative version control system – to support a distributed community of developers in accessing and developing the OMDOC format, software, and documentation, see [Mat] for a general introduction to the setup. The resources for version 1.2 of the OMDOC format which is described in this book are accessible on the web at `https://svn.mathweb.org/repos/`

mathweb.org/branches/omdoc-1.2 via a regular web browser[1]. The SVN server allows anonymous read access to the general public. To check out the OMDoc distribution, use

svn co https://svn.mathweb.org/repos/mathweb.org/branches/omdoc-1.2

This will create a directory omdoc, with the sub-directories

directories	content
bin, lib, oz, thirdParty	programs and third-party software used in the administration and examples
css, xsl	style sheets for displaying OMDoc documents on the web, see Chapter 25 for a discussion.
doc	The OMDoc documentation, including the specification, papers about a the OMDoc format and tools.
dtd, rnc	The OMDoc document type definition and the RELAXNG schemata for OMDoc
examples	Various example documents in OMDoc format.
projects	various contributed developments for OMDoc. Documentation is usually in their doc sub-directory

After the initial check out, the OMDoc distribution can be kept up to date by the command svn -q update in the top-level directory from time to time. To obtain write access contact svnadmin@mathweb.org.

23.3   The OMDoc Bug Tracker

MathWeb.org supplies a BugZilla bug-tracker [Bug05] at http://bugzilla.mathweb.org:8000 to aid the development of the OMDoc format. BugZilla is a server-based discussion forum and bug tracking system. We use it to track, archive and discuss tasks, software bugs, and enhancements in our project. Discussions are centered about threads called "bugs" (which need not be software bugs at all), which are numbered, can be searched, and can be referred to by their URL. To use BUGZILLA, just open an account and visit the OMDoc content by querying for the "product" OMDoc. For offline use of the bug-tracker we recommend the excellent DESKZILLA application [Des05], which is free for open-source projects like OMDoc.

Further development of the OMDoc format will be public and driven by the discussions on BUGZILLA, the OMDoc mailing list, and the OMDoc Wiki (see Section 23.1).

---

[1] Ongoing development of the OMDoc format can be accessed via the head revision of the repository at https://svn.mathweb.org/repos/mathweb.org/trunk/omdoc.

23.4 An XML Catalog for OMDoc

Many XML processes use system IDs (in practice URLs) to locate supporting files like DTDs, schemata, style sheets. To make them more portable, OM-Doc documents will often reference the files on the `mathweb.org` web server, even in situations, where they are accessible locally e.g. from the OMDoc distribution. This practice not only puts considerable load on this server, but also slows down or even blocks document processing, since the XML processors have to retrieve these files over the Internet.

Many processors can nowadays use XML catalogs to remap public identi-fiers and URLs as an alternative to explicit system identifiers. A catalog can convert public IDs like the one for the OMDoc DTD (`-//OMDoc//DTD OMDoc V1.2//EN`) into absolute URLs like `http://www.mathweb.org/omdoc/omdoc.dtd`. Moreover, it can replace remote URLs like this one with local URLs like `file:///home/kohlhase/omdoc/dtd/omdoc.dtd`. This offers fast, reliable access to the DTDs and schemata without making the documents less portable across systems and networks.

To facilitate the use of catalogs, the OMDoc distribution provides a catalog file `lib/omdoc.cat`. This catalog file can either be imported into the system's catalog[2] using a `nextCatalog` element of the form

---
`<nextCatalog xml:id="omdoc.cat" catalog="file:///home/kohlhase/omdoc/lib/omdoc.cat"/>`

---

or by making it known directly to the XML processor by an application-specific method. For instance for `libxml2` based tools like `xsltproc` or `xmllint`, it is sufficient to include the path to `omdoc.cat` in the value of the `XML_CATALOG_FILES` environment variable (it contains a whitespace-separated list of FILES).

23.5 External Resources

The OMDoc format has been used on a variety of projects. Chapter 26 gives an overview over some of the projects (use the project home pages given there for details), a up to date list of links to OMDoc projects can be found at `http://www.mathweb.org/omdoc/projects.html`. These projects have contributed tools, code, and documentation to the OMDoc format, often stressing their special vantage points and applications of the format.

---

[2] This catalog is usually at `file:///etc/xml/catalog` on UNIX systems; unfortu-nately there is no default location for WINDOWS machines.

Validating OMDoc Documents

In Chapter 1 we have briefly discussed the basics of validating XML documents by document type definitions (DTDs) and schemata. In this chapter, we will instantiate this discussion with the particulars of validating OMDOC documents.

Generally, DTDs and schemata are context-free grammars for trees[1], that can be used by a **validating parser** to reject XML documents that do not conform to the constraints expressed in the OMDOC DTD or schemata discussed here.

Note that none of these grammars can enforce all constraints that the OMDOC specification in Part III of this book imposes on documents. Therefore grammar-based validation is only a necessary condition for OMDOC-**validity**. Still, OMDOC documents should be validated to ensure proper function of OMDOC tools, such as the ones discussed in Chapters 25 and 26. Validation against multiple grammars gives the best results. With the current state of validation technology, there is no clear recommendation, which of the validation approaches to prefer for OMDOC. DTD validation is currently best supported by standard XML applications and supports default values for attributes. This allows the author who writes OMDOC documents by hand to elide implicit attributes and make the marked-up text more readable. XML- and RELAXNG schema validation have the advantage that they are namespace-aware and support more syntactic constraints. Neither of these support mnemonic XML entities, such as the ones used for UNICODE characters in Presentation-MATHML, so that these have to be encoded as UNICODE code points. Finally RELAXNG schemata do not fully support default val-

---

[1] Actually, a recent extension of the XML standard (XLINK) also allows to express graph structures, but the admissibility of graphs is not covered by DTD or current schema formalisms.

ues for attributes, so that OMDoc documents have to be normalized[2] to be
RelaxNG-valid.

We will now discuss the particulars of the respective validation formats.
As the RelaxNG schema is the most expressive and readable for humans
we consider it as the normative grammar formalism for OMDoc, and have
included it in Appendix D for reference.

24.1   Validation with Document Type Definitions

The OMDoc document type definition [Kohc] can be referenced by the public
identifier "-//OMDoc//DTD OMDoc V1.2//EN" (see Section 23.4). The DTD
driver file isomdoc.dtd, which calls various DTD modules.

DTD-validating XML parsers are included in almost all XML processors. The author uses the open-source RXP [Tob] and XMLLINT [Veia] as
stand-alone tools. If required, one may validate OMDoc documents using
an SGML parser such as nsgmls, rather than a validating XML parser. In
this case an SGML declaration defining the constraints of XML applicable
to an SGML parser must be used (see [Cla97] for details).

To allow DTD-validation, OMDoc documents should contain a document
typedeclaration of the following form:

<!DOCTYPE omdoc PUBLIC "-//OMDoc//DTD OMDoc V1.2//EN"
                "http://www.mathweb.org/omdoc/dtd/omdoc.dtd" >

The URI may be changed to that of a local copy of the DTD if required, or
it can be dropped altogether if the processing application has access to an
XML catalog (see Section 23.4). Whether it is useful to include document
type declarations in documents in a production environment depends on the
application. If a document is known to be DTD- or even OMDoc-valid, then
the validation overhead a DOCTYPE declaration would incur from a validating
parser[3] may be conserved by dropping it.

24.1.1   Parametrizing the DTD

The OMDoc DTD makes heavy use of parameter entities, so we will briefly
discuss them to make the discussion self-contained. Parameter entity declarations are declarations of the form

---

[2] An OMDoc document is called **normalized**, iff all required attributes are
present. Given a DTD or XML schema that specifies default values, there are
standard XML tools for XML-normalization that can be pipelined to allow RelaxNG validation, so this is not a grave restriction.
[3] The XML specification requires a validating parser to perform validation if a
DOCTYPE declaration is present

```
<!ENTITY % assertiontype "theorem|proposition|lemma|%otherassertiontype;">
```

in the DTD. This one makes the abbreviation %assertiontype; available for the string "theorem|proposition|lemma|observation" (in the DTD of the document in Listing 24.1). Note that parameter entities must be fully defined before they can be referenced, so recursion is not possible. If there are multiple parameter entity declarations, the first one is relevant for the computation of the replacement text; all later ones are discarded. The internal subset of document type declaration is pre-pended to the external DTD, so that parameter entity declarations in the internal subset overwrite the ones in the external subset.

The (external) DTD specified in the DOCTYPE declaration can be enhanced or modified by adding declarations in square brackets after the DTD URI. This part of the DTD is called the internal subset of the DOCTYPE declaration, see Listing 24.1 for an example, which modifies the parameter entity %otherassertiontype; supplied by the OMDOC DTD to extend the possible values of the type attribute in the assertion element for this document. As a consequence, the assertion element with the non-standard value for the type attribute is DTD-valid with the modified internal DTD subset.

Listing 24.1. A document type declaration with internal subset

```
<!DOCTYPE omdoc PUBLIC "-//OMDoc//DTD OMDoc V1.2//EN"
               "http://www.mathweb.org/omdoc/omdoc.dtd"
    [<!ENTITY % otherassertiontype "observation">]>
...
5  <assertion type="observation">...</assertion>
...
```

24.1.2 DTD-Based Normalization

Note that if a OMDOC fragment is parsed without a DTD, i.e. as a well-formed XML fragment, then the default attribute values will not be added to the XML information set. So simply dropping the DOCTYPE declaration may change the semantics of the document, and OMDOC documents should be normalized[4] first. Normalized OMDOC documents should carry the standalone attribute in the XML processing instruction, so that a normalized OMDOC document has the form given in Listing 24.2.

Listing 24.2. A normalized OMDOC document without DTD

```
<?xml version="1.0" standalone="true"?>
<omdoc xml:id="something" version="1.2" xmlns="http://www.mathweb.org/omdoc">
...
</omdoc>
```

---

[4] The process of DTD-normalization expands all parsed XML entities, and adds all default attribute values

The attribute `version` and the namespace declaration `xmlns` are fixed by the DTD, and need not be explicitly provided if the document has a `DOCTYPE` declaration.

24.1.3   Modularization

In OMDoc 1.2 the DTD has been modularized according to the W3C conventions for DTD modularization [ABD+01]. This partitions the DTD into **DTD modules** that correspond to the OMDoc modules discussed in Part III of this book.

These DTD modules can be deselected from the OMDoc DTD by changing the **module inclusion entities** in the local subset of the document type declaration. In the following declaration, the module PF (see Chapter 17) has been deselected, presumably, as the document does not contain proofs.

```
<!DOCTYPE omdoc PUBLIC "-//OMDoc//DTD OMDoc V1.2//EN"
                       "http://www.mathweb.org/omdoc/dtd/omdoc.dtd"
   [<!ENTITY % omdoc.pf.module "IGNORE">]>
```

Module inclusion entities have the form `%omdoc.`⟨*ModId*⟩`.module;`, where ⟨*ModId*⟩ stands for the lower-cased module identifier. The OMDoc DTD repository contains DTD driver files for all the sub-languages discussed in Section 22.3, which contain the relevant module inclusion entity settings. These are contained in the files `omdoc-`⟨*SlId*⟩`.dtd`, where ⟨*SlId*⟩ stands for the sub-language identifier.

Except for their use in making the OMDoc DTD more manageable, DTD modules also allow to include OMDoc functionality into other document types, extending OMDoc with new functionality encapsulated into modules or upgrading selected OMDoc modules individually. To aid this process, we will briefly describe the module structure. Following [ABD+01], DTD modules come in two parts, since we have inter-module recursion. The problem is for instance that the `omlet` element can occur in mathematical texts (`mtext`), but also contains `mtext`, which is also needed in other modules. Thus the modules cannot trivially be linearized. Therefore the DTD driver includes an entity file ⟨*ModId*⟩`.ent` for each module ⟨*ModId*⟩, before it includes the grammar rules in the **element modules** ⟨*ModId*⟩`.mod` themselves. The entity files set up parameter entities for the qualified names and content models that are needed in the grammar rules of other modules.

24.1.4   Namespace Prefixes for OMDoc Elements

Document type definitions do not natively support XML namespaces. However, clever coding tricks allow them to simulate namespaces to a certain extent. The OMDoc DTD follows the approach of [ABD+01] that parametrizes namespace prefixes in element names to deal gracefully with syntactic effects of namespaced documents like we have in OMDoc.

Recall that element names are **qualified name**s, i.e. pairs consisting of a namespace URI and a local name. To save typing effort, XML allows to abbreviate qualified names by namespace declarations via `xmlns` pseudo-attribute: the element and all its descendants are in this namespace, unless they have a namespace attribute of their own or there is a namespace declaration in a closer ancestor that overwrites it. Similarly, a namespace abbreviation can be declared on any element by an attribute of the form `xmlns:nsa="nsURI"`, where `nsa` is a name space abbreviation, i.e. a simple name, and `nsURI` is the URI of the namespace. In the scope of this declaration (in all descendants, where it is not overwritten) a qualified name `nsa:n` denotes the qualified name `nsURI:n`.

The mechanisms described in [ABD+01] provide a way to allow for namespace declarations even in the (namespace-agnostic) DTD setting simply by setting a parameter entity. If `NS.prefixed` is declared to be `INCLUDE`, using a declaration such as `<!ENTITY % NS.prefixed "INCLUDE">` either in the local subset of the `DOCTYPE` declaration, or in the DTD file that is including the OMDOC DTD, or the DTD modules presented in this appendix, then all OMDOC elements should be used with a prefix, for example `<omdoc:definition>`, `<omdoc:CMP>`, etc. The prefix defaults to `omdoc:` but another prefix may be declared by declaring in addition the parameter entity `omdoc.prefix`. For example, `<!ENTITY % omdoc.prefix "o">` would set the prefix for the OMDOC namespace to `o:`.

Note that while the Namespaces Recommendation [Bra99] provides mechanisms to change the prefix at arbitrary points in the document, this flexibility is not provided in this DTD (and is probably not possible to specify in any DTD). Thus, if a namespace prefix is being used for OMDOC elements, so that for example the root element is:

```
<omdoc:omdoc xmlns:omdoc="http://www.mathweb.org/omdoc">
...
</omdoc:omdoc>
```

then the prefix must be declared in the local subset of the DTD, as follows:

```
<!DOCTYPE omdoc:omdoc PUBLIC "-//OMDoc//DTD OMDoc V1.2//EN"
                "http://www.mathweb.org/omdoc/dtd/omdoc.dtd"
  [<!ENTITY % NS.prefixed "INCLUDE"><!ENTITY % omdoc.prefix "omdoc">]>
```

The OMDOC DTD references six namespaces:

language	namespace	prefix
MATHML	`http://www.w3.org/1998/Math/MathML`	`m:`
OPENMATH	`http://www.openmath.org/OpenMath`	`om:`
XSLT	`http://www.w3.org/1999/XSL/Transform`	`xsl:`
Dublin Core	`http://purl.org/dc/elements/1.1/`	`dc:`
Creative Commons	`http://creativecommons.org/ns`	`cc:`
OMDOC	`http://www.mathweb.org/omdoc`	`omdoc:`

These prefixes can be changed just like the OMDOC prefix above.

24.2   Validation with RelaxNG Schemata

RELAXNG [Vli03] is a relatively young approach to validation developed outside the W3C, whose XML schema paradigm was deemed overburdened. As a consequence, RELAXNG only concerns itself with validation, and leaves out typing, normalization, and entities. RELAXNG schemata come in two forms, in XML syntax (file name extension .rng) and in compact syntax (file name extension .rnc). We provide the RELAXNG schema [Kohe] as the normative validation schema for OMDOC. As compact syntax is more readily understandable by humans, we have reprinted it as the normative grammar for OMDOC documents in Appendix D. Just as in the case for the OMDOC DTD, we provide schemata for the OMDOC sub-languages discussed in Section 22.3. These are contained in the driver files omdoc-⟨⟨SlId⟩⟩.rnc, where ⟨⟨SlId⟩⟩ stands for the sub-language identifier.

There is currently no standard way to associate a RELAXNG schema with an XML document[5]; thus validation tools (see http://relaxng.org for an overview) have to be given a grammar as an explicit argument. One consequence of this is that the information that an OMDOC document is intended for an OMDOC sub-languages must be managed outside the document separately from the document.

There are various validators for RELAXNG schemata, the author uses the open-source XMLLINT [Veia] as a stand-alone tool, and the nXML mode [Cla05] for the EMACS editor [Sta02] for editing XML files, as it provides powerful RELAXNG-based editing support (validation, completion, etc.).

24.3   Validation with XML Schema

For validation[6] with respect to XML schemata (see [XML]) we provide an XML schema for OMDOC [Kohf], which is generated from the RELAXNG schema in Appendix D via the TRANG system described above. Again, schemata for the sub-languages discussed in Section 22.3 are provided as omdoc-⟨⟨SlId⟩⟩.rnc, where ⟨⟨SlId⟩⟩ stands for the sub-language identifier.

To associate an XML schema with an element, we need to decorate it with an xsi:schemaLocation attribute and the namespace declaration for XML schema instances. In Listing 24.3 we have done this for the top-level omdoc element, and thus for the whole document. Note that this mechanism makes mixing XML vocabularies much simpler than with DTDs, that can only be associated with whole documents.

---

[5] In fact this is not an omission, but a conscious design decision on the part of the RELAXNG developers.

[6] There are many schema-validating XML parsers, the author uses the open-source XMLLINT [Veia].

Listing 24.3. An XML document with an XML schema

```
<?xml version="1.0"?>
<omdoc xml:id="example.with.schema"
       xmlns:xsi="http://www.w3.org/2001/XMLSchema-instance"
       xsi:schemaLocation="http://www.mathweb.org/omdoc
                           http://www.mathweb.org/omdoc/xsd/omdoc.xsd">
   ...
</omdoc>
```

Transforming OMDoc by XSLT Style Sheets

In the introduction we have stated that one of the design intentions behind OMDoc is to separate content from presentation, and leave the latter to the user. In this section, we will briefly touch upon presentation issues. The technical side of this is simple: OMDoc documents are regular XML documents that can be processed by XSLT style sheets [XSL99] to produce the desired output formats. There are several high-quality XSLT transformers freely available including `saxon` [Kay], `xalan` [Fou], and `xsltproc`[Veib]. Moreover, XSLT is natively supported by the newest versions of the browsers MS Internet Explorer [Cor] and MOZILLA [Org].

XSLT style sheets can be used for several tasks in maintaining OM-Doc, such as for instance converting other XML-based input formats into OMDoc (e.g. `cd2omdoc.xsl` for converting OPENMATH content dictionaries into OMDoc format), or migrating between different versions of OMDoc e.g. the style sheet `omdoc1.1adapt1.2.xsl` that operationalizes all the syntax changes from Version 1.1 of OMDOC to version 1.2 (see Appendix A for a tabulation). We will now review a set of XSLT style sheets for OMDOC, they can be found in the OMDoc distribution (see Section 23.2) or on the web at [Kohg].

25.1 Extracting and Linking XSLT Templates

One of the main goals of content markup for mathematical documents is to be independent of the output format. In Chapter 19, we have specified the conceptual infrastructure provided by the OMDoc language, in this section we will discuss the software infrastructure needed to transform OMDoc documents into various formats.

The `presentation` elements for symbols in OPENMATH or Content-MATHML formulae allow a declarative specification of the result of transforming expressions involving these symbols into various formats. To use this information in XSLT style sheets, the content of `presentation` elements must be transformed into XSLT templates, and these must be linked into the generic transformation style sheet. The OMDoc distribution provides two meta-style-sheets for these tasks.

The first one — `expres.xsl` — compiles the content of the `presentation` and `omstyle` elements in the source file into XSLT templates. The style sheet takes the parameter `report-errors`, which is set to 'no' by default; setting it to 'yes' will enable more verbose error reports. The OMDoc distribution provides UNIX `Makefiles` that specify the target ⟨*base*⟩`-tmpl.xsl` for each OMDoc file ⟨*base*⟩`.omdoc`, so that the templates file can be generated by typing `make` ⟨*base*⟩`-tmpl.xsl`. Note that `expres.xsl` follows the references in the `ref` elements (it `ref`-normalizes the document see Section 11.5) before it generates the templates[1].

The second style sheet — `exincl.xsl` — generates link table for a specific OMDoc document. This style sheet `ref`-normalizes the document and outputs an XSLT style sheet that includes all the necessary template files. `expres.xsl` takes two parameters: `self` is the name of the source file name itself[2]. The `Makefiles` in the OMDoc distribution specify the target ⟨*base*⟩`-incl.xsl`, so that the link table can be generated by typing `make` ⟨*base*⟩`-incl.xsl`.

Let us now consider the example scenario in Figure 25.1: Given an OM-Doc document ⟨*document*⟩`.omdoc` that uses symbols from theories a, b, c, d, and e which are provided by the OMDoc documents ⟨*background*⟩`.omdoc`, ⟨*special*⟩`.omdoc` and ⟨*local*⟩`.omdoc`, we need to generate the template files ⟨*background*⟩`-tmpl.xsl`, ⟨*special*⟩`-tmpl.xsl`, and ⟨*local*⟩`-tmpl.xsl` (via `expres.xsl`) as well as ⟨*document*⟩`-incl.xsl` (via `exincl.xsl`). Now it is only necessary to include the link table ⟨*document*⟩`-incl.xsl` into a generic transformation style sheet to specialize it with the notation information specified in the `presentation` elements in theories a, b, c, d, and e.

The transformation architecture based on the `Makefiles` provided with the OMDoc distribution does the linking by creating a specialized style sheet ⟨*document*⟩`2html.xsl` that simply includes the generic OMDoc transformation style sheet `omdoc2html.xsl` (see Section 25.3), and the style sheet ⟨*document*⟩`-incl.xsl`. Changing this simple control style sheet allows to add site- or language-specific templates (by adding them directly or including respective style sheets). An analogous processing path leads ⟨*document*⟩`.tex` using `omdoc2tex.xsl` and from there to PDF using tools like `pdflatex`.

---

[1] In the current implementation, `expres.xsl` generates one large template that combines the XSLT code for all target formats. This simplifies the treatment of the default presentations as requested by the specification in Section 19.3, but hampers mixing presentation information from multiple sources. An implementation based on modes would probably have advantages in this direction in the long run.

[2] For some reason XSLT processors do not provide access to this information portably.

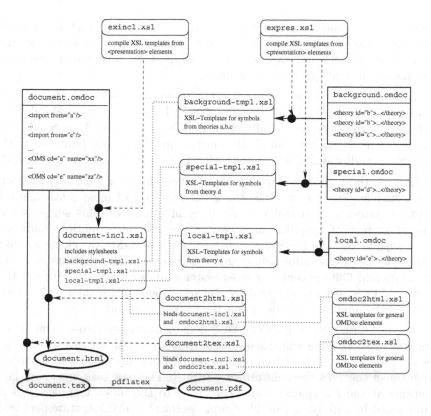

Fig. 25.1. The OMDOC presentation process

Other processing architectures may be built up using on-demand technologies, e.g. servlets, mediators, or web services, but will be able to follow the same general pattern as our simpleminded implementation in `Makefiles`.

We will make use of this general architecture based on extraction and linking via XSLT style sheets in the transformation of OMDOC documents below.

25.2 OMDOC Interfaces for Mathematical Software Systems

One of the original goals of the OPENMATH, Content-MATHML and OMDOC languages is to provide a communication language for mathematical software systems. The main idea behind this is to supply systems with interfaces to a universally accepted communication language standard (an interlingua), and so achieve interoperability for n systems with only $2n$ translations instead of n^2. As we have seen in Section 2.1, OPENMATH and Content-MATHML provide a good solution at the level of mathematical objects, which is sufficient for systems like computer algebra systems. OMDOC adds the level of mathematical statements and theories to add support for automated reasoning systems and formal specification systems.

To make practical use of the OMDOC format as an interlingua, we have to support building OMDOC interfaces. An XSLT style sheet is a simple way to come up with (the input half) of an OMDOC interface. A more efficient way would be to integrate an XML directly into the system (suitable XML parsers are readily available for almost all programming languages nowadays).

Usually, the task of writing an XSLT style sheet for such a conversion is a relatively simple task, since the input language of most mathematical software system is isomorphic to a subset of OMDOC. This suggests the general strategy of applying the necessary syntax transformations (this has to be supplied by the style sheet author) on those OMDOC elements that carry system-relevant information and transforming those that are not (e.g. Metadata and CMP elements for most systems) into comments. Much of the functionality is already supplied by the style sheet omdoc2sys.xsl, which need only be adapted to know about the comment syntax.

The task of translating an OMDOC document into system-specific input has two sub-tasks. We will discuss them using the concrete example of the omdoc2pvs.xsl style sheet that transforms OMDOC documents to the input language of the PVS theorem prover [ORS92]: The first task is to translate elements at the statement- and theory level to the input language this is hand-coded by supplying suitable templates for the OMDOC statement and theory elements in an extension of the omdoc2sys.xsl style sheet. The second task is to translate the formulae to the input language. Here, the system usually has a particular way of expressing complex formulae like function applications and binding expressions; in the concrete case of PVS, function application uses a prefix function argument syntax, and n-ary binding expressions, where the scope is separated by a colon from the variable list. This information must also be encoded in respective templates for the om:OMA, om:OMBIND, om:OMV elements from OPENMATH and the m:apply and m:ci from Content-MATHML. For the symbol elements, we have to distinguish two cases: the predefined symbols of the system language and the object symbols that are introduced by the user to formalize a certain problem. In both cases, the transformation procedure needs input on how these symbols are to be represented in the system language. For the object symbols we assume that there are suitable theory structures available, which declare them in symbol elements, thus we can assume that these theory structures also contain use elements with appropriate format attribute in the presentation elements for those symbols that need special representations in the system language. For the predefined symbols of the system language, we assume the same. To be able to transform an OMDOC document into system input, we need a language definition theory, i.e. an OMDOC document that contains a theory which provides **symbols** for all the predefined words of the system language. This theory must also contain **presentation** elements with **use** children specialized the input formats of all systems targeted for communication.

Listing 25.2. A **symbol** in a language definition theory

```
<symbol name="sigmatype">
 <metadata>
  <dc:description>
    The dependent function type constructor is a binding operator. The source type is
    the type of the bound variable X, the target type is represented in the body.
  </dc:description>
 </metadata>
</symbol>

<presentation xml:id="pr-sigmatype" for="#sigmatype" role="binding">
 <style format="pvs">
  <text>[</text>
  <recurse select="*[2]/*"/><text> -&gt; </text><recurse select="*[3]"/>
  <text>]</text>
 </style>
 <style format="nuprl">
  <recurse select="*[2]/*"/><text> -&gt; </text><recurse select="*[3]"/>
 </style>
</presentation>
```

The other direction of the translation needed for communication is usually much more complicated, since it involves parsing the often idiosyncratic output of these systems. A better approach is to write specialized output generators for these systems that directly generate OMDOC representations. This is usually a rather simple thing to do, if the systems have internal data structures that provide all the information required in OMDOC. It is sometimes a problem with these systems that they only store the name of a symbol (logical constant) and not its home theory. At other times, internal records of proofs in theorem provers are optimized towards speed and not towards expressivity, so that some of the information that had been discarded has to be recomputed for OMDOC output.

One of the practical problems that remains to be solved for interfaces between mathematical software systems is that of semantic standardization of input languages. For mathematical objects, this has been solved in principle by supplying a theory level in the form of OPENMATH or OMDOC content dictionaries that define the necessary mathematical concepts. For systems like theorem provers or theory development environments we need to do the same with the logics underlying these systems. For an effort to systematize logics into a hierarchy that fosters reuse and communication of systems, based on a series of experiments of interfacing with the theorem proving systems ΩMEGA [BCF+97], INKA [HS96], Pvs [ORS92], λClam [RSG98b], TPS [ABI+96] and CoQ [Tea] see Section 26.18

25.3   Presenting OMDoc to Humans

We will now discuss the software infrastructure needed to transform OMDoc documents into human-readable form in various formats. We speak of of OMDoc **presentation** for this task.

Due to the complex nature of OMDoc presentation, only part of it can actually be performed by XSLT style sheets. For instance, sub-tasks like reasoning about the prior knowledge of the user, or her experience with certain proof techniques is clearly better left to specialized applications. Our processing model is the following: presenting an OMDoc is a two-phase process.

The first phase is independent of the final output format (e.g. HTML, MathML, or LaTeX) and produces another OMDoc representation specialized to the respective user or audience, taking into account prior knowledge, structural preferences, bandwidth and time constraints, etc. This phase usually generates a narrative-structured document from a knowledge-centered one.

The second phase is a formatting process that can be extracted by XSLT style sheets that transforms the resulting specialized document into the respective output format with notational- and layout preferences of the audience. We will only discuss the second one and refer the reader for ideas about the first process to systems like P.rex [Fie01a, FH01].

The presentation of the OMDoc document elements and statements is carried out by the style sheets `omdoc2html.xsl` for XHTML, `omdoc2html.xsl` for XHTML+MathML and `omdoc2tex.xsl` for LaTeX. These style sheets are divided into files according to the OMDoc modules and share a large common code base `omdoc2share.xsl`, basically the first two include the latter and only redefine some format-specific options. For instance, `omdoc2share.xsl` supplies an infrastructure for internationalization introduced in Section 14.1. This allows to generate localized presentations of the OMDoc documents, if enough information is present in the multilingual groups of `CMP` elements. `omdoc2share.xsl` takes a parameter `TargetLanguage`, whose value can be a whitespace-separated preference list of ISO 639 norm two-letter country codes. If `TargetLanguage` consists of a single entry, then the result will only contain this language with gaps where the source document contains no suitable `CMP`. Longer `TargetLanguage` preference lists will generally result in more complete, but multilingual documents. Apart from the language-specific elements in the source document, localization also needs to know about the presentation of certain keywords used in OMDoc markup, e.g. the German "Lemma" and the French "Lemme" for `<assertion type="lemma">`. This information is kept in the keyword table `lib/locale.xml` in the OMDoc distribution, which contains all the keywords necessary for presenting the OMDoc elements discussed so far. An alternative keyword table can be specified by the parameter `locale`.

OMDoc Applications and Projects

This chapter presents a variety of applications and projects that use the OMDoc format or are related to it in a substantive way.

Apart from the projects directly reported here, the OMDoc format is used by the new research field of Mathematical Knowledge Management (MKM; cf. http://www.mkm-ig.org/), which combines researchers in mathematics, computer science, and library science. We refer the reader to the proceedings of the annual MKM conference [BC01b, Asp03, ABT04, Koh05b, BF06].

26.1 Introduction

The text in the project descriptions has been contributed[1] by the authors marked in the section headings, for questions about the projects or systems, please visit the web-sites given or contact the authors directly. Note that the material discussed in this chapter is under continuous development, and the account here only reflects the state of mid-2006, see http://www.mathweb.org/omdoc for more and current information.

26.1.1 Overview

The OMDoc format as a whole and the applications mentioned above are supported by a variety of tools for creating, manipulating, and communicating OMDoc documents. We can distinguish four kinds of tools:

Interfaces for Mathematical Software Systems like automated theorem provers. These system are usually add ins that interpret the internal representation of formalized mathematical objects in their host systems and

---

[1] If your OMDoc project is not represented here, please contact kohlhase@mathweb.org to arrange for inclusion in later editions of this book.

recursively generate formal OMDoc documents as output and communication streams. Some of these systems also have input filters for OMDoc like the √eriFun described in Section 26.20, but most rely on the OMDoc transformation to their native input syntax described in Section 25.2.

Invasive Editors i.e. are add-ins or modes that "invade" common general-purpose editing systems and specialize them to deal with the OM-Doc format. The OMDoc mode for the EMACS editor presented in Section 26.16, the CPOINT add-in for MS PowerPoint (Section 26.14), the MATHEMATICA® notebook converter (Section 26.17), the SENTIDO plugin for MOZILLA-based browsers, and the plugin for TEXMACS (Section 26.19) are examples for this kind of editor. They differ from simple output filter in providing editing functionality for OMDoc specific information.

Human-Oriented Frontend Formats for instance the QMATH project described in Section 26.2 defines an interface language for a fragment of OMDoc, that is simpler to type by hand, and less verbose than the OMDoc that can be generated by the qmath parser. STEX defines a human-oriented format for OMDoc by extending the TEX/LATEX with content markup primitives, so that it can be transformed to OMDoc. See Section 26.15 for details.

Mathematical Knowledge Bases mathematical knowledge base The MBASE and MAYA systems described in Sections 26.4 and 26.12 are web-based mathematical knowledge bases that offer the infrastructure for a universal, distributed repository of formalized mathematics represented in the OMDoc format.

26.1.2   Application Roles of the OMDoc Format

The applications above support the utilization of the OMDoc format in several roles. Generally, OMDoc can used of as a

Communication Standard between mechanized reasoning systems.

Data Format for Controlled Refinement from informal presentation to formal specification of mathematical objects and theories. Basically, an informal textual presentation can first be marked up, by making its structure explicit (classifying text fragments as definitions, theorems, proofs, linking text, and their relations), and then formalizing the textually given mathematical knowledge in logical formulae (by adding FMP elements; see Chapter 14).

Document Preparation Language. The OMDoc format makes the large-scale document- and conceptual structures explicit and facilitates maintenance on this level. Individual documents can be encoded as lightweight narrative structures, which can directly be transformed to e.g. XHTML+MATHML or LATEX, which can in turn be published on the Internet.

Basis for Individualized (Interactive) Documents. Personalized narrative structures can be generated from MBASE content making use of the conceptual structure encoded in MBASE together with a user model. For instance, the MMiSS, MATHDOX, and ACTIVEMATH projects described in Sections 26.6 to 26.8 use the OMDOC infrastructure in an educational setting. They make use of the content-orientation and the explicit structural markup of the mathematical knowledge to generate on the fly specialized learning materials that are adapted to the students prior knowledge, learning goals, and notational tastes.

Interface for Proof Presentation. As the proof part of OMDOC allows small-grained interleaving of formal (FMP) and textual (CMP) presentations in multiple languages (see e.g. [HF97, Fie99]).

26.2   QMath: A Human-Oriented Language and Batch Formatter for OMDoc

Project Home	http://www.matracas.org/qmath/index.en.html
Authors	Alberto González Palomo Toledo, Spain[2]

QMATH is a batch processor that produces an OMDOC file from a plain UNICODE text document, in a similar way to how TEX produces a DVI file from a plain text source. Its purpose is to allow fast writing of mathematical documents, using plain text and a straightforward syntax (like in computer algebra systems) for mathematical expressions.

The "Q" was intended to mean "quick", since QMATH began in 1998 as an abbreviated notation for MATHML. The first version (0.1) just expanded the formulas found enclosed by "$" signs, which were abbreviated forms of the MATHML element names, and added the extra markup such as <mrow> and the like. The second (0.2) did the same thing, but this time allowing an infix notation that was fixed in the source code. Finally, version 0.3 allowed the redefinition of symbols while parsing, but it was not capable of expanding formulas embedded in XML documents like the previous ones did until version 0.3.8.[3] For a more detailed history see [Palb].

QMATH is very simple: it just parses a text (UTF-8) file according to a user-definable table of symbols, and builds an XML document from that. The symbol definitions are grouped in files called "contexts". The idea is that when you declare a context, its file is loaded and from then on these symbol definitions take precedence over any previous one, thus setting the context for parsing of subsequent expressions.

The grouping of symbols in the context files is arbitrary. However, the ones included with QMATH follow the OPENMATH Content Dictionaries hierarchy so that, for instance, the English language syntax for the symbols in the "arith1" CD is defined in the context "Mathematics/OpenMath/arith1".

Figure 26.1 shows a minimal QMATH document, and the OMDOC document generated from it. The first line (`"QMATH 0.3.8"`) in the QMATH document is required for the parser to recognize the file. The lines beginning with ":" are metadata items, the first of which, `:en`, declares the primary language for the document, in this case English. Specifying the language is required, as it sets the basic keywords accordingly, and there is no default (privileged) language in QMATH. For example, the English keyword "`Context`" is written "`Contexto`" if the language is Spanish. (Similarly, the arithmetic context is "`Matemáticas/Aritmética`"). Then, the "OMDoc" context is loaded, defining the XML elements to be produced by the metadata items and the different kinds of paragraphs: plain text, theorem, definition, proof, example, etc.

---

[2] The author is currently employed part-time in the ACTIVEMATH project, developed by Saarland University and the DFKI, but this work was done on his own, without their supervision or support.

[3] This offers an alternative to the OQMATH wrapper mentioned in Section 26.8.

```
QMATH 0.3.8
:en
Context: "Mathematics/OMDoc"

:"Diary"
:W. Smith
:1984−04−04 18:43:00+00:00

Context: "Mathematics/Arithmetic"

Theory:[<−thoughtcrime]

:"Down with Big Brother"
Freedom is the freedom to say $2+2=4$.
If that is granted, all else follows.
```

From contexts/en/Mathematics/OpenMath/arith1.qmath:

```
Symbol: plus OP_PLUS "arith1:plus"
Symbol: + OP_PLUS "arith1:plus"
Symbol: sum APPLICATION "arith1:sum"
Symbol: Σ APPLICATION "arith1:sum"
...
```

From contexts/en/Mathematics/OpenMath/relation1.qmath:

```
Symbol: = OP_EQ "relation1:eq"
Symbol: neq OP_EQ "relation1:neq"
Symbol: ¬= OP_EQ "relation1:neq"
Symbol: ≠ OP_EQ "relation1:neq"
...
```

```xml
<?xml version='1.0' encoding='UTF−8' standalone='no'?>
<!DOCTYPE omdoc PUBLIC "−//OMDoc//DTD OMDoc V1.2//EN"
          " ../../../../ dtd/omdoc.dtd">
<omdoc xmlns='http://www.mathweb.org/omdoc' version='1.2'
       xmlns:dc='http://purl.org/dc/elements/1.1/'>
 <metadata>
  <dc:language>en</dc:language>
  <dc:title>Diary</dc:title>
  <dc:creator role='aut'>W. Smith</dc:creator>
  <dc:date>1984−04−04T18:43:00+00:00</dc:date>
 </metadata>
 <theory xml:id='thoughtcrime'>
  <imports from="arith1"/>
  <imports from="relation1"/>
  <omtext>
   <metadata><dc:title>Down with Big Brother</dc:title></metadata>
   <CMP>
    Freedom is the freedom to say
    <OMOBJ xmlns='http://www.openmath.org/OpenMath'>
     <OMA>
      <OMS cd='relation1' name='eq'/>
      <OMA>
       <OMS cd='arith1' name='plus'/>
       <OMI>2</OMI><OMI>2</OMI>
      </OMA>
      <OMI>4</OMI>
     </OMA>
    </OMOBJ>.
    If that is granted, all else follows.
   </CMP>
  </omtext>
 </theory>
</omdoc>
```

Fig. 26.1. A minimal QMATH document (top left) and its OMDOC result (bottom). Some symbol definitions are displayed in the top right.

After that setup come the document title, author name (one line for each author), and date, which form the content of the OMDOC metadata element.

The document is composed of paragraphs (which can be nested) separated by empty lines, and formulas are written enclosed by "$" signs.

There is an Emacs mode included in the source distribution, that provides syntax highlighting and basic navigation based on element identifiers.

It is also possible to use it on an XML document for expanding only the mathematical expressions. QMATH will detect automatically the input format, either QMATH text or XML, and in the later case output everything verbatim except for the QMATH language fragments found inside the XML processing instructions of the form <?QMath ... ?> and the mathematical expressions between "$".

```
<?xml version='1.0' encoding='UTF−8' standalone='no'?>
<!DOCTYPE omdoc PUBLIC "−//OMDoc//DTD OMDoc V1.2//EN"
                       " ../../../../ dtd/omdoc.dtd">
<?QMath
:en
Context: "Mathematics/Arithmetic"
?>
<omdoc xmlns='http://www.mathweb.org/omdoc' version='1.2'
       xmlns:dc='http://purl.org/dc/elements/1.1/'>
 <metadata>
  <dc:language>en</dc:language>
  <dc:title>Diary</dc:title>
  <dc:creator role='aut'>W. Smith</dc:creator>
  <dc:date>1984−04−04T18:43:00</dc:date>
 </metadata>
 <theory xml:id='thoughtcrime'>
  <omtext>
   <metadata><dc:title>Down with Big Brother</dc:title></metadata>
   <CMP>
   Freedom is the freedom to say $2+2=4$.
   If that is granted, all else follows .
   </CMP>
  </omtext>
 </theory>
</omdoc>
```

Fig. 26.2. The same example document, using QMATH only for the formulas

While QMATH was a good improvement over manual typing of the OMDOC XML, it does not scale well: in real documents, with more than a couple of nesting levels, it is difficult to keep track of where the current paragraph belongs.

One solution is to use it only for the mathematical expressions, and rely on some XML editor for the document navigation and organization, such as the OMDOC mode for Emacs described in Section 26.16 or the OQMATH mode for JEDIT in Section 26.9. Another is to use the SENTIDO browser/editor in Section 26.3, which reimplements and extends QMATH's functionality.

QMATH is Free Software distributed under the GNU General Public License (GPL [FSF91]).

26.3   Sentido: An Integrated Environment for OMDoc

Project Home	http://www.matracas.org/sentido/index.en.html
Authors	Alberto González Palomo
	Toledo, Spain[4]

SENTIDO is an integrated environment for browsing, searching, and editing collections of OMDOC documents. It is implemented as an extension for the MOZILLA/FIREFOX browsers to avoid the biggest problems found when using QMATH: the need to compile the program for installing, the batch mode of interaction that made small corrections consume much of the author's time, and the lack of any support for document navigation and search.

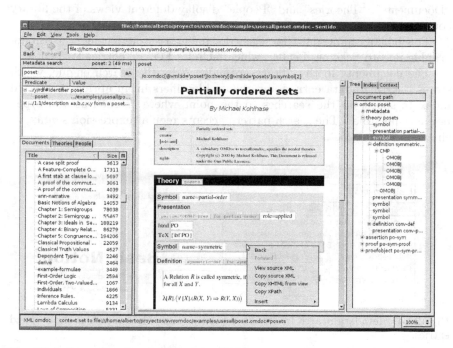

Fig. 26.3. SENTIDO after indexing the OMDoc repository in the library (left) and loading a document from it (center and right).

Figure 26.3 shows a typical session initiated by searching in the document library (described below in more detail) and opening one of the results. The context menu displays the options for browsing back and forward, viewing the XML (OMDOC) source of the selected element, copying it to the system clipboard, copying its MATHML rendering or an XPATH expression that identifies it, and inserting new elements.

---

4   The author is currently employed part-time in the ACTIVEMATH project, developed by Saarland University and the DFKI, but this work was done on his own, without their supervision or support.

26.3.1 The User Interface

The window is made to resemble the web browser, and consists of two main panes: the smaller one on the left contains the interface for the "document library", and the right one the "document view" and associated information like the document tree, element identifiers index, and context at the current cursor position.

The document library is a knowledge base about documents, the theories defined in them, and people mentioned in their metadata as authors, editors, translators, etc. It is implemented as an RDF store with the documents organized in collections called "volumes" with references to documents, so that different volumes can have documents in common. The tabs labelled "Documents", "Theories" and "People" display different views of the library content.

The bibliographic data for each document is stored using the Bibliographic Record Schema [Len04], which includes FOAF[5] entries for people.

The documents in the library are indexed by the search engine, which stores their metadata entries and theory identifiers in an abridged inverted index to speed up the searches to the point where "search as you type" becomes possible[6]. The search pattern accepts regular expression syntax, as shown in Figure 26.4.

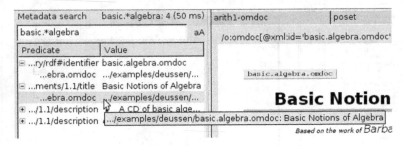

Fig. 26.4. Metadata Search in SENTIDO: tooltips show the content of the cropped entries.

The document view is built using XHTML + MATHML that can be edited normally, with the changes being propagated to the internal OMDOC.

The view is built on demand (using XSLT) as the subparts of the document are unfolded in the document navigation tree found in the right part of the window. This has been found important in practice since many real

---

[5] "Friend of a friend", described in their web page http://www.foaf-project.org as being "about creating a Web of machine-readable homepages describing people, the links between them and the things they create and do."

[6] On the author's 1 GHz laptop computer, the search times in a library of around two thousand documents are usually between 100 and 200 milliseconds.

uses of OMDOC involve documents that contain large lists of elements, like exercises, that are largely independent of each other and thus do not usually require being viewed at the same time, and the biggest delay in opening a medium to large sized document was by far the display of the XHTML view. Another motivation for this approach is to progress towards handling the source document more like a database, and customize its presentation for the task at hand.

SENTIDO adds some options to the context menu in the browser, to allow the user to open links to OMDOC files from web pages (see Figure 26.5).

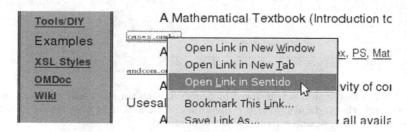

Fig. 26.5. MOZILLA's Context Menu after Installing SENTIDO.

26.3.2 Formula Editing

Mathematical expressions are entered using a selectable linear syntax, translated by a new version of the QMATH parser described in Section 26.2. This is a much more capable implementation based on finite-state cascades [Abn96].

There are five grammars included in the install package, that are used for translating back and forth between OPENMATH and the linear syntax of QMATH and the Computer Algebra Systems MAXIMA, YACAS, MAPLE™ and MATHEMATICA®. More syntaxes can be added by writing new grammars, with a format similar to QMATH "context" files.

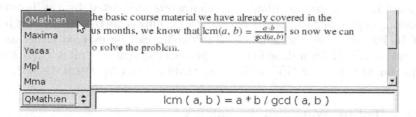

Fig. 26.6. The formula editor under the document view, with the input syntax menu and the text field where the formula is typed, which updates continuously the internal OPENMATH representation and the MATHML view.

When the cursor enters a formula, the linear input field appears at the bottom of the document view, as seen in Figure 26.6. It contains a text field for editing, and a menu button for selecting the syntax, which can be done at any moment: the linear expression is regenerated instantaneously from its OPENMATH form, so it is possible to enter a formula using, for instance, MATHEMATICA® syntax, then select another syntax such as MAPLE™, and get the expression immediately translated, going through its OPENMATH representation (Figure 26.7).

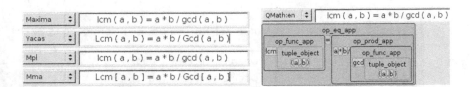

Fig. 26.7. The formula is translated by SENTIDO each time the user selects another syntax (left, the vertical line is the blinking caret), and it is possible to view the parse tree (right), updated as the input is modified.

Insertion of formulae is achieved by typing the dollar symbol "$", which produces an empty formula readily editable so that the sequence of keystrokes is similar to typing TEX or QMATH text: one can type $e^(pi*i)+1=0$ and get $e^{\pi i} + 1 = 0$ without having to look at the formula editor or use the mouse. The changes as the formula is being modified are stored, and the display updated from the OPENMATH form, at each point when there is a complete parse of the formula. This gives immediate feedback on how the program understands the input.

An important difference is that there is no need to care about "context files" any more. In QMATH, specifying a "context" had a double function: putting symbols in scope for disambiguation, and selecting a notation style for them. Those aspects are separated in SENTIDO: the in-scope symbols are automatically determined from the enclosing theory and those imported from it (recursively), and the notation is selectable by the user.

Note that the parser allows any characters supported by the browser rendering engine of MOZILLA/FIREFOX (a big subset of UNICODE), not just ASCII. For example, the number 3.14159265... can be entered either as π or with an ASCII form depending on the selected syntax: "pi" for QMATH, "%pi" for MAXIMA, or "Pi" for YACAS, MAPLE™ and MATHEMATICA®.

26.3.3 Future Work and Availability

SENTIDO is a long term personal project that has been in development for several years (since 2004), entirely in the author's spare time and using his own computing resources, based on experiments[7] and notes collected during the development of QMATH. Therefore, we expect it to continue developing during the foreseeable future unless a better application appears that makes it redundant.

Its components are designed to be reusable, which is tested from time to time by producing spin-off applications that use subsets of its functionality in a self-contained way. One example is the small Computer Algebra System called ALGEBRA [Pala], that contains parts of SENTIDO such as the new parser combined with specific ones like the function plotter and the term rewriting engine.

Future developments will focus on what we consider the two main tasks for a development environment for semantic encoding of mathematical content:

– Ease the tedium of writing all the details needed for an unambiguous encoding of the content. This is where the flexible input parser comes into play: having a syntax redefinable at any point in the content simplifies the expression input, as the syntax can be adapted to the context in which an expression occurs.
– Provide some benefit once we have the semantic encoding which would not be present with an ambiguous encoding such as TEX. Here we need to implement detailed checking and strong search capabilities. A next step would be to assist the writing process by inferring new content and informing the input interface about the context as mentioned above.

Some planned improvements in SENTIDO are:

– Make the browser open OMDOC documents linked from normal pages directly in SENTIDO, by implementing a stream handler for the MIME type `application/omdoc+xml`.
– Integrate ALGEBRA into SENTIDO, to add automated symbolic manipulation to the document editing process.
– Extend the checking being done on the theories: at the time of writing these lines, only the theory import relations are checked for loops and unknown theory references, which was already enough to locate several mistyped theory identifiers in the OMDOC repository.
– Implement useful features found in other projects such as THEOREMA [PB04]. This is strongly related to the two points above since THEOREMA implements many features needed for the task of content checking which are still missing in SENTIDO, and some of them are available in proof-of-concept form in ALGEBRA.

SENTIDO is Free Software distributed under the GNU General Public License (GPL [FSF91]).

---

[7] Some of those early experiments with MOZILLA inspired work done on adapting OPENOFFICE and TEX$_{MACS}$ for OMDOC in collaboration with George Goguadze [GP03]

26.4   MBase, an Open Mathematical Knowledge Base

Project Home	http://www.mathweb.org/mbase
Authors	Andreas Franke[1], Michael Kohlhase[2] [1] Computer Science, Saarland University [2] School of Engineering and Science, International University Bremen

We describe the MBASE system, a web-based mathematical knowledge base. It offers the infrastructure for a universal, distributed repository of formalized mathematics. Since it is independent of a particular deduction system and particular logic, the MBASE system can be seen as an attempt to revive the QED initiative from an infrastructure viewpoint. See [KF01] for the logical issues related to supporting multiple logical languages while keeping a consistent overall semantics. The system is realized as a mathematical service in the MATHWEB system [FK99, Zim04], an agent-based implementation of a mathematical software bus for distributed mathematical computation and knowledge sharing. The content language of MBASE is OMDOC.

We will start with a description of the system from the implementation point of view (we have described the data model and logical issues in [KF01]).

The MBASE system is realized as a distributed set of MBASE servers (see figure 26.8). Each MBASE server consists of a Relational Data Base Management System (RDBMS) connected to a MOZART process (yielding a MATHWEB service) via a standard data base interface. For browsing the MBASE content, any MBASE server provides an http server (see [MBa] for an example) that dynamically generates presentations based on HTML or XML forms.

This architecture combines the storage facilities of the RDBMS with the flexibility of the concurrent, logic-based programming language Oz [Smo95], of which MOZART (see [Moz]) is a distributed implementation. Most importantly for MBASE, MOZART offers a mechanism called **pickling**, which allows for a limited form of persistence: MOZART objects can be efficiently transformed into a so-called pickled form, which is a binary representation of the (possibly cyclic) data structure. This can be stored in a byte-string and efficiently read by the MOZART application effectively restoring the object. This feature makes it possible to represent complex objects (e.g. logical formulae) as Oz data structures, manipulate them in the MOZART engine, but at the same time store them as strings in the RDBMS. Moreover, the availability of "Ozlets" (MOZART functors) gives MBASE great flexibility, since the functionality of MBASE can be enhanced at run-time by loading remote functors. For instance complex data base queries can be compiled by a specialized MBASE client, sent (via the Internet) to the MBASE server and applied to the local data e.g. for specialized searching (see [Duc98] for a related system and the origin of this idea).

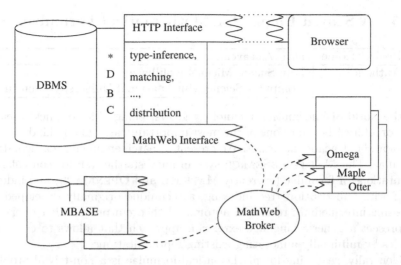

Fig. 26.8. System architecture

MBASE supports transparent distribution of data among several MBASE servers (see [KF01] for details). In particular, an object O residing on an MBASE server S can refer to (or depend on) an object O' residing on a server S'; a query to O that needs information about O' will be delegated to a suitable query to the server S'. We distinguish two kinds of MBASE servers depending on the data they contain: *archive servers* contain data that is referred to by other MBASEs, and *scratch-pad* MBASEs that are not referred to. To facilitate caching protocols, MBASE forces archive servers to be *conservative*, i.e. only such changes to the data are allowed, that the induced change on the corresponding logical theory is a conservative extension. This requirement is not a grave restriction: in this model errors are corrected by creating new theories (with similar presentations) shadowing the erroneous ones. Note that this restriction does not apply to the non-logical data, such as presentation or description information, or to scratchpad MBASEs making them ideal repositories for private development of mathematical theories, which can be submitted and moved to archive MBASEs once they have stabilized.

26.5   A Search Engine for Mathematical Formulae

Project Home	`http://search.mathweb.org/`
Authors	Ioan Sucan, Michael Kohlhase
	Computer Science, International University Bremen

As the world of information technology grows, being able to quickly search data of interest becomes one of the most important tasks in any kind of environment, be it academic or not. We present a search engine for mathematical formulae. The MATHWEBSEARCH system harvests the web for content representations of formulae (currently MATHML and OPENMATH) and indexes them with substitution tree indexing, a technique originally developed for accessing intermediate results in automated theorem provers. For querying, we present a generic language extension approach that allows to construct queries by minimally annotating existing representations.

Generally, searching for mathematical formulae is a non-trivial problem — especially if we want to be able to search occurrences of the query term as sub-formulae — for the following reasons:

1. *Mathematical notation is context-dependent.* For instance, binomial coefficients can come in a variety of notations depending on the context: $\binom{n}{k}$, $_nC^k$, C_k^n, and C_n^k all mean the same thing: $\frac{n!}{k!(n-k)!}$. In a formula search we would like to retrieve all forms irrespective of the notations.
2. *Identical presentations can stand for multiple distinct mathematical objects*, e.g. an integral expression of the form $\int f(x)dx$ can mean a Riemann Integral, a Lebesgue Integral, or any other of the 10 to 15 known anti-derivative operators. We would like to be able to restrict the search to the particular integral type we are interested in at the moment.
3. *Certain variations of notations are widely considered irrelevant*, for instance $\int f(x)dx$ means the same as $\int f(y)dy$ (modulo α-equivalence), so we would like to find both, even if we only query for one of them.

To solve this formula search problem, we concentrate on *content representations of mathematical formulae*, since they are presentation-independent and disambiguate mathematical notions.

A Running Example: The Power of a Signal. A standard use case for MATHWEBSEARCH is that of an engineer trying to solve a mathematical problem such as finding the power of a given signal $s(t)$. Of course our engineer is well-versed in signal processing and remembers that a signal's power has something to do with integrating its square, but has forgotten the details of how to compute the necessary integrals. He will therefore call up MATHWEBSEARCH to search for something that looks like $\int_?^? s^2(t)dt$ (for the concrete syntax of the query see Listing 26.9 in Section 26.5). MATHWEBSEARCH finds a document about Parseval's Theorem, more specifically $\frac{1}{T}\int_0^T s^2(t)dt = \Sigma_{k=-\infty}^{\infty} \mid c_k \mid^2$ where c_k are the Fourier coefficients of the signal. In short, our engineer found the exact formula he was looking for (he had missed the factor in front and the integration limits) and moreover a theorem he may be able to use.

Indexing Mathematical Formulae. For indexing mathematical formulae on the web, we will interpret them as first-order terms. This allows us to use a technique from automated reasoning called *term indexing* [Gra96]. This is the process by which a set of terms is stored in a special purpose data structure (the **index**) where common parts of the terms are potentially shared, so as to minimize access time and storage. The indexing technique we work with is a form of tree-based indexing called *substitution-tree indexing*. A substitution tree, as the name suggests, is simply a tree where substitutions are the nodes. A term is constructed by successively applying substitutions along a path in the tree, the leaves represent the terms stored in the index. Internal nodes of the tree are **generic terms** and represent similarities between terms.

The main advantage of substitution tree indexing is that we only store substitutions, not the actual terms, and this leads to a small memory footprint. Adding data to an existing index is simple and fast, querying the data structure is reduced to performing a walk down the tree. Index building is done in similar fashion to [Gra96]. Once the index is built, we keep the actual term instead of the substitution at each node, so we do not have to recompute it with every search. At first glance this may seem to be against the idea of indexing, as we would store all the terms again, not only the substitutions. However, due to the tree-like structure of the terms, we can in fact store only a pool of (sub)terms and define the terms in our index using pointers to elements of the pool (which are simply other terms). To each of the indexed terms, a data string is attached — a string that represents the exact location of the term. We use XPointer [GMMW03b] to specify this.

Unfortunately, substitution tree indexing does not support subterm search in an elegant fashion, so when adding a term to the index, we add all its subterms as well. This simple trick works well: the increase in index size remains manageable and it greatly simplifies the implementation.

A Query Language for Content Mathematics. When designing a query language for mathematical formulae, we have to satisfy a couple of conflicting constraints. The language should be content-oriented and familiar, but it should not be specialized to a given content representation format. Our approach to this problem is to use a simple, generic extension mechanism for XML based representation formats rather than a genuine query language itself. The query extension is very simple, it adds two new attributes to the respective languages: `mq:generic` and `mq:anyorder`, where the prefix `mq:` abbreviates the namespace URI `http://mathweb.org/MathQuery/` for MATHWEBSEARCH.

In this way, the user need not master a new representation language, and we can generate queries by copy and paste and then make parts of the formulae generic by simply adding suitable attributes. We will use Content

MATHML [ABC⁺03] in the example, but MATHWEBSEARCH also supports
OPENMATH and a shorthand notation that resembles the internal representation we are using for terms (prefix notation). The `mq:generic` attribute
takes string values and can be specified for any element in the query, making
it into a (named) query variable: its contents are ignored and it matches any
term in the search process.

While of searching expressions of the form $A = B$, we might like to find
occurrences of $B = A$ as well. At this point the `mq:anyorder` attribute comes
in. Inside an `apply` tag, the first child defines the function to be applied; if
this child has the attribute `mq:anyorder` defined with the value "yes", the
order of the subsequent children is ignored. If we do not want to specify the
function name, we can use the `mq:generic` attribute again, but this time for
the first child of an `apply` tag. Given the above, the query of our running
example has the form presented in Listing 26.9. Note that we do not know
the integration limits or whether the formula is complete or not.

Listing 26.9. Query for signal power

```
<math xmlns="http://www.w3.org/1998/Math/MathML"
     xmlns:mq="http://mathweb.org/MathQuery">
 <apply><int/>
   <domainofapplication mq:generic="domain"/>
   <bvar> <ci mq:generic="time"/> </bvar>
   <apply><power/>
     <apply><ci mq:generic="fun"></ci><ci mq:generic="time"/></apply>
     <cn>2</cn>
   </apply>
 </apply>
</math>
```

Input Processing. MATHWEBSEARCH can process any kind of XML representation for content mathematics. The system is modular and provides
an interface which allows to add XML-to-index-term transformers. . We will
discuss input processing for Content-MATHML.

Given an XML docu-
ment, we create an index
term for each of its `math` elements. Consider the example on the right: We have the
standard mathematical no-
tation of an equation (1), its
Content MATHML represen-
tation (2), and the term we

1) Mathematical
   expression:
   $f(x) = y$

3) Term
   representation:
   $eq(f(x), y)$

2) Content MATHML:
```
<apply><eq/>
 <apply>
   <ci>f</ci>
   <ci>x</ci>
 </apply>
 <ci>y</ci>
</apply>
```

extract for indexing (3). As previously stated, any mathematical construct
can be represented in a similar fashion.

Search modulo α-renaming becomes available via a very simple input
processing trick: during input processing, we add a `mq:generic` attribute
to every bound variable (but with distinct strings for different variables).
Therefore in our running example the query variable t (`@time` in Listing 26.9)

in the query $\int_?^? s^2(t)dt$ is made generic, therefore the query would also find the variant $\frac{1}{T}\int_0^T s^2(x)dx = \Sigma_{k=-\infty}^{\infty} \mid c_k \mid^2$.

Result Reporting. For a search engine for mathematical formulae we need to augment the set of result items (usually page title, description, and page link) reported to the user for each hit. As typical pages contain multiple formulae, we need to report the exact occurrence of the hit in the page. We do this by supplying an XPOINTER reference where possible. Concretely, we group all occurrences into one page item that can be expanded on demand and within this we order the groups by number of contained references. For any given result, a detailed view is available. This view shows the exact term that was matched (using Presentation MATHML) and the used substitution (a mapping from the query variables specified by the mq:generic attributes to certain subterms) to match that specific term.

Case Studies and Results. We have tested our implementation on the content repository of the CONNEXIONS Project, available via the OAI protocol [OAI02]. This gives us a set of over 3,200 articles with mathematical expressions to work on. The number of terms represented in these documents is approximately 53,000 (77,000 including subterms). The average term depth is 3.6 and the maximal one is 14. Typical query execution times on this index are in the range of milliseconds. The search in our running example takes 14 ms for instance. There are, however, complex searches (e.g. using the mq:anyorder attribute) that internally call the searching routine multiple times and take up to 200 ms but for realistic examples execution time is below 50 ms. We are currently building an index of the 86,000 Content MATHML formulae from http://functions.wolfram.com. Here, term depths are much larger (average term depth 5.7, maximally around 50) resulting in a much larger index: it is just short of 2 million formulae. First experiments indicate that search times are largely unchanged by the increase in index size.

In the long run, it would be interesting to interface MATHWEBSEARCH with a regular web search engine and create a powerful, specialized, full-feature application. This would resolve the main disadvantage our implementation has – it cannot search for simple text. A simple socket-based search API allows to integrate MATHWEBSEARCH into other content-based mathematical software systems.

26.6   Semantic Interrelation and Change Management

Project Home	http://www.mmiss.de
Authors	Bernd Krieg-Brückner, Achim Mahnke
	Computer Science, University of Bremen, Germany

The corpus of electronically available mathematical knowledge increases rapidly. Usually, mathematical objects are embedded in and related to different kinds of documents like articles, books, or lecture material, the domain of which can be different from mathematics, e.g., engineering or computer science. Therefore, maintaining high-quality mathematical knowledge becomes a non-trivial engineering task for teams of authors.

In this scenario, sharing and reuse is the key to efficient development. Unfortunately, while there has been a large body of research concerning the sharing and reuse of program developments, sharing and reuse of documents has until now been mainly done by little more than cut and paste. However, to ensure sustainable development, i.e. continuous long-term usability of the contents, sharing and reuse needs to be supported by tools and methods taking into account the semantic structure of the document. In developing these methods and tools we can benefit from the experience in and associated support tools.

We address this problem by providing a methodology to specify coherence and consistency of documents by interrelation of semantic terms and structural entities, supported by a tool for fine-grained version control and configuration management including change management. Semantic interrelation explicates the meaning lying behind the textual content, and relates the semantic concepts within and across documents by means of an ontology. To allow change management, each document is structured in-the-small. Each document corresponds to a package, and packages may be structured in-the-large using folders and import relations. The ideas and methods explained here have been developed in the MMiSS project which aimed at the construction of a multi-media Internet-based adaptive educational system (see [KBHL+03, KBLL+04, KBKB+04]).

26.6.1   Semantic Interrelation via Ontologies

Ontologies provide the means for establishing a semantic structure. An ontology is a formal explicit description of concepts in a domain of discourse. The MMiSSLATEX package for ontologies provides a set of easy-to-use macros for the declaration of ontologies in LATEX documents. They are used to *declare* the ontology of semantic terms used in a document, in a prelude up front. This *specification* of the document contains at least a rigorous hierarchical structure of the terminology (a taxonomy, the *signature* of the document), and may be seen as an elaborate index structure. Moreover, relations between terms may be defined for more semantic interrelation.

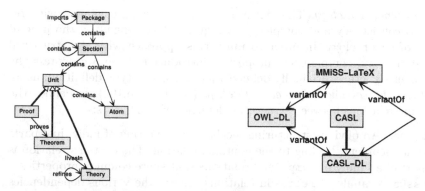

Fig. 26.10. (a) Parts of the system's ontology (b) Formalism variants

The ontology serves a dual purpose — just as the specification of an abstract data type in program development: it specifies the content to be expected in the body of the document in an abstract, yet precise, manner — the content developers requirement specification; and it specifies the content for reference from the outside — the user's perspective, who may then view the body of the document as a black box. The content developer will use the MMiSSLaTeX Def command to specify the *defining* occurrence of a promised term, as for an index. Using the structuring in-the-large facilities via packages, the external user may then refer between documents using various kinds of *reference* commands, as the content developer may within a document.

The next section will show, how we can explore this domain ontology — supplied by the author — in order to capture semantic relations between document parts and use these relations for supporting a management of change for mathematical documents.

26.6.2   Change Management

The notion of *change management* is used for the maintenance and preservation of consistency and completeness of a development during its evolution. More precisely, we want to have a consistent configuration in which all versions are compatible and there are no cyclic definitions or proofs. At the same time, it should be a complete configuration: there should be no dangling forward references.

Such notions are well-known for formal languages. In contrast, natural language used for writing teaching material does not usually possess a well-defined semantics, and the notion of consistency is arguable. Different authors may postulate different requirements on the material in order to regard it as being consistent. The existence of a user-defined ontology helps a great deal to check references. However, we can make even better use of the information contained in the ontology.

The System's Ontology. The aim is to allow change management with regard to consistency and completeness requirements defined by the user in terms of an ontology. In order to unify this approach with the structural consistency and completeness properties introduced above, we express the document structure, originally defined by a document type definition, as an ontology, the so-called *System's Ontology* (see Fig. 26.10a). It defines the following relations between structural elements of documents:

comprises An obvious structuring mechanism is nesting of individual parts of a document, leading to the contains relation. The contains relation is part of a family of **comprises** relations that share common properties.

reliesOn A family of **reliesOn** relations reflects the various dependencies between different parts of a document. For example, a theorem *lives in* a theory, or proof *proves* a theorem.

pointsTo The family of **pointsTo** relations is very similar, and relates references with the defining occurrence of a semantic term.

variantOf Another structuring relation is introduced by variants. Parts of a document may e.g. be written in various languages which gives rise to a **variantOf** relation between these document parts and their constituents; it is an equivalence relation.

It is now rather straightforward to formulate consistency and completeness rules in terms of invariants of these relations. Formulating these invariants as formal rules will enable us to implement a generic and flexible change management that keeps track of the invariants and informs the user about violations when a previously consistent document has been revised, leading to various kinds of error (e.g. for **reliesOn** relations) or warning messages (e.g. for **pointsTo** relations).

Properties of Interactions between Structuring Mechanisms.. This approach also allows us to lift relations to structuring mechanisms allowing more modular and localized change management. For example, relating the **comprises** and **reliesOn** relations allows us to formalize invariants regarding the closure of document parts with respect to the **reliesOn** relation: We can require that there is a proof for each theorem in a package. Furthermore, if two structural entities are related by **reliesOn**, their relation is propagated along the **comprises** relation towards the root of the hierarchy of nested structural entities, such that (for a theorem T a proof P, and sections A, B):

$$B \text{ contains } P \ \& \ A \text{ contains } T \ \& \ P \text{ proves } T \Rightarrow B \textbf{ reliesOn } A.$$

If the user changes section A, the repository will only need to check all sections that A relies on (such as B here) for invariants, and not the whole document. However, in contrast to formal developments as in e.g. the MAYA system [AH05], there is no rigorous requirement that a document should obey all the rules. There may be good reasons, for instance, to present first a "light-weight" introduction to all notions introduced in a section before giving the detailed definitions. In this particular case, one would want to introduce forward pointers to the definitions rather than making the definitions rely on the introduction; thus the rules are covered.

In any case, the more structure there is, the better the chances are for preserving consistency and completeness; any investment in introducing more `reliesOn` relations, for example, will pay off eventually. The change management will observe whether revisions by the user will affect these relations and, depending on the user's preferences, emit corresponding warnings.

The aim is to allow users to specify individual notions of consistency by formulating the rules that the relations should obey. This should be possible for the relations between the particular (predefined) structuring mechanisms, but also in general between semantic terms of the user's own ontology. Our work in this direction will rely on the methods and tools provided by the HETS system (see Section 26.13).

26.6.3 Variants

The concept of variants adds a new dimension to hierarchically structured documents. The idea is to maintain and manage different variants of structural entities (document subtrees) which represent the same information in different ways — variants are meant to glue them together.

Managing different natural language variants in parallel is an obvious example. Another one is the formalism variant which denotes the particular formalism in which a formal content part like a theorem or a definition is expressed. Considering ontology development itself, for example, we propose to use variants to maintain different formal representations for the same semantic concept together with its documentation. Figure 26.10b shows the possible variants for declaring ontology components (see [MKB04] for details).

The MMiSS repository provides functions to store and retrieve these structural variants by means of specifications for selecting particular variants for editing or presentation.

26.6.4 Relations to OMDoc

OMDoc provides modules for marking up the knowledge structure and the narrative structure of mathematical documents. MMiSS combines these two viewpoints by giving means for structuring the document contents (which constitutes the narrative structure) and for specifying the incorporated knowledge by use of ontologies. Therefore, we have implemented an export of MMiSS documents to (content and narrative) OMDoc documents and vice versa.

26.7   MathDox: Mathematical Documents on the Web

Project Home	http://www.mathdox.org
Authors	A.M. Cohen, H. Cuypers, E. Reinaldo Barreiro
	Department of Mathematics and Computer Science,
	Eindhoven University of Technology

Abstract. The MATHDOX system provides an infrastructure for interactive mathematical documents that make use of the World Wide Web. These documents take input from various sources, users, and mathematical services. Communication between these different entities can be realized using OPEN-MATH. But, such communication and the interactivity inside the mathematical document take place in a specific, dynamic context. In this paper we discuss our approach to such a dynamic mathematical context: MATHDOX. It consists of both an XML-based markup language for interactive mathematical contents and a set of software tools realizing the interactivity. cl

26.7.1   Introduction

Although the notion of an interactive mathematical document has been around for several years, cf. [CM98], its realization is nowhere near the final stage. For instance, recent progress in web technologies has enabled a much smoother communication of mathematics than ever before. The use of an interactive mathematical document (IMD) can provide a window to the world of mathematical services on the Internet, and a mathematical service on the Internet can be created by the building of an interactive mathematical document. MATHDOX is an ensemble of software tools for creating IMDs, it includes

1. an XML based language that offers markup support for the source texts of IMDs;
2. a document server, rendering interactive mathematical documents from source text and interactively obtained information;
3. mathematical services, providing connections with CASs like MATHEMATICA® and GAP via OPENMATH phrasebooks (cf. [OM]).

The creation of MATHDOX is a project at the Technische Universiteit Eindhoven (the RIACA institute). Several people at RIACA have helped creating it; here we mention Manfred Riem, Olga Caprotti, Hans Sterk, Henny Wilbrink, Mark Spanbroek, Dorina Jibetean. The system is mainly built with Java and related technology. The products are available via the project web site and will be licensed under the Lesser Gnu Public License [FSF99].

26.7.2   The Language

The MATHDOX source is an XML document. We have derived our own document type definitions (DTD) for these source texts. We have been influenced by both DOCBOOK [WM99] and OMDOC. The former is a fairly general standard for electronic books, the latter is a very rich, and strongly logic-oriented standard for mathematical documents—the main subject of this book. Both OMDOC and MATHDOX use OPENMATH [BCC⁺04], the difference being that OMDOC focuses on representing mathematical knowledge whereas MATHDOX focuses on interactivity. The connections with both Doc-Book and OMDOC are of importance to us because we expect several authoring tools for it to emerge in the coming few years, and we want to profit from their presence.

The mathematics in the MATHDOX source is given by means of OPEN-MATH objects. This feature has clear advantages in terms of portability. The DocBook type grammar sees to it that there are natural scopes, where mathematical objects 'live'. For instance, when a chapter begins with "Let $\mathbb{F}$ be a field", the scope of the variable $\mathbb{F}$ is assumed to be the whole chapter (although, somewhere further down the hierarchy, say in a section of the chapter, this assignment can be overridden).

Interactivity in MATHDOX is taken care of by XML tags representing various programming constructs as well as queries to external mathematical services. These actions take place within part of the context, which fixes the precise semantics of the objects involved. Further constructs are available for handling context and user input. Our notion of context is based on [FHJ⁺99b]. Context is divided into static and dynamic context. The static context may be defined as the set of all XML sources from which a interactive document can be prepared for use. Two extreme forms are OPENMATH Content Dictionaries and a chapter of an ordinary book. The dynamic context behaves more like the state of a CAS. It keeps track of the variables introduced, their properties, their values, and their scopes. The MATHDOX language has constructs for storing and changing this information. For example, the field $\mathbb{F}$ introduced at the beginning of a chapter may be specified to be a finite field of five elements in the context of a particular section of the chapter.

Although semantics is the primary target, some features for presentation have been built into the language. In order to have a flexible presentation, we use presentation-annotated OPENMATH. In MATHDOX we allow style attributes inside OPENMATH objects. By discarding these style attributes, regular OPENMATH is obtained. For instance, by providing the appropriate value for the style attribute, the author has a choice between a slash and a fraction display. In $\frac{3/4 \mid 2/3}{5}$ we have used them both.

Another way of solving presentation issues is illustrated by the statement: $3, 4 \in \mathbb{Z}$. The corresponding OPENMATH expression would be the equivalent of $3 \in \mathbb{Z} \wedge 4 \in \mathbb{Z}$, but our OPENMATH statement reads that the sequence $3, 4$ belongs to $\mathbb{Z}$. So here, the semantics of the element-of symbol has been stretched so as to help out presentation.

26.7.3   The MATHDOX System

An essential component of the MATHDOX software is its document server. It provides a view to the client of the content and manages both the static and the dynamic context. The usage of the MATHDOX document server is shown in Figure 26.11. We explain in some detail the main components shown in this picture.

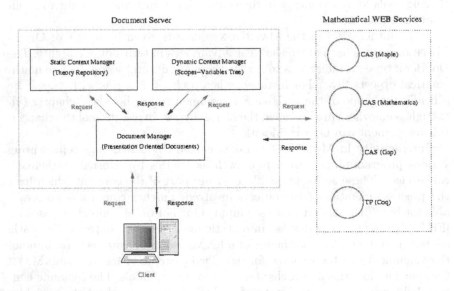

Fig. 26.11. The MATHDOX software

1. The *client*. The client is realized by a math-enabled web browser. It will present views of the documents served to the user, interact with the user, and communicate user input to the document server.

   The communication between client and server takes place via the HTTP request/response mechanism. The responsibility for interaction is mostly on the server side.

2. The documentserver. This server caters for presentation, communication, and context. It supports a wide range of actions ranging from handling queries to searching within documents for mathematical content and from placing (and retrieving) objects into the context, to rendering documents in different views.

   The document server is realized as a Java enhanced web application [JSP] inside a web server. It is not a monolithic entity. As shown in Figure 26.11, it is formed by the system parts. The *document manager* serves views to the client. IMDs can be thought of as programs (scripts) encoding the production of a response. In generating the response, they can make use of the information contained in the static context, and in the dynamic context (scopes and variables), the user input communicated along with request, and the results of computations carried on by one or more mathematical services.

Another part is the *static context manager* which is responsible for managing a repository of MATHDOX mathematical theories.

The final (third) part is the *dynamic context manager* which is responsible for the dynamic information.

3. *mathematical services*. Mathematical services can be very diverse: some may serve as general interfaces to CAS or to Theorem Provers. The MATHDOX software provides ways to access these services via standard protocols, among which those developed under the MONET project [Mon]. The mechanism extends the phrasebook set-up for OPENMATH [CCC+00, CCR00]. For constructing specific OPENMATH services, we employ our Java OPENMATH library ROML [ROM].

26.7.4 Conclusion

Now that MATHDOX is close to a complete working version, trial applications are in the make. We mention

- a server for providing designs of experiments on command to statisticians,
- an exercise repository for the EU funded LeActiveMath project,
- a mathematics course on calculus, with automated natural language production from a formal-mathematical source for the EU funded project WebALT,
- interactive lecture notes (the successor of [CCS99]) for an Abstract Algebra course within a mathematically oriented Bachelor curriculum,
- educational material for highschool mathematics in the Netherlands.

26.8   OMDoc in ActiveMath

Project Home	http://www.activemath.org/
Authors	The ACTIVEMATH group: Erica Melis, Giorgi Goguadse, Alberto Gonzales-Palomo, Adrian Frischauf, Martin Homik, Paul Libbrecht, Carsten Ullrich DFKI GmbH and Universität des Saarlandes

ACTIVEMATH is a mature web-based intelligent learning environment for mathematics that has been developed since 2000 at the University of Saarland and at the German Research Institute of Artificial Intelligence (Intelligent Learning Environments Group headed by Erica Melis). Its learning objects are encoded in an extension of OMDOC.

26.8.1   The ACTIVEMATH System

In addition to presenting pre-defined interactive materials, it adaptively generates courses according to the learner's goals, learning scenarios, competencies, and preferences. For this, **Tutorial Component** requests **learning**objects [8], related to the learning goal to be retrieved from several repositories. The retrieval of object-IDs is realized by a mediator taking into account structures and meta data of learning objects, and then the Tutorial Component assembles them to a course skeleton depending on a **Learner Model**. For details see [Ull05, Ull04].

In several stages a **Presentation Component** fills and transforms this skeleton to a material in the requested output format. In the interactive browser formats dummies can represent Learning Objects that can be instantiated dynamically — depending on the learning progress or on requests by the user.

This Learner Model stores the learning history, the user's profile and preferences, and a set of beliefs that the systems holds about the cognitive and meta-cognitive competencies and the motivational state of the learner. The domain model that underlies the structure of the learner model is inferred from the content for that domain and its meta data represented in the OM-DOC source.

ACTIVEMATH is internationalized and 'speaks' German, English, French, Spanish, Russian, and Chinese by now. Its mathematical notation rendering can as well be adapted to national' standards.

To realize a smooth and efficient cooperation of all components and in order to integrate further internal and external services, ACTIVEMATH has adopted a modular service-oriented architecture displayed in Figure 26.12. It

---

[8] Following the classical definitions, learning objects are any resources that are used the learning activity. When in OMDOC, learning objects considered are such as a `definition`, an `omtext` or an interactive exercise.

includes the XML-RPC web communication protocol for its simplicity and support. In addition, an event framework enables the asynchronous messaging for any changes.

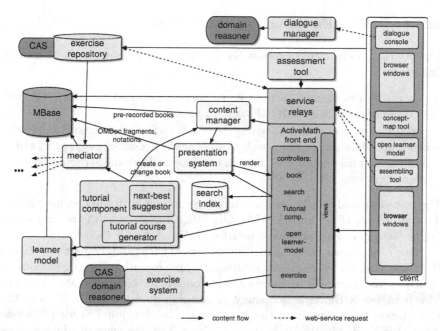

Fig. 26.12. The components, services and information flow in ACTIVEMATH

A complex subsystem in its own right is ACTIVEMATH's exercise subsystem [GPE05] that plays interactive exercises, computes diagnoses and provides feedback to the learner in a highly personalized way. It reports events to inform the other components about the user's actions.

In 2005, large educational contents exist in ACTIVEMATH's repositories for Fractions (German), Differential Calculus (German, English, Spanish) at high school and first year university level, operations research (Russian, English), Methods of Optimization (Russian), Statistics and Probability Calculus (German), Matheführerschein (German), and a Calculus course from University of Westminster in London.

ACTIVEMATH's Service-Approach. The encoding of content in OMDOC is an advantage for ACTIVEMATH's Web-service approach. If available, the services – including Web repositories – can communicate more semantic information than just meta data. However, the interoperability of the content encoding is only one side of the Semantic Web coin. Hence, the developments for ACTIVEMATH also include the reuse and interoperability of components and tools [MGH+05].

External services that are being connected currently are the SIETTE assessment tool [CGM+04] and one or more repository of interactive exercises and interactive content.

26.8.2 OMDoc **Extensions for** ACTIVEMATH

The ACTIVEMATH DTD extends the OMDoc DTD version 1.1 in several directions:

- new types of items such as `misconceptions`, additional types of items such as types of exercises (`MCQ`, `FIB`, `map`, `problem`),
- additional several relations with types such as `for` or `prerequisite-of`,
- other additional meta data such as `difficulty`, `competency`, or `field`,
- additional infrastructure as, e.g., in exercises, additional structure such as content packages [GUM+04].

The metadata and relation extensions are compliant with the Learning Metadata Standards IEEE and IMS LOM [IEE02, Con01]. Most of the extensions are pedagogically/educationally motivated. Some details follow.

The educational metadata include `competency` and `competencylevel` that are used for assessment, evaluation, and for adaptive suggestions of examples and exercises in course generation. As for competencies, ACTIVEMATH supports Bloom's taxonomy of learning goal levels [Blo56] and the more recent taxonomy from the Program for International Student Assessment (PISA) [KAB+04] and National Council of Teachers of Mathematics (NCTM).

ACTIVEMATH educational metadata include `learning_context` which was in first versions of LOM. Metadata values, such as `difficulty`, `abstractness`, and "typical learning time" have been annotated with the corresponding learning context (allowing to say that an example is hard for an undergraduate but not for a higher class). The ACTIVEMATH DTD introduced some educational relation types which facilitate adaptive course generation and concept map exercises, among others.

The OMDoc format has been refactored in ACTIVEMATH in order to represent metadata in a form that is separable from the representation of the knowledge item. For example, some metadata represented in form of attributes of an item is moved inside the metadata element. The purpose of such a separation is to facilitate the management of learning materials in ACTIVEMATH. Components such as Tutorial Component and Learner Model do not deal with the content of the knowledge items but rather with their metadata only and hence it is convenient to have a way to extract metadata records from the content.

For the internationalization each OMDoc item may have sub-elements in several languages since ACTIVEMATH does not translate learning objects on the fly.

ACTIVEMATH extends the OMDOC `example` element. A detailed explanation can be found in [MG04]. In case of a worked-out example, the micro-structure of this element is enriched with a solution that has a structure similar to a proof in OMDOC. It differs from the proof element since the solution might not only prove a statement, but also calculate the value of some expression or explore the properties of a particular structure (e.g. curve discussion). This representation allows for different presentations, and serves as a basis for the automatic generation of exercises by fading some parts of the structure of a worked-out example (see [MG04]).

The new exercise representation of ACTIVEMATH was the basis for extending the Math QTI standard [MP04]. Even though its origin can be traced to OMDOC originally not much is left from the QUIZ representation of OMDOC which supports only very limited types of exercises and did not have enough infrastructure. The micro-structure of an interactive exercise has to allow for different kinds of interactivity, checking the correctness of the answer, providing feedback, etc. This interaction graph can be automatically filled with information by the exercise subsystem components that can communicate with external systems in order to generate feedback to the user.

A description of ACTIVEMATH language for exercises can be found in [GPE05].

26.8.3 Usage of Semantic Representation in ACTIVEMATH

The fact that the Tutorial Component employs metadata to search for appropriate learning objects and assemble them has been sketched above. In addition, other tools and components of ACTIVEMATH make use of the semantics of OPENMATH, the ACTIVEMATH metadata and OMDOC more generally.

Computer Algebra Services. Computer algebra system (CAS) — currently YACAS [Yac], MAXIMA [Max], and WIRIS [Wir] — are integrated as external services. Via a broker, a CAS receives queries (partially Monet queries) to evaluate OPENMATH expressions. This enables the exercise system to evaluate user input, e.g., for numerical or semantic equivalence with a particular expression. The service CAS has to translate in- and output via phrasebooks.

Presentation Component. The naive approach to rendering OMDOC documents would be to fetch the items from a data base, assemble them (or parts of them) and then run several style-sheets on the resulting sequence; Those style-sheets would depend on the requested output format (HTML+ Unicode, XHTML+ MATHML, PDF via LaTeX, SVG, or slides), the target browser (we support MOZILLA, FIREFOX, Internet Explorer) and the personalization.

This approach turned out to be infeasible for complex, real-world applications. Therefore ACTIVEMATH includes a multi-stage presentation process as described in [ULWM04]. It has many advantages, among them a much better performance and even better perceived performance through multiple caching, a clear separation of different concerns which provides more flexibility for the various adaptivity dimensions that ACTIVEMATH supports, including selection of learning objects, link annotations language, specific presentations of pages, exercises etc, and of mathematical expressions, target output format, browser.

The final rendering maintains the references to mathematical symbols but renders them invisible. This information can then be used by copy-and-paste and for tool tips that indicate the name of a symbol on the page.

For an even more specialized presentation of mathematical notation which is often requested by authors and users we developed a complex presentation tag representation and an authoring facility for it [MLUM05]. These special presentations are integrated into the presentation process upon request.

Copy and Paste. The rendering includes an invisible reference to the unique identifier of mathematical symbols and expressions. This provides a basis for copying the reference to an OPENMATH expression, i.e., the semantics of the expression to a computer algebra system, to the input editor (in dictionary and exercises), and into exercise blanks. The actual transfer mechanism is, because of security limitations and because of resource management, a drag-and-drop operation which allows immediate negotiation between the recipient and source. This allows to transform to the appropriate encoding on demand. Alternate encodings include OPENMATH with a restricted set of content-dictionaries, HTML with embedded presentation and content MATHML. Reference to OMDOC items and to pages of a book in ACTIVEMATH are exchangeable using the same paradigm.

Interactive Concept Map Tool ICMAP. ICMAP utilizes the OMDOC encoding and relations for generating feedback to users' inputs [MKH05]. The tool visualizes (parts of) a domain and relations between concepts and between concepts and satellites.

Semantic Search. ACTIVEMATH's search facility has been upgraded to enable not only approximate search results but also to search semantically for (OPENMATH) mathematical expressions, for certain types of learning objects and objects with particular metadata. The implementation of the search uses Jakarta Lucene with its high-performance and easy deployment.

OMDOC-Related Components and Tools of ACTIVEMATH. Many of the tools described above have not been sufficient for the purposes of a complex and mature educational application such as ACTIVEMATH. Therefore, we had to improve them or implement some from scratch. In particular, these include authoring tools (for which improvement is still ongoing), transformation tools, validation tools, and style sheets. Moreover, new tools have been developed or integrated into ACTIVEMATH, e.g., an input editor that returns OPENMATH.

The conversion of OMDOC source to presentation code is done using XSLT style sheets. We started with the style sheets available in OMDOC repository and added to them the ACTIVEMATH linking schemata. These style sheets needed more polishing since they were too big and the management of notations was not feasible. Moreover, the TEX oriented style sheets had to be refurbished in order to work well with big documents.

Further tools have been realized within the authoring tools which are covered in Section 26.9.

26.9   Authoring Tools for ACTIVEMATH

Project Home	http://www.activemath.org/projects/jEditOQMath
Authors	Paul Libbrecht
	DFKI GmbH and Universität des Saarlandes

The OMDOC content to be delivered by ACTIVEMATH are OMDOC documents with OPENMATH formulae. Experience has shown that writing the XML-source by hand is feasible and even preferred if the author wants to follow the evolution of content's structure. It is similar to HTML editing. However, the complexity of XML makes it hard to keep an overview when writing mathematical expressions. Therefore, the OQMATH processor has been implemented: it uses QMATH for formulae and leaves the rest of the OMDOC written as usual XML.

OQMATH has been integrated in a supporting XML-editor, jEdit. This editor provides structural support at writing XML-documents. Authors, even with no XML-knowledge, can easily write valid document JEDITO-QMATH. This package includes, in a one-click installer, QMATH, OQMATH, JEDIT, and Ant-scripts for publication of the content in ACTIVE-MATH knowledge bases. These scripts validate the references in the content. These scripts also provide authors with short cycles edit-in-JEDITOQMATH-and-test-in-ACTIVEMATH. More about JEDITOQMATH can be seen from http://www.activemath.org/projects/jEditOQMath at [Lib04]

JEDITOQMATH provides search facilities as well as contextual drops from items presented in an ACTIVEMATH window. This way the testing of content in the target environment and the authoring experience are bound tighter together, thus making JEDITOQMATH closer to the WYSIWYG paradigm without being limited to its simple visual incarnation.

To date, more than 10'000 *items* of OMDOC content has been written using these authoring tools in Algebra and Calculus. This experience with authors considerably improved our understanding of what today's authors need and what different classes of authors can cope with.

Among the greatest difficulties of authoring content for ACTIVEMATH was the art of properly choosing mathematical semantic encoding: the mathematical discourse is made of very fine notation rules along with subtle digressions to these rules... formalizing them, as is needed when writing OPENMATH or the QMATH formulae for them, turns out to often be overwhelming. The usage of the ellipsis in such a formula as $1, \ldots, k, \ldots, n$ is a simple example of semantic encoding challenge. The knowledge organization of OMDOC that makes it possible to define one's own OPENMATH symbols has been a key ingredient to facing this challenge.

Among the features most requested by authors, which we have tried to answer as much as possible, are a short edit-and-test cycle and validation facilities taking in account the overall content.

Validation Tools. Automated validation of OMDoc content has many facets. XML-validation with a DTD and Schema is a first step. However there are still many structure rules mentioned only as human readable forms in the OMDoc specifications. References between OMDoc items is another important facet which has been answered by ActiveMath knowledge bases and publishing scripts. Experience has proved that ignoring such errors has lead repeatedly to authors complaining about the weirdest behaviours of the overall learning environment. Many other simple validations could be done in order to support the author, for example the validation of a picture embedding, or of fine grained typing of relations (for example, that a definition should only be *for* a symbol).

Further validation tools are being investigated, for example, those tuned to particular pedagogical scenarios.

Further Authoring Tools for ActiveMath. jEditOQMath clearly remains for users who feel comfortable with source editing. Experience has shown that authors having written HTML or TeX earlier did not find this paradigm problematic. It is, however, a steep learning slope for beginner authors. A more visual component is being worked upon, able to display and edit visually the children of a CMP, including formulae.[9] This component, along with forms and summaries for metadata, should provide a visual environment to edit OMDoc content for ActiveMath in a relatively accessible way.

Another area where source editing has shown difficulties is in the process of authoring exercises with many steps... the rich structure of the exercises, along with the non-neglectable space taken by the display of XML-source has challenged several authors, having difficulties to overview such sources as 600 Kb of OQMath source for a single exercise. A web-based visual authoring environment is under work within the ActiveMath group.

---

[9] More about the component for OMDoc micro-structure can be read from http://www.activemath.org/projects/OmdocJdomAuthoring/.

26.10   SWiM – An OMDoc-Based Semantic Wiki

Project Home	http://kwarc.eecs.iu-bremen.de/projects/swim
Authors	Christoph Lange, Michael Kohlhase
	Computer Science, International University Bremen

SWiM is a semantic wiki for collaboratively building, editing and browsing a mathematical knowledge base of OMDoc theories. Our long-term objective is to develop a software that facilitates the creation of a shared, public collection of mathematical knowledge and serves work groups of mathematicians as a tool for collaborative development of new theories. Even though the work reported here was initially motivated by solving the MKM author's dilemma [KK04], we contend that the new application area MKM can also contribute to the development of semantic wikis.

Technically, SWiM is based on the semantic wiki engine IkeWiki [Sch06], which was chosen because of its modular design, its rich semantic web infrastructure, its user assistance for annotations, and its orientation towards learning [SBB+06].

26.10.1   Semantic Wikis

A wiki [LC01] is a web server application that allows users to browse, create, and edit hyperlinked pages in a web browser, usually using a simple text syntax. In contrast to most content management systems, wiki pages are accessible via an URL containing their title. A new page can be created by linking from an existent page to the page to be created. This link will then lead to an edit form. Usually, anyone is allowed to edit pages on a wiki, but access can be restricted. Other characteristics of wikis include permanent storage of old page versions (with facilities to display differences between two versions and to restore a certain version), notification about recent changes, and full-text search.

Semantic wikis [VKS+06, TB06] enhance wikis by Semantic Web technologies, such as RDF [LS99] or ontologies. Usually one page represents one concept from a real-world domain, which has a type, possibly some metadata, and typed links to other concepts. For example, a link from a wiki page about "Life, the Universe and Everything" to another page about Douglas Adams could be typed as "is author of". In terms of RDF, this can be expressed by the following subject–predicate–object triple,

("Douglas Adams", isAuthorOf, "Life, the Universe and Everything")

where the *isAuthorOf* relation would be defined in an ontology. These links are usually displayed in a navigation box next to the page contents. Semantic wikis only deal with wiki text, not with mathematics, though some allow to embed mathematical formulae as presentational-only TeX.

SWiM encourages users to collaborate: Non-mathematicians can collaborate in creating a "Wikipedia of mathematics" by compiling the knowledge available so far, while scientists can collaboratively develop new theories. Users get an immediate reward for many of their contributions: Once they specify the type of a page or relations of one page to another, this information will be displayed in a box of navigation links. We intend to make the data created in SWiM usable for external services by offering an export facility for OMDoc documents and by integrating them into SWiM. Mathematicians developing theories will be assisted to retain an overview of theory dependencies in order not to break them. Social software services will further utilize the semantic information available from the theories and from tracking the user interaction log ("Who did what on which page when?"). User feedback to pages can be extended to social bookmarking, which is "the practice of saving bookmarks [of Internet resources] to a public web site and 'tagging' them with keywords." [Lom05] The more users tag a certain resource, the higher a social bookmarking service will rank it.

The enhancements of the data model semantic wikis bring along — compared to traditional wikis — are already present in the OMDoc format, so that an OMDoc-based wiki only needs to operationalize their underlying meaning. For example, typed links, which are implemented via an extension to the wiki syntax in SEMANTIC MEDIAWIKI [VKVH06] or editable through a separate editor in IKEWIKI [Sch06], are implemented by means of the for attribute to OMDoc's elements (e.g. `<example for="#id-of-assertion">`). SWiM makes them editable easily and visualizes them adequately. A semantic wiki targeted at mathematics must ensure that dependencies between concepts are preserved. Results in this area will be interesting for non-mathematical semantic wikis as well, especially when they support higher levels of formalization such as ontologies.

26.10.2 Design of SWiM

Concepts and Relations. The smallest unit that can be displayed, edited, linked to, or archived in a wiki is a page. In a semantic wiki, it usually describes one *concept*, including its properties and its relations to other concepts. While standalone OMDoc documents can contain more than one theory, is is important to keep pages small in a wiki to improve the effectivity of usage. Furthermore, usual semantic wikis only store and display metadata and typed links per page; SWiM does too.[10] Users are strongly encouraged to define at most one theory per wiki page and to roll out non-constitutive statements (see Section 15.1) to separate pages, referencing their context theory. As constitutive statements cannot exist without an enclosing theory, but

---

[10] Semantic information will only be considered on the theory and statement levels of OMDoc — directly or through reasoning in the case of transitive closures —, not on the object level.

as, on the other hand, we want each wiki page to form a valid document, we introduced a new element `swim:page`, which can be a child of an `omdoc` element and which has the same content model as a `theory` element — in particular, it can hold several theory-constitutive statements and connect them to their context theory.

OMDOC's system on-
tology has been partly
coded in OWL-DL
and imported to the
wiki's RDF store,
which is implemented
using the Jena Se-
mantic Web Frame-
work for Java [JEN].
Theories as well as
statements of any type
form concepts, and
the most important
relations between
those concepts are ex-
tracted from the OM-

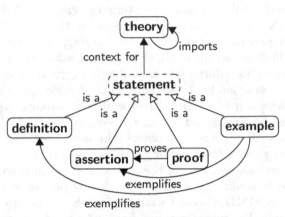

Fig. 26.13. Subset of OMDOC's system ontology

DOC pages on saving and then stored as RDF triples. These relations include:

- The import relation between theories
- The relation of a statement to its context theory
- The relation of an example to the statement it exemplifies
- The relation of a proof to the assertion it proves

It is planned to also take relations given by user interaction into consideration, such as "Who edited which page when?", and to combine ontology-defined relations and user relations. For example, a metric estimating the *degree of difficulty* of a page, calculated by counting the questions on the discussion page, could be implemented. Furthermore, the user can specify taxonomic relations, which cannot be stated explicitly in OMDOC, such as ("all differentiable functions are continuous"), as annotations in an ontology language like RDF Schema or OWL.

User Interface and Interaction Model. Pages can be rendered to XHTML plus presentational MathML using the transformations described in Chapter 25. There is also a browsable source code view, which is useful for documents that are not written in textbook style.

Not only will the user be able to navigate along the dependency graph, she will also be able to *interact* with the system: she will be asked whether she wants to explore the theories required as dependencies in further detail.

Suppose that the user is currently reading the page containing the theory `ring` from the elementary algebra example from Chapter 7. In this case the wiki will not only display navigation links to the direct dependencies **group** and **monoid**, but it will also provide unobtrusive buttons that allow the user to give one of the commands in Figure 26.14. Not only the last case will be recorded — the others are interesting as well for *social bookmarking*. For example, if many users requested a theory t to be explained, the system could default to display not only the direct dependencies but also the level-two dependencies, for it seems that t is too difficult for only being explained shallowly.

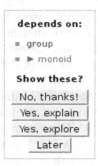

No, thanks! *"I already know group and monoid."*

Explain *"Please show me group and monoid, I want to learn about ring's prerequisites."* — group and monoid will be displayed.

Explore *"Please show me all prerequisites for ring."* — group, monoid, and semigroup, are opened in separate windows or serialized into one page.

Suspend *"I want to know about group and monoid, but only later."* — SWiM keeps a notice in the user's profile that she wants to read group and monoid sometime. Reminder links to suspended theories are shown on a separate navigation bar.

Fig. 26.14. The command buttons to navigate along the dependencies

Further Work. Further work on SWiM will concentrate on integrating a lightweight management of change process. Second, while the wiki is yet a user-friendly *browser*, there is still a demand for assisting users to *edit* OM-DOC. To this end, the QMATH preprocessor (see Section 26.2) will be integrated into SWiM. Mathematical objects entered as QMATH will be kept in this syntax for display in the edit form, but they will be converted to OM-DOC for rendering for presentation and when pages are exported to another application.

26.11   Induction Challenge OMDoc Manager (ICOM)

Project Home	`http://www.cs.nott.ac.uk/~lad/research/` `challenges/challenge_manager.html`
Authors	Thomas D. Attfield, Monica C. Duarte, Lin Li, Ho-Ying Mak, Adam M. Neal, Lewis M. Toft, Zixuan Wang, Louise A. Dennis School of Computer Science and Information Technology, University of Nottingham

We describe work in progress to create a system for organising and presenting a set of challenge problems collected by the Induction Theorem Proving community. These challenge problems come from a number of sources and are presented in different logics using different presentation conventions.

The intention is to provide a system which will allow these problems to be stored in a unified format and will support the collection, browsing and extraction of the problems.

OMDOC is an obvious choice for representing such problems and the system is able to take advantage of much existing work on the manipulation of XML documents.

26.11.1   The Induction Challenge Problems

Inductive Theorem proving is a small field. The main theorem provers within this field are NQTHM [BM79] (now re-engineered as ACL2 [KM96]), INKA [AHMS99], the CLAM series [BvHHS90, RSG98a] and RRL [KZ95]. TWELF [PS99] also looks at the automation of inductive proof in the context of logical frameworks. Within the field it is hard to assess claims for the superiority of any given system since there is naturally a tendency to report "successes" – difficult or challenging problems automatically proved. There is also a desire within the community to develop a store of shared knowledge about the challenges that face the automation of proof by mathematical induction.

TPTP (Thousands of Problems for Theorem Proverss) [SS98] is a library of test problems for first-order ATP systems. They provide the ATP community with a comprehensive library complete with unambiguous names and references. All the problems are stated in a standardised formulation of first-order logic and are widely used to benchmark first-order systems. They are also used as the test set for the CASC competition [Sut01] which compares such systems. One of the benefits of the TPTP library to the ATP community is the existence of a common set of problems by which comparisons can be made.

It is not practical for inductive theorem provers to follow the pattern of the TPTP library. Various attempts have been made to build a similar corpus of problems requiring inductive reasoning. The most mature of these was based on the Boyer-Moore [BM79] corpus[11]. This corpus was unpopular partly because there was repetition within the problem set and partly because many problems depended on a few particular function definitions. But the major objection was that induction theorem provers use a number of different logics, some of which are typed and some of which are not, which made it difficult to agree on a standard format. The use of other logics also raised translation issues and a fully automated process for converting the theorems, even into an agreed typed language was never produced.

A group of researchers within the community[12] agreed that instead of a large set of benchmarks in a standard logic they would each put forward a number of "Challenge Problems". These should present interesting challenges to the automation of inductive proof or illustrate important features which an inductive prover should be able to handle. A set of these problems would be collected which would remain sufficiently small that an individual could represent them within their own theorem proving system as they saw fit[13]. These challenge problems are currently described in a high-level way and written up in an ad hoc fashion. The descriptions contain both mathematical notation and commentary. They are difficult to read, navigate or use in any particular system.

OMDOC seems ideally suited as a format for representing these challenge problems: it can represent both text and formulae; it is not tied to any particular logic and it supports the extraction of data into a number of different formats. As an added benefit its hyper-text features would potentially allow definitions to be stored separately and shared between problems. Individual theorem provers can then concentrate on translations between OPENMATH content dictionariesnd their own logics and individuals submitting problems can specify the appropriate content dictionary for the problem.

26.11.2   System Description

The Induction Challenge OMDOC Manager (ICOM) is designed to be a system which will ease the submission and extraction process for the problems. Our intention is to provide a submission interface that will create a simple OMDOC markup for the problems which can subsequently be edited by a user and to provide browsing and extraction capabilities.

Each challenge problem description contains six distinct sections (e.g. Summary, Definitions, Comments). Currently a user who wishes to enter a problem into our system is presented with the form shown on the right with a field for each section.

---

[11] This has become known as the Dmac corpus after David McAllester who translated a fragment of the NQTHM corpus into a simpler language.

[12] At the 2000 CADE Workshop on the Automation of Proof by Mathematical Induction.

[13] The current set can be found at http://www.cs.nott.ac.uk/~lad/research/challenges.

Each section, once entered by a user, is placed in a CMP tag. These tagged fragments are wrapped in standard OMDOC headers and footers to produce a valid OMDOC. This completed document is then written to disk and stored. We are currently working on a simple parser to translate equations into om:OMOBJ structures which a user will then be able to edit (for instance to specify the appropriate content dictionaries). We hope this will be easier than adding all the OPEN-MATH tags by hand.

An existing document can be displayed as a tree and from this tree the document can be directly manipulated. This tree display also allows the user to see the structure of the document more clearly. It is also possible to extract an HTML view of the contents of the document so it can be displayed in a web browser and read by a human.

Our implementation language is JAVA and we use its JAXP DOM API. DOM [DOM] is a W3C standard which uses a tree-based model (storing data in hierarchies of nodes). This means that once an OMDOC has been created or opened all the document's data is in memory and so data can be accessed rapidly. DOM also enables simple modification of documents by adding or deleting nodes. Although Sax (an alternative model) achieves better performance and less memory overhead than DOM, it is easier to traverse and modify XML documents using a DOM tree structure. Since

we anticipate that users may wish to modify the initial OMDOC produced by our system we adopted the DOM model instead.

26.11.3   Further Work

ICOM is still in the early stages of development. Currently our most pressing aim is to provide improved support for entering equations. Once this is in place we hope to add searching facilities and provide better mechanisms for links to be created between challenge problems. We would also like to experiment with the automatic extraction of problems into a theorem prover via an MBASE [KF00] and a MATHWEB [FHJ+99a].

26.12   Maya: Maintaining Structured Developments

Project Home	www.dfki.de/~inka/maya.html
Authors	Serge Autexier[1], Dieter Hutter[1], Till Mossakowski[2], Axel Schairer[1] [1] DKFI GmbH, Stuhlsatzenhausweg 3, D 66123 Saarbrücken [2] Computer Science, University of Bremen, Germany

26.12.1   Overview

The MAYA-system was originally designed to maintain and utilize the structuring mechanisms incorporated in various specification languages when evolving and verifying formal software developments. In this setting, a software system as well as their requirement specifications are formalised in a textual manner in some specification language like CASL [CoF04] or VSE-SL [AHL+00]. All these specification languages provide constructs similar to those of OMDOC to structure the textual specifications and thus ease the reuse of components. Exploiting this structure, e.g. by identifying shared components in the system specification and the requirement specification, can result in a drastic reduction of the proof obligations, and hence of the development time which again reduces the overall project costs.

However, the logical formalisation of software systems is error-prone. Since even the verification of small-sized industrial developments requires several person months, specification errors revealed in late verification phases pose an incalculable risk for the overall project costs. An *evolutionary formal development* approach is absolutely indispensable. In all applications so far development steps turned out to be flawed and errors had to be corrected. The search for formally correct software and the corresponding proofs is more like a *formal reflection* of partial developments rather than just a way to assure and prove more or less evident facts.

The MAYA-system supports an evolutionary formal development since it allows users to specify and verify developments in a structured manner, incorporates a uniform mechanism for verification *in-the-large* to exploit the structure of the specification, and maintains the verification work already done when changing the specification. MAYA relies on *development graphs* [AH05, Hut00] as a uniform (and institution independent[14]) representation of structured specifications, and which provide the logical basis for the *Complex theories* and *Development graphs* of OMDOC[15]. Relying on development graphs enables MAYA to support the use of various (structured) specification languages like OMDOC, CASL [CoF04], and VSE-SL [AHL+00]

---

[14] This includes, for instance, that it does not require a particular logic (see e.g. [MAH06] for more details).

[15] These are the modules CTH and DG, respectively.

to formalise mathematical theories or formal software developments. To this end MAYA provides a generic interface to plug in additional parsers for the support of other specification languages. Moreover, MAYA allows the integration of different theorem provers to deal with the actual proof obligations arising from the specification, i.e. to perform verification *in-the-small*.

Textual specifications are translated into a structured logical representation called a development graph, which is based on the notions of consequence relations and morphisms and makes arising proof obligations explicit. The user can tackle these proof obligations with the help of theorem provers connected to MAYA like Isabelle [Pau94] or INKA [AHMS99].

A failure to prove one of these obligations usually gives rise to modifications of the underlying specification. MAYA supports this evolutionary process as it calculates minimal changes to the logical representation readjusting it to a modified specification while preserving as much verification work as possible. If necessary it also adjusts the database of the interconnected theorem prover. Furthermore, MAYA communicates explicit information how the axiomatization has changed and also makes available proofs of the same problem (invalidated by the changes) to allow for a reuse of proofs inside the theorem provers. In turn, information about a proof provided by the theorem provers is used to optimise the maintenance of the proof during the evolutionary development process.

26.12.2   From Textual to Logical Representation

The specification of a formal development in MAYA is always done in a textual way using specification languages like CASL , OMDOC or VSE-SL. MAYA incorporates parsers to translate such specifications into the MAYA-internal specification language DGRL ("Development Graph Representation Language"). DGRL provides a simply-typed λ-calculus to specify the local axiomatization of a theory in a higher-order logic. While unstructured specifications are solely represented as a signature together with a set of logical formulas, the structuring operations of the specification languages are translated into the structure of a development graph. Each node of this graph corresponds to a theory. The axiomatization of this theory is split into a local part which is attached to the node as a set of higher-order formulas and into global parts, denoted by ingoing definition links, which import the axiomatization of other nodes via some consequence morphisms (such as the `imports` element in OMDOC). While a *local* link imports only the local part of the axiomatization of the source node of a link, *global* links are used to import the entire axiomatization of a source node (including all the imported axiomatization of other nodes). In the same way local and global *theorem links* are used to postulate relations between nodes (see [AH05] for details) which correspond to OMDOC's `theory-inclusion` and `axiom-inclusion` elements.

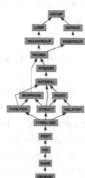

Fig. 26.15. The graphical user interface of MAYA & the development graph for the OMDOC representation of groups in MBASE

On the left hand side, Figure 26.15 shows the graphical user interface of MAYA. The right hand side shows the development graph in MAYA for a formalisation of groups. The formalisation was given in OMDOC and imported into MAYA from the OMDOC database MBASE (see [KF01] and Section 26.4).

26.12.3 Verification In-the-Large

The development graph is the central data-structure to store and maintain the formal (structured) specification, the arising proof obligations and the status of the corresponding verification effort (proofs) during a formal development.

MAYA distinguishes between proof obligations postulating properties between different theories (like the `theory-inclusion` and `axiom-inclusion` elements in OMDOC) and lemmata postulated within a single theory (like `assertion` in OMDOC). As theories correspond to subgraphs within the development graph, a relation between different theories, represented by a global theorem link, corresponds to a relation between two subgraphs. Each change of these subgraphs can affect this relation and would invalidate previous proofs of this relation. Therefore, MAYA decomposes relations between different theories into individual relations between the local axiomatization of a node and a theory (denoted by a local theorem link). Each of these relations decomposes again into a set of proof obligations postulating that each local axiom of the node is a theorem in the target theory with respect to the morphism attached to the link.

While definition links establish relations between theories, theorem links denote lemmata postulated about such relations. Thus, the reachability between two nodes establishes a formal relation between the connected nodes (i.e. the theory of the source node is part of the theory of the target node wrt.

the morphisms attached to the connecting links). MAYA uses this property to prove relations between theories by searching for paths between the corresponding nodes (instead of decomposing the corresponding proof obligation in the first place).

26.12.4   Verification In-the-Small

When verifying a local theorem link or proving speculated lemmata, the conjectures have to be tackled by some interconnected theorem prover. In both cases the proofs are done *within* a specific theory. Thus, conceptually each theory may include its own theorem prover. In principle, there is a large variety of integration types. The tightest integration consists of having a theorem prover for each node wrt. which theory conjectures must be proven, and the theorem prover returns a proof object generated during the proof of a conjecture. Those are stored together with the conjecture and can be used by MAYA to establish the validity of the conjecture if the specification is changed. The loosest integration consists in having a single generic theorem prover, which is requested to prove a conjecture within some theory and is provided with the axiomatization of this theory. The theorem prover only returns whether it could prove a conjecture or not, without any information about axioms used during the proof. For a detailed discussion of the advantages and drawbacks of the different integration scenarios see [AM02].

Currently, MAYA supports two integration types: One where information about used axioms is provided by the theorem prover, and one where no such information is provided. In the first case, MAYA stores the proof information and the axioms used during the proof. In the second case, MAYA assumes there is a proof for the proof obligation, as there is no information about the proof. In both scenarios, MAYA makes use of generic theorem provers which are provided with the axiomatization of the current theory. Currently MAYA provides all axioms and lemmata located at theories that are imported from the actual theory by definition links to the prover. Switching between different proof obligations may cause a change of the current underlying theory and thus a change of the underlying axiomatization. MAYA provides a generic interface to plug in theorem provers (based on an XML-RPC protocol) that allows for an incremental update of the database of the prover.

26.12.5   Evolution of Developments

The user executes changes to specifications in their textual representation. Parsing a modified specification results in a modified DGRL-specification. In order to support a management of change, MAYA computes the differences of both DGRL-specifications and compiles them into a sequence of basic operations in order to transform the development graph corresponding to the original DGRL-specification to a new one corresponding to the modified

DGRL-specification. Examples of such basic operations are the insertion or deletion of a node or a link, the change of the annotated morphism of a link, or the change of the local axiomatization of a node. As there is currently no optimal solution to the problem of computing differences between two specifications, MAYA uses heuristics based on names and types of individual objects to guide the process of mapping corresponding parts of old and new specification. Since the differences of two specifications are computed on the basis of the internal DGRL-representation, new specification languages can easily be incorporated into MAYA by providing a parser for this language and a translator into DGRL.

The development graph is always synthesised or manipulated with the help of the previously mentioned basic operations (insertion/deletion/change of nodes/links/axiomatization) and MAYA incorporates sophisticated techniques to analyse how these operations will affect proof obligations or proofs stored within the development graph. They incorporate a notion of monotonicity of theories and morphisms, and take into account the sequence in which objects are inserted into the development graph. Furthermore, the information about the decomposition and subsumption of global theorem links obtained during the verification *in-the-large* is explicitly maintained and exploited to adjust them once the development graph is altered. Finally, the knowledge about proofs, e.g. the used axioms, provided by the interconnected theorem provers during the verification *in-the-small* is used to preserve or invalidate the proofs.

26.12.6 Conclusion and System Availability

The MAYA-system is mostly implemented in Common Lisp while parts of the GUI, shared with the ΩMEGA-system [SHB+99], are written in Mozart. The CASL-parser is provided by the CoFI-group in Bremen (see Section 26.13). The MAYA-system is available from the MAYA-web-page at http://www.dfki.de/~inka/maya.html.

The Heterogeneous Tool Set (HETS, see Section 26.13) extends MAYA with a treatment of hiding [MAH06], and a uniform treatment of different logics based on the notion of heterogeneous development graphs [Mos02]. Furthermore, it is planned to extend this with the maintenance of theory-specific control information for theorem provers. The latter comprises a management for building up the database of theorem provers by demand rather than providing all available axioms and lemmata at once and it comprise the management of meta-level information, like tactics or proof plans, inside MAYA.

26.13   Hets: The Heterogeneous Tool Set

Project Home	www.tzi.de/cofi/hets
Authors	Till Mossakowski, Christian Maeder, Klaus Lüttich
	Computer Science, University of Bremen, Germany

26.13.1   Motivation

> *"There is a population explosion among the logical systems used in computer science. Examples include first order logic, equational logic, Horn clause logic, higher order logic, infinitary logic, dynamic logic, intuitionistic logic, order-sorted logic, and temporal logic; moreover, there is a tendency for each theorem prover to have its own idiosyncratic logical system. We introduce the concept of* institution *to formalize the informal notion of* 'logical system'*." [GB92]*

In the area of formal specification and logics used in computer science, numerous logics are in use:

— logics for specification of data types,
— process calculi and logics for the description of concurrent and reactive behaviour,
— logics for specifying security requirements and policies,
— logics for reasoning about space and time,
— description logics for knowledge bases in artificial intelligence and for the Semantic Web,
— logics capturing the control of name spaces and administrative domains (e.g. the ambient calculus), etc.

Indeed, at present, it is not imaginable that a combination of all these (and other) logics would be feasible or even desirable — even if it existed, the combined formalism would lack manageability, if not become inconsistent. Often, even if a combined logic exists, for efficiency reasons, it is desirable to single out sublogics and study translations between these (cf. e.g. [Sch04]). Moreover, the occasional use of a more complex formalism should not destroy the benefits of *mainly* using a simpler formalism.

   This means that for the specification of large systems, heterogeneous multi-logic specifications are needed, since complex problems have different aspects that are best specified in different logics. Moreover, heterogeneous specifications additionally have the benefit that different approaches being developed at different sites can be related, i.e. there is a formal interoperability among languages and tools. In many cases, specialized languages and tools often have their strengths in particular aspects. Using heterogeneous specification, these strengths can be combined with comparably small effort.

   OMDOC deliberately refrains from a full formalization of mathematical knowledge: it gains its flexibility through avoiding the specification of a formal semantics of the logic(s) involved. By contrast, the Heterogeneous Tool Set

(HETS, [MMLW]) is based on a rigorous formal semantics. HETS gains its flexibility by providing *formal interoperability*, i.e. integration of different formalisms on a clear semantic basis. Hence, HETS is a both flexible, multilateral *and* formal (i.e. based on a mathematical semantics) integration tool. Unlike other tools, it treats logic translations (e.g. codings between logics) as first-class citizens.

26.13.2   Institutions, Entailment Systems and Logics

Heterogeneous specification is based on individual (homogeneous) logics and logic translations [Mos05]. To be definite, the terms 'logic' and 'logic translation' need to be formalized in a precise mathematical sense. We here use the notions of *institution* [GB92] and *entailment system* [Mes89], and of *comorphism* [GR02] between these.

Logical theories are usually formulated over some (user-defined) vocabulary, hence it is assumed that an institution provides a notion of *signature*. Especially for modular specification, it is important to be able to relate signatures, which is done by *signature morphisms*. These can be composed, and hence form a *category of signatures and signature morphisms*.

Furthermore, an institution provides notions of *sentences* and *models* (over a given signature Σ). Models and sentences are related by a *satisfaction relation*, which determines when a given sentence holds in a model. An entailment system also provides an *entailment (provability) relation*, which allows to infer sentences (conclusions) from given sets of sentences (premises, or axioms).

Finally, it is assumed that each signature morphism leads to translations of sentences and models that preserve satisfaction and entailment.

A *institution comorphism* is a translation between two institutions. It maps signatures to signatures, sentences to sentences and models to models, such that satisfaction is preserved (where models are mapped contravariantly, i.e. against the direction of the comorphism).

We refer the reader to the literature [GB92, Mes89, MGDT05] for formal details of institutions and comorphisms. Subsequently, we use the terms "institution" and "logic" interchangeably, as well as the terms "institution comorphism" and "logic translation".

26.13.3   The Architecture of the Hets System

HETS is a tool for parsing, static analysis and proof management combining various such tools for individual specification languages, thus providing a tool for heterogeneous multi-logic specification (see Fig. 26.16). The graph of currently supported logics and logic translations is shown in Fig. 26.17. However, syntax and semantics of heterogeneous specifications as well as their implementation in HETS is parametrized over an arbitrary such logic graph.

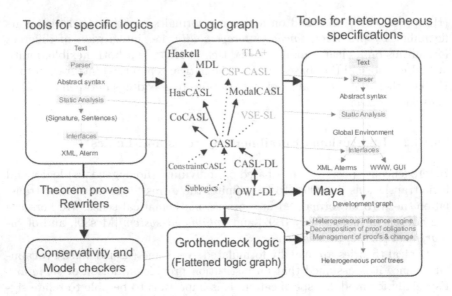

Fig. 26.16. Architecture of the heterogeneous tool set

Indeed, the HETS modules implementing the logic graph can be compiled independently of the HETS modules implementing heterogeneous specification, and this separation of concerns is essential to keep the tool manageable from a software engineering point of view.

Heterogeneous CASL (HETCASL; see [Mos04]) includes the structuring constructs of CASL, such as union and translation. A key feature of CASL is that syntax and semantics of these constructs are formulated over an arbitrary institution (i.e. also for institutions that are possibly completely different from first-order logic resp. the CASL institution). HETCASL extends this with constructs for the translation of specifications along logic translations.

Like MAYA (see Section 26.12), HETS provides a representation of structured specifications which are the logical basis for the *Complex theories* and *Development graphs* of OMDOC[16].

For proof management, MAYA's calculus of development graphs has been extended with hiding and adapted to heterogeneous specification. Development graphs provide an overview of the (heterogeneous) specification module hierarchy and the current proof state, and thus may be used for monitoring the overall correctness of a heterogeneous development.

HETS also provides a translation of CASL to and from a subset of OM-DOC (namely some formal first-order subset). Future work aims at a deeper integration of HETS and OMDOC that provides a translation to and from OMDOC for each of the logics integrated in HETS. Moreover, OMDOC itself

---

[16] These are the modules CTH and DG, respectively.

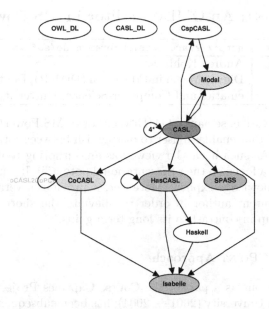

Fig. 26.17. Graph of logics currently supported by HETS. The more an ellipse is filled, the more stable is the implementation of the logic.

will become a "logic" (but only with syntax, without model theory) within HETS, such that also informal OMDoc documents (or formal OMDoc documents written in a logic currently not available in HETS) will be manageable for HETS. In this way, the data formats of OMDoc and HETS will converge, such that tools e.g. for searching, versioning or management of change can be implemented uniformly for both.

26.14   CPOINT: An OMDOC Editor in MS PowerPoint

Project Home	http://kwarc.eecs.iu-bremen.de/software/CPoint/
Authors	Andrea Kohlhase
	Digital Media in Education (DiMeB), Dept. of Math-
	ematics and Computer Science, University Bremen

CPOINT is an invasive, semantic OMDOC editor in MS PowerPoint (with an OMDOC outlet) that enables a user to distinguish between form and content in a document. As such it can be viewed as an authoring tool for OMDOC documents with a focus on their presentational potential. It enables a user to make implicit knowledge explicit. Moreover, it provides several added-value services to a content author in order to alleviate the short term costs of semantic mark up in contrast to its long term gains.

26.14.1   The CPOINT Approach

CPOINT started out as a part of the Course Capsules Project (CCAPS) at Carnegie Mellon University (2001 — 2004), has been subsequently supported by the International University Bremen (2004) and is now developed further at the 'Digital Media in Education' group at Bremen University. CPOINT is distributed under the Gnu Lesser General Public License (LGPL) [FSF99]. The newest version can be downloaded from the project homepage.

PowerPoint (PPT) slides address exclusively the issue of presentation — the placement of text, symbols, and images on the screen, carefully sequenced and possibly animated or embellished by sound. This directly leads to the question: *What exactly is the content in a PPT presentation?*

Obviously, the text and the pictures carry content as does the textual, presentational, and placeholder structure. For instance the ordering of information by writing it in list form, grouping information bubbles in one slide, or marking text as title by putting it into a 'title' placeholder can be mapped directly onto the OMDOC `omgroup` and `metadata` elements. Unfortunately though, this content exploits neither OMDOC's theory level nor the statement or formula level in more than a very superficial way.

The 'real' content is hidden beneath the presentation form: the authors, lecturers, and audience know or learn this real content by **categorizing** what they see, and **combining** it with what they already know and presently hear. CPOINT stands for 'Content in PowerPoint'. It models this by providing the author with a tool to explicitly store the additional implicit knowledge with the PPT show itself and from within the PPT environment without destroying the presentational aspects of the PPT document. Moreover, CPOINT **converts** the additional content to the appropriate OMDOC levels, so that the resulting OMDOC document captures all content. For an author the semantic markup process is a long-term investment. In order to alleviate the author's costs, CPOINT has implemented several **added-value services**.

26.14.2   The CPOINT Application

CPOINT extends PPT's presentational functionalities by semantic ones to get a handle on its visible and invisible content. As an invasive editor (see [Koh05a]) CPOINT makes these semantic authoring tools available through a toolbar in the PPT menu (see Figure 26.18) where they are available whenever PPT is running. CPOINT is written in Visual Basic for Applications and can be distributed as a PPT add-in.

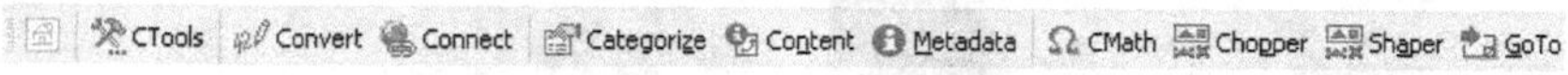

Fig. 26.18. The CPOINT menu bar

The top-level structure of a PPT presentation is given by slides. Each slide contains **PPT objects**, e.g. text boxes, shapes, images, or tables. These objects carry certain properties like text structure (e.g. ordered lists), document structure (e.g. being a title in the text hierarchy), or presentational structure (e.g. color, bold font, italic font, or symbol font). CPOINT enables the author to attach additional information to each PPT object. In particular, the author is empowered to transform implicit into explicit knowledge by categorizing, combining and enhancing these objects semantically.

Categorizing. The semantic annotation process typically starts with understanding an object's role in the to be transmitted knowledge and a subsequent categorization. The author selects the respective PPT object and assigns a suitable (didactic) role and category from a pre-defined list ranging from hard core mathematical categories like "Theory", "Definition", or "Assertion" to didactic elements like "Question" or "Comment". If a PPT object is part of a multi-part presentation (e.g. ranging over multiple slides) of a semantic entity, it can be marked as a sequel and inherits all information from previous parts. This way the PPT dependant linearity of the objects can be overcome.

Combining. For categorized PPT objects the author can input category specific content via the respective details form (see Figure 26.19 as an example for a PPT group categorized as "Axiom"). In particular, PPT objects can be assigned a relation via CPOINT's reference system. For instance, the axiom in Figure 26.19 sits in the theory called 'taxonomy of shapes'. A more sophisticated example would be a proof *for* an assertion that is constructed out of several, individual proof steps succeeding one another. Frequently, an author wants to reference implicit knowledge (e.g. theories can comprise entire concepts and as such are typically not explicitly presented in a lecture). Here, she can use CPOINT to create abstract PPT objects called **abstract objects** that are invisible in the actual PPT show but can be dealt with like all other PPT objects.

The information annotated in these processes can be exploited for added-value services.

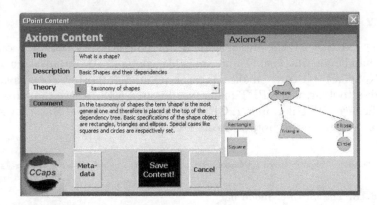

Fig. 26.19. The CPOINT content form for an axiom object

OMDOC *Conversion.* The heart of CPOINT is the functionality for converting a fully (CPOINT-)edited presentation into a valid OMDOC document. This generated OMDOC document can for instance be read into computer-supported education systems like ACTIVEMATH (see [MAF+01] and Section 26.8).

Added-Value Services. As author support is essential for the motivation doing the semantic markup process, CPOINT offers the following added-value services:

Content Search and Navigation. CPOINT's GOTO facility makes use of the additional semantic quality of PPT objects by offering content search. For instance if an author remembers the existence of a definition of "equivalence" in some (older) PPT presentation, she might look up all PPT objects in a collection of several PPT presentations that are categorized as "Definition" and whose title contain the word "equivalence". The author is offered a list of all these objects and by selecting one she is directed to the specific PPT object.

Dependency Graphs. CPOINTGRAPHS enables the user to view graph based presentations of the annotated knowledge on distinct detail levels.

Semantics-Induced Presentation. The module CPOINTAUTHOR offers the presentation of the underlying semantics. Whenever the author selects a PPT object basic semantic information (like category, title, and main references) is presented to her. With CPOINT's Visualize Mode semantic labels for annotated PPT objects are generated.

Creation of Pre-Categorized PPT Objects. Based on an individually designed CSS style sheet categorized, styled PPT objects can be *created* with CPOINTAUTHOR. The layout is determined in the CSS file by the respective category (e.g. proposition) or superordinate classification (e.g. assertion, content, general).

Math Glyphs in PPT. Based on the PPT add-in TEXPOINT, the CMath functionalities empower an author to define individual symbol presentations. CPOINT introduces a mathematical user interface, which fully integrates mathematical symbols into PowerPoint presentations based on the semantics of the underlying objects rather than simply generating appropriate ink marks. For instance, the author might categorize a PPT object as a symbol with the name 'reals' for the real numbers. The specific Unicode character to represent the real numbers can be declared with CPOINT. Subsequently, whenever the author writes the text '\reals' and activates the math mode, then this sequence of characters is replaced by the previously declared presentation. The symbol presentation may also be given in LATEX form so that TEXPOINT can transform the LATEX code into PPT glyphs. Note that this feature is not limited to math glyphs but can be used for handy abbreviations (macros) as well.

Editorial Notes. Treating PPT presentations as content documents requires more editing, therefore CPOINTNOTES add editorial functionalities like grouped editorial notes and navigation within these.

OMDOC To PPT. The CPOINTIMPORT module enables the import of OM-DOC documents into the PPT application. According to an individual underlying CSS style sheet PPT objects in a newly created PPT presentation are generated.

ActiveMath. Integrated development environment for ActiveMath content and specific ActiveMath book creation for a selected PPT object.

26.14.3 Future Work

In the future the addition of other added-value services for users is planned. We want to shift the focus from the authoring role to the recipient role of a PPT presentation, e.g. in form of a CPOINTSTUDENT module in accordance with the CPOINTAUTHOR module. Furthermore, a new, more basic and therefore more user-friendly interface for CPOINT novices will be implemented. This CPOINTBASIC module will try to overcome the heavily form-oriented format of CPOINT. In a next step the growing of a CPOINT user will be supported by offering advanced CPOINT utilities that will extend CPOINTBASIC. Additionally, the success of "social software" under the Web 2.0 paradigm like "social bookmarking" gives rise to the idea of a new personal and sharable PPT objects management where the predefined categories in CPOINT are replaced by "social tags". Another CPOINT project is its extension for usage by teachers in school, which usefulness has already been established in [And06]. The newest project at the International University of Bremen is the implementation of a CPOINT-like editor for MS Word.

26.15   sTeX: A LaTeX-Based Workflow for OMDoc

Project Home	http://kwarc.iu-bremen.de/projects/stex/
Authors	Michael Kohlhase
	Computer Science, International University Bremen

One of the reasons why OMDoc has not been widely employed for representing mathematics on the web and in scientific publications, may be that the technical communities that need high-quality methods for publishing mathematics already have an established method which yields excellent results — the TeX/LaTeX system. A large part of mathematical knowledge is prepared in the form of TeX/LaTeX documents.

We present sTeX (Semantic TeX) a collection of macro packages for TeX/LaTeX together with a transformation engine that transforms sTeX documents to the OMDoc format. sTeX extends the familiar and time-tried LaTeX workflow until the last step of Internet publication of the material: documents can be authored and maintained in sTeX using a simple text editor, a process most technical authors are well familiar with. Only the last (publishing) step (which is fully automatic) transforms the document into the unfamiliar XML world. Thus, sTeX can serve as a conceptual interface between the document author and OMDoc-based systems: Technically, sTeX documents are transformed into OMDoc, but conceptually, the ability to semantically annotate the source document is sufficient.

26.15.1   Recap of the TeX/LaTeX System

TeX [Knu84] is a document presentation format that combines complex page-description primitives with a powerful macro-expansion facility, which is utilized in LaTeX (essentially a set of TeX macro packages, see [Lam94]) to achieve more content-oriented markup that can be adapted to particular tastes via specialized document styles. It is safe to say that LaTeX largely restricts content markup to the document structure[17], and graphics, leaving the user with the presentational TeX primitives for mathematical formulae. Therefore, even though LaTeX goes a great step into the direction of a content/context markup format, it lacks infrastructure for marking up the functional structure of formulae and mathematical statements, and their dependence on and contribution to the mathematical context.

But the adaptable syntax of TeX/LaTeX and their tightly integrated programming features have distinct advantages on the authoring side:

– The TeX/LaTeX syntax is much more compact than OMDoc, and if needed, the community develops LaTeX packages that supply new functionality with a succinct and intuitive syntax.

---

[17] supplying macros e.g. for sections, paragraphs, theorems, definitions, etc.

– The user can define ad-hoc abbreviations and bind them to new control
  sequences to structure the source code.
– The TEX/LATEX community has a vast collection of language extensions
  and best practice examples for every conceivable publication purpose. Additionally, there is an established and very active developer community
  that maintains these.
– A host of software systems are centered around the TEX/LATEX language
  that make authoring content easier: many editors have special modes for
  LATEX, there are spelling/style/grammar checkers, transformers to other
  markup formats, etc.

In other words, the technical community is heavily invested in the whole
workflow, and technical know-how about the format permeates the community. Since all of this would need to be re-established for an OMDoc-based
workflow, the community is slow to take up OMDoc over TEX/LATEX, even
in light of the advantages detailed in this book.

26.15.2 A LATEX-Based Workflow for XML-Based Mathematical Documents

An elegant way of sidestepping most of the problems inherent in transitioning
from a LATEX-based to an XML-based workflow is to combine both and take
advantage of the respective values.

The key ingredient in this approach is a system that can transform
TEX/LATEX documents to their corresponding XML-based counterparts.
That way, XML-documents can be authored and prototyped in the LATEX
workflow, and transformed to XML for publication and added-value services.

There are various attempts to solve the TEX/LATEX to XML transformation problem; the most mature is probably Bruce Miller's LATEXML
system [Mil]. It consists of two parts: a re-implementation of the TEX analyzer with all of its intricacies, and an extensible XML emitter (the component that assembles the output of the parser). Since LATEX style files are
(ultimately) programmed in TEX, the TEX analyzer can handle all TEX extensions[18], including all of LATEX. Thus the LATEXML parser can handle all
of TEX/LATEX, if the emitter is extensible, which is guaranteed by the LA-
TEXML binding language: To transform a TEX/LATEX document to a given
XML format, all TEX extensions must have "LATEXML bindings", i.e. directives to the LATEXML emitter that specify the target representation in
XML.

The STEX system that we present here supplies a set of TEX/LATEX packages and the respective LATEXML bindings that allow to add enough structural information in the TEX/LATEX sources, so that the LATEXML system
can transform them into documents in OMDoc format.

---

[18] i.e. all macros, environments, and syntax extensions used int the source document

26.15.3   Content Markup of Mathematical Formulae in TEX/LATEX

The main problem here is that run-of-the-mill TEX/LATEX only specifies the presentation (i.e. what formulae look like) and not their content (their functional structure). Unfortunately, there are no universal methods (yet) to infer the latter from the former. Consider for instance the following "standard notations"[19] for binomial coefficients: $\binom{n}{k}$, $_nC^k$, C_k^n, and $\mathcal{C}_n^k$ all mean the same thing: $\frac{n!}{k!(n-k)!}$. This shows that we cannot hope to reliably recover the functional structure (in our case the fact that the expression is constructed by applying the binomial function to the arguments n and k) from the presentation alone short of understanding the underlying mathematics.

The apparent solution to this problem is to dump the extra work on the author (after all she knows what she is talking about) and give her the chance to specify the intended structure. The markup infrastructure supplied by the STEX collection lets the author do this without changing the visual appearance, so that the LATEX workflow is not disrupted. We speak of **semantic preloading** for this process. For instance, we can now write

$$\texttt{\textbackslash CSum\{k\}1\textbackslash infty\{\textbackslash Cexp\{x\}k\}} \quad \text{instead of} \quad \texttt{\textbackslash sum\_\{k = 1\}\^{}\textbackslash infty x\^{}k} \quad (26.1)$$

for the mathematical expression $\sum_{k=1}^{\infty} x^k$. In the first form, we specify that we are applying a function (CSumLimits $\hat{=}$ sum with limits) to four arguments: (*i*) the bound variable k (*ii*) the number 1 (*iii*) ∞ (*iv*) \Cexp{x}k (i.e. x to the power k). In the second form, we merely specify hat LATEX should draw a capital sigma character (Σ) whose subscript is the equation $k = 1$ and whose superscript is ∞. Then it should place next to it an x with an upper index k.

Of course human readers (who understand the math) can infer the content structure from the expression $\sum_{k=1}^{\infty} x^k$ of the right-hand representation in (26.1), but a computer program (who does not understand the math or know the context in which it was encountered) cannot. However, a converter like LATEXML can infer this from the left-hand LATEX structure with the help of the curly braces that indicate the argument structure. This technique is nothing new in the TEX/LATEX world, we use the term "**semantic macro**" for a macro whose expansion stands for a mathematical object. The STEX collection provides semantic macros for all Content-MATHML elements together with LATEXML bindings that allow to convert STEX formulae into MATHML.

26.15.4   Theories and Inheritance of Semantic Macros

Semantic macros are traditionally used to make TEX/LATEX code more portable. However, the TEX/LATEX scoping model (macro definitions are

---

[19] The first one is standard e.g. in Germany and the US, the third one in France, and the last one in Russia

scoped either in the local group or to the end of the document), does not mirror mathematical practice, where notations are scoped by mathematical environments like statements, theories, or such (see [Koh05c] for a discussion and examples). Therefore the SₜₑX collection provides an infrastructure to define, scope, and inherit semantic macros.

In a nutshell, the SₜₑX symdef macro is a variant of the usual newcommand, only that it is scoped differently: The visibility of the defined macros is explicitly specified by the module environment that corresponds to the OMDoc theory element. For this the module environment takes the optional KeyVal arguments id for specifying the theory name and uses for the semantic inheritance relation. For instance a module that begins with

---

\begin{module}[id=foo,uses={bar,baz}]

---

restricts the scope of the semantic macros defined by the \symdef form to the end of this module given by the corresponding \end{module}, and to any other module environment that has [uses={...,foo,...}] in its declaration. In our example the semantic macros from the modules bar and baz are inherited as well as the ones that are inherited by these modules.

We will use a simple module for natural number arithmetics as an example. It declares a new semantic macro for summation while drawing on the basic operations like $+$ and $-$ from LᴬTₑX. \Sumfromto allows us to express an expression like $\sum_{i=1}^{n} x^i$ as \Sumfromto{i}1n{2i-1}. In this example we have also made use of a local semantic symbol for n, which is treated as an arbitrary (but fixed) symbol (compare with the use of \arbitraryn below, which is a new — semantically different — symbol).

```
\begin{module}[id=arith]
  \symdef{Sumfromto}[4]{\sum_{#1=#2}^{#3}{#4}}
  \symdef[local]{arbitraryn}{n}
  What is the sum of the first $\arbitraryn$ odd numbers, i.e.
  $\Sumfromto{i}1\arbitraryn{2i-1}?$
\end{module}
```

---

is formatted by SₜₑX to

What is the sum of the first n odd numbers, i.e. $\sum_{i=1}^{n} 2i - 1?$

Moreover, the semantic macro Sumfromto can be used in all module environments that import it via its uses keyword. Thus SₜₑX provides sufficient functionality to mark up OMDoc theories with their scoping rules in a very direct and natural manner. The rest of the OMDoc elements can be modeled by LᴬTₑX environments and macros in a straightforward manner.

The SₜₑX macro packages have been validated together with a case study [Koh05c], where we semantically preloaded the course materials for a two-semester course "General Computer Science I&II" at International University Bremen and transform them to the OMDoc, so that they can be used in the ActiveMath system (see Section 26.8).

26.16   An Emacs Mode for Editing OMDoc Documents

Project Home	http://www.cs.cmu.edu/~ccaps
Authors	Peter Jansen
	School of Computer Science, Carnegie Mellon University
	versity

We describe an EMACS major mode for editing OMDOC documents, developed by the COURSE CAPSULES project group at the CMU School of Computer Science. This mode extends the EMACS editor [Sta02] with functionality intended to help visualize, edit, and create documents written in OMDOC format.

The mode is part of the OMDOC distribution (see Section 23.2), it is provided under the conditions specified in the Library Gnu Public License [FSF99].

26.16.1   Introduction

The CCaps project has developed tools to convert legacy materials written in a variety of formats (POWERPOINT, MATHEMATICA®, etc.) into the OMDOC format (see Sections 26.14 and 26.17). In many cases the output generated by such tools needs to be post-processed or otherwise modified.

To this end, a user must open the file, read and understand its contents and perform the appropriate modifications. Though an OMDOC document is a regular text file, most of its content consists of markup, which is hard to read and tedious to type. It is therefore important to support the user with tools that make a document easier to read and modify, either in the form of a separate editor, or as an extension of an existing editor.

One approach to this is to build a *visual* OMDOC *editor*, which presents the document in a form resembling conventional mathematical documents (i.e., without showing the markup explicitly, and with appropriate formatting for mathematical formulae), and offers the user functionality to modify or annotate its content.

While this is ideal for user understanding of document content, it presupposes consistent syntactic correctness, makes it more difficult to inspect or change markup directly, and may present challenges as to resolving user action ambiguities.

We have taken this approach in the CPOINT and MATHEMATICA® add-ins (see Sections 26.14 and 26.17). But we also wanted a tool that would maintain full control of all the textual information, while offering support for readability and editing functionality. We chose for this tool to extend the EMACS editor [Sta02], which lends itself very well to this task (as well as being the editor for general use for several of our group).

26.16.2   OMDoc Mode Functionality

We now look at the different categories of functionality in slightly more detail.

Visualization. is currently provided by the use of the EMACS *font-lock* mechanism to give different categories of tags and content different fonts and colors to make them easily recognizable. Element categories currently recognized correspond to the OMDoc 1.2 modules: Document structure, Math, Theories, Auxiliaries, Presentation, OPENMATH, and the Dublin Core elements.

A customizable *indentation function* allows for intelligible layout, which is helpful both in hand-coding and the editing of the output of a legacy transformation process. There are key bindings for line, region, and enclosing element indentation.

Editing. Functionality consists mainly of automated insertion of *templates* for each of the OMDoc elements, both via mode-dependent menu options and key bindings, grouped by element category (the same categories as given above).

The template insertion mechanism is based on `tempo.el`, which allows for the maintenance of a *list of insertion points* the user can navigate in between to supply or change the values of certain attributes.

The main function currently available for completing incomplete elements is the equivalent of the standard `electric-/` function. We are planning to add several other completion functions in the near future (for tags, tag sets, attribute names, and symbol and theory names).

The mode also provides for *validation*: either internally (as a simple local syntax check to check well-formedness) or externally (via an external xml validation validation engine). Internal validation builds an abbreviated parse tree, and highlights discrepancies, suggesting possible modifications of insertions of element opening or closing tags. External validation runs an external xml validation engine (RXP or nsgmls, depending on the configuration variables), and shows the output in a separate buffer.

Document Creation. is supported by automatic insertion of a basic OMDoc *skeleton* in new buffers as well as a *time-stamp updating* mechanism and some smaller functions that extract information from the user's environment variables to supply information for some of the metadata slots (see the example in Figure 26.20 below).

26.16.3   Examples

We illustrate some of the above by means of a few screen shots. The example in Figure 26.21 is taken while editing a document that was semi-automatically generated from part of a MATHEMATICA® notebook ([Sut06]). Here, the user has already run an automated indentation function, for example by activating `omdoc-indent-enclosing-main` by typing C-c C-q, and is now about to use the OMDoc menu to enter a new construct.

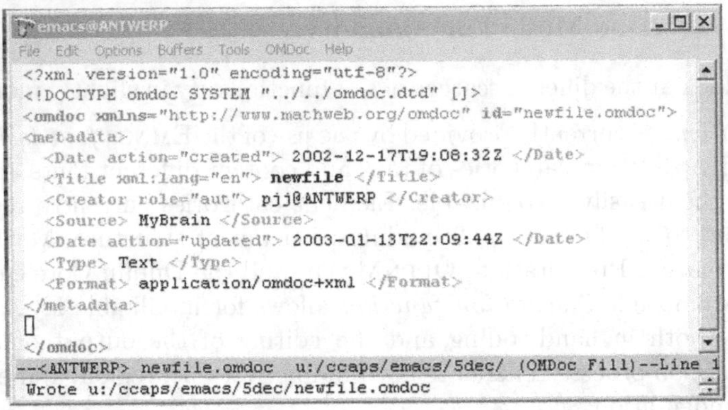

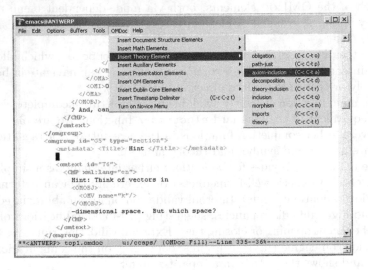

Fig. 26.20. Opening a new buffer in OMDoc mode

Fig. 26.21. Editing an OMDoc document

After this operation, which could also have been performed by typing the key sequence C-c C-t a), EMACS inserts the following text at the point (i.e. cursor position).

```
<axiom−inclusion xml:id="" to=""> </axiom−inclusion>
```

The second example shows the skeleton template that is automatically inserted when the user opens a new file: Figure 26.20. Note that the file name has been used as id and title automatically, and the user's address appears in the Author field. Timestamps are inserted in Date fields for both creation and update, and the latter is adjusted automatically every time changes are saved to the file.

26.17   Converting Mathematica Notebooks to OMDoc

Project Home	`http://www.cs.cmu.edu/~ccaps`
Authors	Klaus Sutner School of Computer Science, Carnegie Mellon University

We describe a tool that converts MATHEMATICA® notebooks to OMDOC. The program is implemented entirely in MATHEMATICA® and easily extensible.

Creating an editor for general mathematical documents is notoriously difficult, in particular when input methods are required that mimic the traditional two-dimensional layout of many formulae. Thus, it seems natural to use an existing high-quality system such as the MATHEMATICA® notebook front end as an authoring tool for mathematical documents. A considerable amount of effort has gone into the design of this front end, see for example [Wol00], resulting in a surprisingly versatile system. The notebook front end provides a rich set of palettes that allow inexperienced users to construct complicated expressions almost instantaneously. For more advanced users there is a well-thought-out set of keyboard operations that make it possible to create, navigate and edit two-dimensional expressions with relative ease and without recourse to time-consuming mouse-based operations. Unlike with TEX, the results are immediately visible and corrections are easy to make. Nonetheless, the quality of the typeset expression approaches that of TEX. Last, but not least, the MATHEMATICA® kernel can be used to generate complicated expressions and even whole notebooks automatically.

MATHEMATICA® provides significant support for import, export and manipulation of XML documents and expressions, see [Wol02]. Thus, one can export a notebook in MATHML format, or in a special NOTEBOOKML format. Unfortunately, these export mechanisms cannot be modified directly to produce highly marked-up documents in OMDOC format.

The NB2OMDOC converter uses a recursive descent parser, that scans the given notebook document and generates corresponding OMDOC. As far as structured text is concerned this is a fairly straightforward operation. However, special care needs to be taken to deal with mathematical text elements, such as definitions, theorems, proofs and such like, and mathematical expressions, in inline format as textual elements as well as in evaluatable format (as input for the MATHEMATICA® kernel). We comment on both issues in turn.

MATHEMATICA® notebooks provide reasonable support for the creation of well-structured documents, but enforce no particular discipline. A fragment of a typical notebook, showing some section headers and a bit of text with inline mathematical formulae is shown in Figure 26.22.

In order to facilitate the translation process it is advisable to front-load the process: the author of the notebook is encouraged to use a special notebook stylesheet, `OMDocStyle.nb`, that defines a number of syntactic cate-

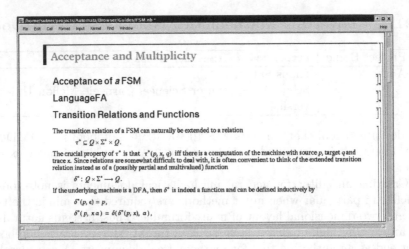

Fig. 26.22. A MATHEMATICA® notebook

gories normally absent in a notebook. These categories are implemented as a combination of the cell types and cell labels. As a typical example, consider a proof of some assertion such as a theorem. Ordinarily, a sequence of plain text cells would be used to express a proof since none of the standard MATHEMATICA® stylesheets provide a special proof style–though some have a theorem style. The elements defined in `OMDocStyle.nb` are easily accessible via pulldown menus or via keyboard shortcuts in the notebook front end. Moreover, the special styles are color-coded in the notebook, so that it is easy for the author to see which elements are present and which might be missing.

The conversion of mathematical expressions in the notebook is accomplished in a two-step procedure. First, we use the built-in MATHEMATICA® operation `ExpressionToSymbolicMathML` that produces a symbolic expression representing a MathML term that corresponds to the original notebook expression. In a second, post-processing step this expression is then transformed into an OPENMATH expression. The post-processing relies heavily on the sophisticated pattern matching mechanism in MATHEMATICA® and uses a special collection of rewrite rules. The rules are based on fairly simple-minded heuristics but do produce adequate results so long as the starting expression is not too complicated. As an example, consider the simple polynomial expression $ax^2 + bx + c$ whose internal representation in MATHEMATICA® looks like so (we assume here the expression appears inline within a block of text, the situation for an input expression is entirely similar):

```
Cell [ BoxData[FormBox[RowBox[{
    RowBox[{"a", " ", SuperscriptBox["x", "2"]}], "+", " ",
    RowBox[{"b", " ", "x"}], " ", "+", " ", "c" }],
    TraditionalForm]]]
```

The first conversion step produces the following MATHEMATICA® expression, shortened here to save space:

```
XMLElement["math",
  {"xmlns" −>"http://www.w3.org/1998/Math/MathML"},
  {XMLElement[ "apply", {}, {XMLElement["plus", {}, {}],
   XMLElement[ "apply", {}, {XMLElement["times", {}, {}],
    XMLElement["ci", {}, {"a"}],
    XMLElement[ "apply", {}, {XMLElement["power", {}, {}],
     XMLElement["ci", {}, {"x"}],
     XMLElement["cn", {"type"−>"integer"}, {"2"}]}]}], ... }]}]
```

The post-processing finally yields the following expression, again shown only in part.

```
XMLElement["OMOBJ", {}, {XMLElement["OMA", {},
  {XMLElement["OMS",
     {"cd"−>"arith1", "name"−>"plus"}, {}],
   XMLElement["OMA", {}, {XMLElement[ "OMS",
     {"cd"−>"arith1", "name"−>"times"}, {}],
   XMLElement["OMV", {"name"−>"a"}, {}],
   XMLElement["OMA", {}, {XMLElement["OMS",
     {"cd"−>"arith1", "name"−>"power"}, {}], ...}]}]}]}]
```

The content dictionary was properly guessed in this instance. Judging from the limited experiments we have undertaken so far, it seems reasonable to expect that a fair amount of the translation can be automated given that the field of discourse is limited, and that the author is willing to customize the rewrite rules that control the post-processing step. Fortuitously, very little knowledge of MATHEMATICA® programming beyond some basic syntax is necessary for the creation of these rules; mathematicians are likely to find these rules fairly intuitive and natural.

At present, the conversion program is somewhat limited in its ability to deal with arbitrarily structured notebooks. It works well with a suite of notebooks developed specifically for the OMDocStyle.nb, but requires modification for other types of notebooks. While it is not our goal to provide a truly general conversion tool with a scope comparable to, say, the built-in conversion to MathML, some generalizations are still needed at this point.

Another crucial issue is the extension of the rewrite rules used in the post-processing step leading from MathML to OPENMATH. No effort has been made so far to systematically generate a set of rules suitable for a large class of documents. At the very least, an extension mechanism is needed that makes it easy for non-expert users to create the necessary rule tables.

Lastly, it is desirable to create a MATHEMATICA® palette-based tool that focuses more narrowly on the authoring and conversion of mathematical expressions only rather than whole notebooks. The generated raw OPENMATH expressions can be fed directly into a low-level editor such as EMACS using the special OMDoc mode created as part of the CCAPS project, see elsewhere in this volume for a description.

26.18   Standardizing Context in System Interoperability

Project Home	http://www.mathweb.org/omdoc/examples/logics
Authors	Michael Kohlhase
	Computer Science, International University Bremen

In this project the OMDOC format is used as a content language for the protocol-based integration of mathematical software systems, where the systems offer mathematical services by publishing service descriptions and interoperate by exchanging computation requests and results. The mechanics of the communication and domain-independent part of meaning of these messages is given by a standardized 'interlingua' (which will not concern us here), a possible implementation of the transport layer we have seen in Chapter 9. Here we are interested in the mathematical objects contained in the messages

OMDOC can help with the task of making mathematical objects interoperable, as we have seen in series of experiments of connecting the theorem proving systems ΩMEGA [BCF+97], INKA [HS96], PVS [ORS92], λClam [RSG98b], TPS [ABI+96], and CoQ [Tea] to the MBASE system by equipping them with an OMDOC interface. As expected, OPENMATH and Content-MATHML solve the problem of syntactically standardizing the representation of mathematical objects. For a semantic interoperability we also need to capture their context. This is not a problem for Content-MATHML, as the context is already standardized in the MATHML recommendation. For OPENMATH, the context is given by the set of content dictionaries in use for representing the mathematical objects. Nevertheless mathematical software systems — such as computer algebra systems, visualization systems, and automated theorem provers — come with different conceptualizations of the mathematical objects (see [KK05] for a discussion). This has been in principle solved by supplying a flexible and structured theory level in the form of OMDOC content dictionaries that define necessary mathematical concepts (see Subsection 26.18.1 for practical considerations). For systems like theorem provers or theory development environments, where the mathematical objects are axioms, definitions, assertions, and proofs there is another problem: that of standardizing the logical language, which we will discuss in Subsection 26.18.2.

26.18.1   Context Interoperability via Theory Morphisms

As an example for the integration of two mathematical software systems we look at the task of integrating the PVS and ΩMEGA set theory libraries. This is simpler than e.g. integrating the computer algebra systems MAPLE™ and MATHEMATICA®, since all the conceptualizations and assumptions are explicitly given, but gives an intuition for the difficulties involved. We summarize the situation in Figure 26.23, where we compare symbol names for

PVS	ΩMEGA	PVS	ΩMEGA
set		subset?	subset
member	in		subset2
empty?	empty	strict_subset?	proper-subset
emptyset	emptyset		superset
nonempty?	not-empty	union	union
full?			union2
fullset			union-over-collection
singleton?	singleton	intersection	intersection
singleton			intersection-over-coll.
complement	set-complement	disjoint?	misses
difference	setminus	meets	
symmetric_difference		add	add-one
	exclunion	remove	

Fig. 26.23. Set theories in ΩMEGA and PVS

set theory concepts in the two systems. The general problem in such an integration of mathematical software systems consists in their independent growth over time, leading to differing names, definitions, theory boundaries, and possibly conceptualizations. Most of these particulars are artefacts of constraints imposed by the system (e.g. file lengths). In this situation theory interpretations suggest themselves as a means for theory integration: We can use theory interpretations to establish inclusion into a suitably constructed **integration theory**. In Figure 26.24 we have executed this for the set theory libraries of the systems PVS, ΩMEGA, TPS, and IMPS; we provide an 'integration theory' mbase:sets — it provides rationally reconstructed versions of all concepts encountered in the system's libraries — and a set of theory inclusions ρ_* that interpret the system concepts in terms of mbase:sets. Note that since the ρ_* are monomorphisms, we can factor any existing theory inclusion (e.g. pvs:sets to pvs:funcs highlighted in Figure 26.23) via the integration theory, using the partial inverse ρ_*^{-1} of ρ_*. For an integration of a set of software systems this refactoring process is repeated recursively from terminal- to initial nodes in the imports relation.

Note that *technically* we do not need to change the interface language of the mathematical software systems[20], we only rationally reconstruct their meaning in terms of the new integration theory, which can act as a gold standard for the integration. *Socially* the existence of the new standard theory may prompt a migration to the nomenclature and coverage of the integration theory. Note furthermore, that we have only treated the simple case, where the mathematical conceptualizations underlying the software systems are already explicitly given in a library. For many mathematical software systems the underlying conceptualizations and assumptions are only documented in scientific papers, user manuals, or inscribed into the code. For such systems, **interface theories** that make them explicit have to be developed to pursue the integration strategy presented above. Of course, this process needs a lot

---

[20] This is important if we want to integrate proprietary software systems, where we have no control over the interfaces.

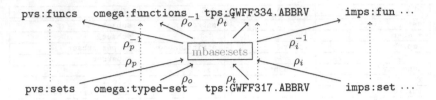

Fig. 26.24. Theory translations for system integration

of manual labor, but leads to true interoperability of mathematical software systems, which can now re-use the work of others.

Finally note that the integration only works as smoothly as in our scenario, if the systems involved make assumptions about mathematical objects that are compatible with each other. In most cases, incompatibilities can be resolved by renaming concepts apart, e.g. one system considers set union to be a binary operation, while the other considers it as n-ary. Here, the integration theory would supply two distinct (though possibly semantically related) concepts; The theory-based integration approach allows to explicitly disambiguate the concepts and thus prevent confusion and translation errors. In very few cases, systems are truly incompatible e.g. if one assumes an axiom which the other rejects. In this case the theory based integration approach breaks down — indeed a meaningful integration seems impossible and unnecessary.

26.18.2   A Hierarchy of Logical Languages

In the example above we made use of the fact that theorem proving systems are simpler to deal with than other mathematical software systems, since they encode the underlying assumptions explicitly into mathematical libraries. Unfortunately though, this is only partially true the underlying base logics are usually not treated in this way. Fortunately, logical concepts are treated in OMDoc just like ordinary ones: by content markup in Content-MathML or OpenMath, so that there is no fundamental barrier to treating them as the theory contexts above; we only have to come up with interface theories for them. We have done just that when we equipped various logic-based systems with OMDoc interfaces observing that even though the systems are of relatively different origin, their representation languages share many features:

- TPS and Pvs are based on a simply typed λ-calculus and only use type polymorphism in the parsing stage, whereas ΩMEGA and λ*Clam* allow ML-style type polymorphism.
- ΩMEGA, InKa and Pvs share a higher sort concept, where sorts are basically unary predicates that structure the typed universe.
- Pvs and CoQ allow dependent- and record types as basic representational features.

but also differ on many others: for instance INKA, PVS, and CoQ explicitly support inductive definitions, but by very different mechanisms and on differing levels. CoQ uses a constructive base logic, whereas the other systems are classical. The similarities are not that surprising, all of these systems come from similar theoretical assumptions (most notably the Automath project [dB80]), and inherit the basic setup (typed λ-calculus) from it. The differences can be explained by differing intuitions in the system design and in the intended applications.

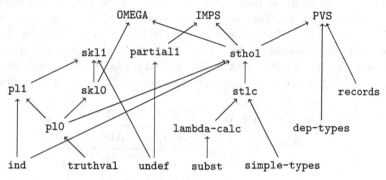

Fig. 26.25. A Hierarchy of Logical Languages

We have started to provide a standardized, well-documented set of content dictionaries for logical languages in the OMDOC distribution. These are organized hierarchically, as depicted in Figure 26.25. In essence, the structured theory mechanism in OMDOC is used to create a language hierarchy that inter-relates the various representation formats of existing theorem provers. For instance the simply typed λ-calculus can be factored out (and thus shared) of the representation languages of all theorem proving systems above. This makes the exchange of logical formulae via the OMDOC format very simple, if they happen to be in a suitable common fragment: In this case, the common (OPENMATH/OMDOC) syntax is sufficient for communication.

26.18.3 Logic Interoperability via Logic Morphisms

In theoretical accounts of the integration of logical languages, one finds categorical accounts like the one described in Section 26.13 or proof-theoretic ones based on definitions like the one below. Both mesh well with the OMDOC representation format and its theory level; we will show this for the proof-theoretic account here.

Definition 26.18.1. *A **logical system** $\mathcal{S} = (\mathcal{L}, \mathcal{C})$ consists of a language $\mathcal{L}$ (i.e. a set of well-formed formulae) and a calculus $\mathcal{C}$ (i.e. a set of inference rules). A calculus gives us a notion of a $\mathcal{C}$-derivation of $\mathbf{A}$ from $\mathcal{H}$, which* *we will denote by* $\mathcal{D}: \mathcal{H} \vdash_{\mathcal{C}} \mathbf{A}$. *Let* S *and* S' *be logical systems, then a* **logic morphism** $\mathcal{F}: S \rightarrow S'$ *consists of a* **language morphism** $\mathcal{F}^{\mathcal{L}}: \mathcal{L} \rightarrow \mathcal{L}'$ *and a* **calculus morphism** $\mathcal{F}^{\mathcal{D}}$ *from* $\mathcal{C}$*-derivations to* $\mathcal{C}'$*-derivations, such that for any* $\mathcal{C}$*-derivation* $\mathcal{D}: \mathcal{H} \vdash_{\mathcal{C}} \mathbf{A}$ *we have* $\mathcal{F}^{\mathcal{D}}(\mathcal{D}): \mathcal{F}^{\mathcal{L}}(\mathcal{H}) \vdash_{\mathcal{C}'} \mathcal{F}^{\mathcal{L}}(\mathbf{A})$.

The intuition behind this is that logic morphisms transport proofs between logical systems. Logic morphisms come in all shapes and sizes, a well-known one is the relativization morphism from sorted logics to unsorted ones, for instance the morphism $\mathcal{R}$ from sorted first-order logic ($\mathbb{S}FOL$) to unsorted first-order logic (FOL). For every sorted constant $\mathcal{R}$ introduces an axiom e.g. $\mathcal{R}([+: \mathbb{N} \rightarrow \mathbb{N} \rightarrow \mathbb{N}]) = \forall X, Y.\mathbb{N}(X) \wedge \mathbb{N}(Y) \Rightarrow \mathbb{N}(X + Y)$. On formulae sorted quantifications are translated into unsorted ones guarded by sort predicates, e.g. $\mathcal{R}(\forall X_{\mathbb{B}}.\mathbf{A}) = \forall X.\mathbb{B}(X) \Rightarrow \mathcal{R}(\mathbf{A})$. Finally, for proofs we have the correspondence given in Figure 26.26, where $\mathbb{A}, \mathbb{B}, \ldots$ are sort symbols.

$$
\mathcal{R} \left(\frac{\mathbf{A}: \mathbb{B} \rightarrow \mathbb{C} \quad \mathbf{B}: \mathbb{B}}{\mathbf{AB}: \mathbb{C}} \right) = \frac{\dfrac{\forall X.\mathbb{B}(X) \Rightarrow \mathbb{C}(\mathbf{A}X)}{\mathbb{B}(\mathbf{B}) \Rightarrow \mathbb{C}(\mathbf{AB})} \quad \mathbb{B}(\mathbf{B})}{\mathbb{C}(\mathbf{AB})}
$$

$$
\mathcal{R} \left(\frac{\forall X_{\mathbb{B}}.\mathbf{A} \quad \mathbf{B}: \mathbb{B}}{[\mathbf{B}/X]\mathbf{A}} \right) = \frac{\dfrac{\forall X.\mathbb{B}(X) \Rightarrow \mathcal{R}(\mathbf{A})}{\mathbb{B}(\mathcal{R}(\mathbf{B})) \Rightarrow \mathcal{R}([\mathbf{B}/X]\mathbf{A})} \quad \mathbb{B}(\mathbf{B})}{\mathcal{R}([\mathbf{B}/X]\mathbf{A})}
$$

Fig. 26.26. Relativization Morphism on Proofs

In Definition 26.18.1 a logical system is a two-partite object consisting of a language and a calculus. In the ontologically promiscuous OMDOC format both parts are represented largely like ordinary mathematically concepts. The notable exception is that proofs have a slightly dual representation, but inference rules of a calculus are still represented as symbols via the Curry-Howard isomorphism (see Chapter 17). Thus a logical system can be represented as an OMDOC theory as we did above, moreover, the logic morphism $\mathcal{R}$ can simply be encoded as a theory inclusion from $\mathbb{S}FOL$ to FOL mapping $\mathbb{S}FOL$ constants for inference rules to FOL terms for proofs. The condition on the form of derivations in Definition 26.18.1 now simply takes on the form of a type compatibility condition.

26.19 Integrating Proof Assistants as Plugins in a Scientific Editor

Project Home	`http://www.ags.uni-sb.de/~omega/projects/` `verimathdoc`
Authors	Serge Autexier, Christoph Benzmüller, Armin Fiedler, and Henri Lesourd Computer Science, Saarland University, Saarbrücken, Germany

In contrast to computer algebra systems (CASs), mathematical proof assistance systems have not yet achieved considerable recognition and relevance in mathematical practice. One significant shortcoming of the current systems is that they are not fully integrated or accessible from within standard mathematical text-editors and that therefore a duplication of the representation effort is typically required. For purposes such as tutoring, communication, or publication, the mathematical content is in practice usually encoded using common mathematical representation languages by employing standard mathematical editors (e.g., LATEX and EMACS). Proof assistants, in contrast, require fully formal representations and they are not yet sufficiently linked with these standard mathematical text editors. Therefore, we have decided to extend the mathematical text editor TEX$_{MACS}$ [dH01] in order to provide direct access from it to the mathematics assistance system ΩMEGA [SBB⁺02, SBA05]. Generally, we aim at an approach that is not dependent on the particular proof assistant system to be integrated [ABFL06].

TEX$_{MACS}$ [dH01] is a scientific WYSIWYG text editor that provides professional typesetting and supports authoring with powerful macro definition facilities like in LATEX. The internal document format of TEX$_{MACS}$ is a SCHEME S-expression composed of TEX$_{MACS}$ specific markup enriched by definable macros. The full access to the document format together with the possibility to define arbitrary SCHEME functions over the S-expressions makes TEX$_{MACS}$ an appropriate text editor for an integration with a mathematical assistance system.

The mathematical proof assistance system ΩMEGA [SBB⁺02, SBA05] provides proof development at a high level of abstraction using knowledge-based proof planning and the proofs developed in ΩMEGA can be verbalized in natural language via the proof explanation system P.rex [Fie01b]. As the base calculus of ΩMEGA we use the CORE calculus [Aut03, Aut05], which supports proof development directly at the *assertion level* [Hua96], where proof steps are justified in terms of applications of definitions, lemmas, theorems, or hypotheses (collectively called *assertions*).

Now, consider a teacher, student, engineer, or mathematician who is about to write a new mathematical document in TEX$_{MACS}$. A first crucial step in our approach is to link this new document to one or more mathematical theories provided in a mathematical knowledge repository. By providing such a link the document is initialized and T_EX_{MACS} macros for the relevant mathematical symbols are automatically imported; these macros overload the pure syntactical symbols and link them to formal semantics. In a T_EX_{MACS} display mode, where this additional semantic information is hidden, the user may then proceed in editing mathematical text as usual. The definitions, lemmas, theorems and especially their proofs give rise to extensions of the original theory and the writing of some proof goes along with an interactive proof construction in ΩMEGA. The semantic annotations are used to *automatically* build up a corresponding formal representation in ΩMEGA, thus avoiding a duplicated encoding effort of the mathematical content. Altogether this allows for the development of mathematical documents in professional type-setting quality which in addition can be formally validated by ΩMEGA, hence obtaining *verified mathematical documents*.

Using T_EX_{MACS}'s macro definition features, we encode theory-specific knowledge such as types, constants, definitions and lemmas in macros. This allows us to translate new textual definitions and lemmas into the formal representation, as well as to translate (partial) textbook proofs into formal (partial) proof plans.

Rather than developing a new user interface for the mathematical assistance system ΩMEGA, we adapt ΩMEGA to serve as a mathematical service provider for T_EX_{MACS}. The main difference is that instead of providing a user interface only for the existing interaction means of ΩMEGA, we extend ΩMEGA to support requirements that arise in the preparation of a semi-formal mathematical document. In the following we present some requirements that we identified to guide our developments.

The mathematical document should be prepared directly in interaction with ΩMEGA. This requires that (1) the semantic content of the document is accessible for a formal analysis and (2) the interactions in either direction should be localized and aware of the surrounding context.

To make the document accessible for formal analysis requires the extraction of the semantic content and its encoding in some semi-formal representation suitable for further formal processing. Since current natural language analysis technology cannot yet provide us with the support required for this purpose, we use semantic annotations in the T_EX_{MACS} document instead. Since these semantic annotations must be provided by the author, one requirement is to keep the burden of providing the annotations as low as possible.

Due to their formal nature the representations of mathematical objects, for instance, definitions or proofs, in existing mathematical assistance systems are very detailed, whereas mathematicians omit many obvious or easily inferable details in their documents: there is a big gap between common mathematical language and formal, machine-oriented representations. Thus another requirement to interfacing T_EX_{MACS} to ΩMEGA is to limit the details that must be provided by the user in the T_EX_{MACS} document to an acceptable amount.

In order to allow both the user and the proof assistance system to manipulate the mathematical content of the document we need a common representation format for this pure mathematical content implemented both in T_EX_{MACS} and in ΩMEGA. To this end we define a language S, which includes many standard notions known from other specification languages, such as terms, formulas, symbol declarations, definitions, lemmas and theorems. The difference to standard specification languages is that our language S (i) includes a language for proofs, (ii) provides means to indicate the logical context of different parts of a document by fitting the narrative structure of documents rather than imposing a strictly incremental description of theories as used in specification languages, and (iii) accommodates various aspects of underspecification, that is, formal details that the writer purposely omitted. Given the language S, we augment the document format of T_EX_{MACS} by the language S. Thus, if we denote the document format of T_EX_{MACS} by T, we define a **semantic document format** $T + S$ as a document format still accepted by T_EX_{MACS}.

Ideally this format of the documents and especially the semantic annotations should not be specific to ΩMEGA in order to enable the combination of the T_EX_{MACS} extension with other proof assistance systems as well as the development of independent proof checking tools. However, an abstract language for proofs that is suitable for our purposes and that allows for underspecification is not yet completely fixed. So far we support the assertion-level proof construction rules provided by CORE [Aut03, Aut05]. Thus, instead of defining a fixed language S, we define a language $S(P)$ parametrized over a language P for proofs and define the document format based on $S(C)$, where C denotes the proof language of CORE. This format supports the *static representation* of semantically annotated documents, which can be professionally typeset with T_EX_{MACS}.

The T_EX_{MACS} document $T + S(C)$ and the pure semantic representation $S(C)$ in the proof assistant must be synchronized. The basic idea here is to synchronize via a diff/patch mechanism tailored to the tree structure of the T_EX_{MACS} documents. The differences between two versions ts_i and ts_{i+1} of the document in $T + S(C)$ are compiled into a patch description p of the corresponding document s_i in $S(C)$ for ts_i, such that the application of p to s_i results in s_{i+1} which corresponds to ts_{i+1}. An analogous diff/patch technique is used to propagate changes performed by the proof assistant tool to documents in $S(C)$ towards the T_EX_{MACS} document in $T + S(C)$. In order to enable the translation of the patch descriptions, a key-based protocol is used to identify the corresponding parts in $T + S(C)$ and $S(C)$.

Beyond this basic synchronization mechanism, we define a language that allows for the description of specific interactions between T_EX_{MACS} and the proof assistant. This language M is a language for structured menus and actions with an evaluation semantics which allows to flexibly compute the necessary parameters for the commands and directives employed in interaction with the proof assistants. The $\text{T}_{\text{E}}\text{X}_{\text{MACS}}$ document format $T + S(C)$ is finally extended to $T + S(C) + M$, where the menus can be attached to arbitrary parts of a document and the changes of the documents performed either by the author or by the proof assistants are propagated between $T + S(C) + M$ and $S(C) + M$ via the diff/patch mechanism. Note that this includes also the adaptation of the menus, which is a necessary prerequisite to support context-sensitive menus and actions contained therein.

The goal of the proposed integration is to use ΩMEGA as a context-sensitive reasoning and verification service accessible from within the first-class mathematical text editor $\text{T}_{\text{E}}\text{X}_{\text{MACS}}$, where the proof assistant adapts to the style an author would like to write his mathematical publication, and to hide any irrelevant system peculiarities from the user. The communication between $\text{T}_{\text{E}}\text{X}_{\text{MACS}}$ and ΩMEGA is realized by an OMDOC-based interface language.

Although so far the proofs are tailored to the rules of the CORE system, the representation language in principle is parametrized over a specific language for proofs. We currently replace the CORE specific proof languages by some generic notion of proofs, in order to obtain a generic format for formalized mathematical documents. Thereby we started from a language for assertion-level proofs with underspecification [ABF+03, AF05], which we developed from previous experiences with tutorial dialogs about mathematical proofs between a computer and students [PSBKK04].

26.20 OMDOC as a Data Format for ✓eriFun

Project Home	http://www.verifun.de/
Authors	Normen Müller
	School of Engineering and Science, International
	University Bremen

✓eriFun (<u>Veri</u>fication of <u>Fun</u>ctional programs) is a semi-automated system for the verification of programs written in a simple functional programming language $\mathcal{FP}$. The system has been developed since 1998 at the university of Darmstadt for use in education and research. The main design goals are a clearly structured, didactically suited system interface (Figure 26.27), an easily portable implementation (JAVA) and an easily but also powerful proof calculus [WS02]. The system's object language consists of a simple definition principle for free data structures, called *sorts* (see Chapter 16), a recursive definition principle for *functions*, and finally a definition principle for statements, called *lemmas*, about the data structures and the functions. To prove a statement ✓eriFun supports the user with a couple of inference rules aggregated in *tactics*. A collection of *sorts*, *functions*, *lemmas*, and *proofs* is called a ✓eriFun *program*. Common file commands, which are based on the JAVA binary serialization mechanism, are provided to save and reload intermediate work.

The OMDOC interface for ✓eriFun described here (see [Mül05] for details) was introduced to alleviate the following drawbacks of the former I/O mechanism based on JAVA binary serialization:

- Files are only machine-readable. Thus, e.g. if the files became corrupted by any circumstance, there is no change of a manual repair.
- Files are strongly bound to the version of the system. Thus any internal system modifications make the files unreadable.

Fig. 26.27. A ✓eriFun session

– Files are not interchangeable with other theorem provers or other mathematical software systems. Thus the information inside the files are only accessible by √eriFun.

√eriFun's interface to OMDoc can be divided into two parts: Encoding and decoding of √eriFun programs to and from OMDoc respectively.

Encoding. In a typical session with the system, a user defines a *program* by stipulating the *sorts* and the *functions* of the program, defines *lemmas* about the sorts and the functions of the program, and finally verifies these lemmas and the termination of the functions.

In general a *program* is mapped to two OMDoc files: The first one consists of the user-defined elements[21] and in the second one √eriFun's logic comprising the predefined symbols, the type system and the proof tactics is defined. At each case one √eriFun-generated-OMDoc file is composed of one `theory` element. The name of a user-defined theory can be set by the user, whereas the name of the theory √eriFun is based on is fixed to VAFP.

Functions are declared by `symbol` elements that also introduces the type of the function (Subsection 15.2.3). The body of a function is encoded as an OpenMath object inside a `definition` element. The corresponding `symbol` element is referenced by the `definition` element in the `for` and relating termination assertions in the `existence` attribute.

Note that instead of using `name` attributes, which only allow XML simple names, we generate a unique ID. The actual √eriFun names are represented in `presentation` elements or rather their `use` elements (Listing 26.30). By using this technique we can use any character string[22] for element names. To cover the whole set of √eriFun fixities (`prefix` (the default), `infix`, `postfix`, `infixl`, `infixr`, and `outfix`) we had to extend the OMDoc format by `infixl` and `infixr`. However, it was not necessary to also add the `outfix` value, but encoding of `outfix` functions is treated slightly different: The name of the function is encoded in the `lbrack` and `rbrack` attribute respectively of the relating `presentation` element and the `use` element is left empty[23].

Lemmata are mapped to `assertion` elements, the value "`lemma`" being assigned to the `type` attribute. The formula of a lemma, analogous to function bodies, is encoded as an OpenMath object inside an `assertion` element.

Particularly convenient is the direct mapping of √eriFun proofs to the OMDoc presentation of proofs. Verifications of lemmas and termination analysis of functions are represented in `proof` elements. The assertion to be proven is referenced in the `for` attribute. VAFP-tactics used inside a proof to achieve the various proof steps (encoded in `derive` elements) are denoted by

---

[21] Actually there are also automatically system-generated elements included, but we may neglect those at this point.

[22] √eriFun has full UNICODE [Inc03] support

[23] As a consequence the previous mentioned special encoding feature does not hold for outfix functions

method elements. Parameters heuristically computed by the system or manually annotated by the user are encoded as OPENMATH objects and appended to each proof step. Furthermore each proof step in ✔eriFun is annotated with a sequence of the form $h_1, \ldots h_n, \forall \ldots ih_1, \ldots, \forall \ldots ih_l \vdash goal$ whereas the expressions h_i are the hypotheses, the expressions $\forall \ldots ih_k$ are the induction hypotheses, and the expression $goal$ is the goal-term of the sequence. Such a sequent is represented by assumption and conclusion child-elements respectively of the relating derive element.

Listing 26.28. A polymorphic ✔eriFun sort

```
structure list [@value] <=
  ∅,
  [ infixr ,100] :: (hd : @value, tl : list [@value])
```

Sorts are wrapped inside adt elements. At this point this integration process provoked two further adaptations of the OMDoc standard. On the one hand, in contrast to OMDoc, sorts in ✔eriFun could be polymorphic (Listing 26.28). This led to the additional, optional parameters attribute of an adt element (Listing 26.29). Within this new attribute one can declare by a comma separated list the names of type variables of the abstract data type.

Listing 26.29. A polymorphic OMDoc ADT

```
   <adt xml:id="vf7b9f3e59−e78e−4221−8064−7fa0c5689f5d.adt" parameters="value">
     <sortdef name="vf7b9f3e59−e78e−4221−8064−7fa0c5689f5d" type="free">
       <constructor name="vf8a6673ac−c1d9−4698−b6ee−90213539a984"/>
       <constructor name="vf38164505−4983−417f−8bdc−6a42b046e933">
 5       <argument>
           <type system="simpletypes">
             <OMOBJ xmlns="http://www.openmath.org/OpenMath">
               <OMV name="value"/>
             </OMOBJ>
10         </type>
           <selector name="vf9fc4c672−207f−45c0−ae61−1f675fde7aed" total="yes"/>
         </argument>
         <argument>
           <type system="simpletypes">
15           <OMOBJ xmlns="http://www.openmath.org/OpenMath">
               <OMA>
                 <OMS cd="VeriFun" name="vf7b9f3e59−e78e−4221−8064−7fa0c5689f5d"/>
                 <OMV name="value"/>
               </OMA>
20           </OMOBJ>
           </type>
           <selector name="vf55767f3a−b019−4308−88f9−d68oo0db505c" total="yes"/>
         </argument>
       </constructor>
25   </sortdef>
   </adt>
```

On the other hand, the child elements of a constructor element had to be expanded by an additional type element to specify the type of the formal parameter of the parent constructor element. Listing 26.30 illustrates the corresponding presentation elements of the ADT in Listing 26.29.

Listing 26.30. Representation of **✔eriFun** names to OMDOC

```
<presentation for="#vf7b9f3e59−e78e−4221−8064−7fa0c5689f5d" role="applied">
  <use format="VeriFun">list</use>
</presentation>
<presentation for="#vf8a6673ac−c1d9−4698−b6ee−90213539a984" role="applied"
              bracket−style="math" precedence="1" fixity="prefix" lbrack="(" rbrack=")">
  <use format="VeriFun">∅</use>
</presentation>
<presentation for="#vf38164505−4983−417f−8bdc−6a42b046e933" role="applied"
              bracket−style="math" precedence="100" fixity="infixr" lbrack="(" rbrack=")">
  <use format="VeriFun">::</use>
</presentation>
<presentation for="#vf9fc4c672−207f−45c0−ae61−1f675fde7aed" role="applied"
              bracket−style="math" precedence="1" fixity="prefix" lbrack="(" rbrack=")">
  <use format="VeriFun">hd</use>
</presentation>
<presentation for="#vf55767f3a−b019−4308−88f9−d68ee0db595e" role="applied"
              bracket−style="math" precedence="1" fixity="prefix" lbrack="(" rbrack=")">
  <use format="VeriFun">tl</use>
</presentation>
```

Decoding. The decoding of a **✔eriFun** program represented in OMDOC is reverse to the encoding mechanism. First we create an empty program and then start the sequential decoding of each **adt**, **symbol** and its relating **definition**, and **assertion** element back into the $\mathcal{FP}$ syntax. After a successful reconstruction of an element it is appended to the current program. Right after such an insertion we check for a **proof** element containing a reference to this new program element. If a proof exists, we re-play all the proof steps and associate the recreated **✔eriFun** proof to the corresponding program element.

One aspect of this decoding exercise is worth mentioning here. The **✔eriFun** system also benefited by the development of the OMDOC standard: Revelation of bugs deep in the system! Especially $\mathcal{FP}$ parser errors and inconsistencies in proof tactics applications could be discovered. Maybe those errors would never have been detected, because in most cases the user is not able to produce them manually, but this errors are automatically generated by the system. So with the assistance of the strict encoding and decoding to and from OMDOC respectively we were able to achieve a much more robust verification system.

By the integration of the open content Markup language OMDOC into the semi-automated theorem prover **✔eriFun**, we made the system more reliable and facilitate the participation in the mathematical network to serve as yet another service. Functional programs and especially proof of statements created in **✔eriFun** are now open to the public. The data is human-readable, machine-understandable, no longer subjected to a particular version of the system. Thus, **✔eriFun** generated knowledge became accessible, robust, interchangeable and transparent.

Part V

Appendix

In this appendix, we document the changes of the OMDoc format over the versions, provide quick reference tables, and discuss the validation helps.

A

Changes to the Specification

After about 18 Months of development, Version 1.0 of the OMDOC format was released on November 1st 2000 to give users a stable interface to base their documents and systems on. It was adopted by various projects in automated deduction, algebraic specification, and computer-supported education. The experience from these projects uncovered a multitude of small deficiencies and extension possibilities of the format, that have been subsequently discussed in the OMDOC community.

OMDOC 1.1 was released on December 29th 2001 as an attempt to roll the uncontroversial and non-disruptive part of the extensions and corrections into a consistent language format. The changes to version 1.0 were largely conservative, adding optional attributes or child elements. Nevertheless, some non-conservative changes were introduced, but only to less used parts of the format or in order to remedy design flaws and inconsistencies of version 1.0.

OMDOC 1.2 is the mature version in the OMDOC 1 series of specifications. It contains almost no large-scale changes to the document format, except that Content-MATHML is now allowed as a representation for mathematical objects. But many of the representational features have been fine-tuned and brought up to date with the maturing XML technology (e.g. ID attributes now follow the XML ID specification [MVW05], and the Dublin Core elements follow the official syntax [DUB03a]). The main development is that the OMDOC specification, the DTD, and schema are split into a system of interdependent modules that support independent development of certain language aspects and simpler specification and deployment of sub-languages. Version 1.2 of OMDOC freezes the development so that version 2 can be started off on the modules.

In the following, we will keep a log on the changes that have occurred in the released versions of the OMDOC format. We will briefly tabulate the changes by element name. For the state of an element we will use the shorthands "dep" for deprecated (i.e. the element is no longer in use in the new OMDOC version), "cha" for changed, if the element is re-structured (i.e. some additions and losses), "new" if did not exist in the old OMDOC version, "lib", if it was liberalized (e.g. an attribute was made optional) and finally "aug" for augmented, i.e. if it has obtained additional children or attributes in the new OMDoc version.

All changes will be relative to the previous version, starting out with OMDoc 1.0.

A.1   Changes from 1.1 to 1.2

Most of the changes in version 1.2 are motivated by modularization. The goal was to modularize the specification so that it can be used as a DTD module, and that restricted sub-languages of OMDoc can be identified.

Perhaps the most disruptive change is in the presentation/style apparatus: In version 1.1, OMDoc used the `style` attribute for all elements that have an `id` attribute to specify generic style classes for the OMDoc elements. This was based on a misunderstanding of the XML cascading style sheet (CSS) mechanism [Bos98], which uses the `class` attribute to specify this information and uses the `style` attribute to specify CSS directives that override the class information. This error in Version 1.1 of OMDoc so severely limits the usefulness for styling that we rename the Version 1.1 of OMDoc `style` attribute to `class`, even though it breaks 1.1-compatible implementations. Concretely, the Version 1.2 of OMDoc `class` attribute takes the role of the Version 1.1 of OMDoc `style`. and the Version 1.2 of OMDoc `style` takes CSS directives.

Furthermore, all `xml:id` on non-constitutive (see Section 15.1) elements in OMDoc were made optional.

Version 1.1 of OMDoc files can be upgraded to version 1.2 with the XSLT style sheet `http://www.mathweb.org/omdoc/xsl/omdoc1.1adapt1.2.xsl`.

element	state	comments	cf.
`alternative`	aug	This element can now have `theory`, `generated-from`, and `generated-via` attributes.	145
`argument`	cha	The `sort` has been replaced by a `type` child, so that higher-order sorts can be specified.	156
`assertion`	aug	the `assertion` element now has an optional `for` attribute. Furthermore, an optional attribute `generated-via` has been added to allow generation via a theory morphism. Finally, two new attributes `status` and `just-by` have been added to mark up the deductive status of the assertion.	142

`assumption`	cha	This element can now have an attribute `inductive` for inductive assumptions. The natural langauge description in the optional `CMP` element is no longer allowed, use a `phrase` element in a `CMP` that is a sibling to the `FMP` instead.	145
`adt`	aug	the `adt` loses the `CMP` and `commonname` children, use the Dublin Core metadata elements `dc:description` and `dc:subject` instead. The `type` attribute is now on the `sortdef` element. Furthermore, an optionala attribute `generated-via` has been added to allow generation via a theory morphism. Finally, an attribute `parameters` has been added to allow for parametric ADTs.	156
`answer`	cha	the `answer` element does not allow `symbol` children any more, if these are needed, the exercise should have its own theory.	210
`attribute`	aug	the `attribute` element now has a optional `ns` attribute for the namespace URI of the generated attribute node and an attribute `select` for an XPATH expression that specifies the value of the generated attribute.	191
`axiom`	aug	the `axiom` element now has an optional `for` attribute which can point to a list of symbols. Furthermore, an optional attribute `generated-via` has been added to allow generation via a theory morphism and an attribute `type` is now also allowed.	138
`axiom-inclusion`	lib	the `axiom-inclusion` element can now contain multiple `path-just` children to record multiple justifications. Furthermore, it can now have `theory`, `generated-from`, and `generated-via` attributes. New optional attributes `conservativity` and `conservativity-just` for stating and justifying conservativity.	180
`catalogue`	dep	the catalogue mechanism has been eliminated.	
`choice`	cha	the `choice` element does not allow `symbol` children any more, if these are needed, the exercise should have its own theory	210
`code`	cha	Attributes `classid` and `codebase` are deprecated. The attributes `pto` and `pto-version` have moved to the `data` element. The attribute `type` has been removed and optional attributes `theory`, `generated-from`, and `generated-via` have been added.	202
`commonname`	dep	This element is deprecated in favor of a `metadata/dc:subject` element.	

conclusion	cha	The natural langauge description in the optional CMP element is no longer allowed, use a phrase element in a CMP that is a sibling to the FMP instead.	124
constructor	cha	The role attribute is now fixed to object. The commonname child has been replaced by an initial metadata element.	156
data	aug	new optional attributes original to specify whether the external resource referenced by the href attribute (value external) or the data content is the original (value local). The data element has acquired attributes pto and pto-version from the code and private elements.	203
dc:*	aug	All Dublin Core tags have been lowercased to synchronize with the tag syntax recommended by the Dublin Core Initiative. The tags were capitalized in OMDoc1.1. Furthermore, dc:contributor, dc:creator, dc:publisher have received an optional xml:id attribute, so that they can be cross-referenced by the new who of the dc:date element.	98
decomposition	aug	The for attribute is now optional, it need not be given, if the element is a child of a theory-inclusion element. Furthermore, it can now have a theory, generated-from, and generated-via attributes.	184
dc:description	aug	The dc:description can now have the optional xml:id, and CSS attributes	99
definition	aug	The definition element can now have the type pattern for pattern-defined functions. This is a degenerate case of the type inductive. Furthermore, an optional attribute generated-via has been added to allow generation via a theory morphism.	139
effect	aug	allows an optional xml:id attribute	204
example	aug	The example element now has the optional theory attribute that specifies the home theory. Furthermore, it can now have attributes theory, generated-from, and generated-via.	146
exercise	cha	the exercise element does not allow symbol children any more, if these are needed, the exercise should have its own theory. Furthermore, it can now have a theory, generated-from, and generated-via attributes.	209
extradata	cha	The content of the old extradata element can now be directly in the metadata/dc:subject element.	

`element`	aug	The `element` element now allows the `map` and `separator` elements in the body. Furthermore, it carries the optional attributes `crid` for parallel markup, `cr` for cross-references, and `ns` for specifying the namespace.	190
`hint`	aug	the `hint` element can now appear on top-level and has a `for` attribute. It does not allow `symbol` children any more, if these are needed, the exercise should have its own theory. Furthermore, the `exercise` can now have a `theory`, `generated-from`, and `generated-via` attributes.	209
`hypothesis`	cha	the `discharged-in` attribute has been eliminated. Scoping is now specified in terms of the enclosing `proof` element. Furthermore, the `symbol` child is no longer allowed inside the element. A sibling `symbol` should be used.	163
`inclusion`	aug	allows optional attributes `xml:id`, `conservativity`, and `conservativity-just` for stating and justifying conservativity.	179
`imports`	lib	the `xml:id` is now optional. New optional attributes `conservativity` and `conservativity-just` for stating and justifying conservativity.	150
`input`	aug	allows an optional `xml:id` attribute	204
`legacy`	new	An element for encapsulating legacy mathematics, can be used wherever `m:math` and `om:OMOBJ` are allowed.	120
`loc`	dep	The catalogue mechanism has been eliminated.	
`m:math`	new	Content-MATHML is now allowed wherever OPENMATH objects were allowed before.	114
`map`	new	this element allows to map its-style directives over a list of e.g. arguments	191
`mc`	aug	the `mc` element can now have a `for` attribute. It does not allow `symbol` children any more, if these are needed, the dominating `exercise` element should have its own theory. Furthermore, the `mc` element can now have a `theory`, `generated-from`, and `generated-via` attributes.	210
`measure`	aug	allows an optional `xml:id` attribute	141
`metacomment`	dep	This element is superseded by the `omtext` element.	124

morphism	aug	The `morphism` element now carries the optional attributes `consistency`, `exhaustivity`, `hiding`, and `type`. Furthermore the content model allows optional elements `measure` and `ordering` after the `requation` children to specify termination information like in `definition`.	92
obligation	aug	allows an optional `xml:id` attribute	178
omdoc	aug	This element can now have a `theory`, `generated-from`, and `generated-via` attributes.	90
omgroup	cha	The values `dataset` and `labeled-dataset` are deprecated in Version 1.2 of OMDoc, since we provide tables in module RT; see Section 14.6 for details. Furthermore, the element can now have the attributes, `modules`, `theory`, `generated-from`, and `generated-via`.	93
omlet	cha	`omlet` can no longer occur at top-level (it just does not make sense). The data model for this element has been totally reworked, inspired by the `xhtml:object` element.	205
omstyle	aug	This element can now have `generated-from`, and `generated-via` attributes. New attribute `xref` that allows to inherit the information from another `omstyle` element.	188
om:*	aug	with OpenMath2, the OpenMath elements carry an optional `id` attribute for structure sharing via the `om:OMR` element. Furthermore, in OMDoc, they carry `cref` attributes for parallel markup with cross-references.	108
om:OMFOREIGN	new	The `om:OMFOREIGN` element can be used to encapsulate arbitrary XML data in OpenMath attributions.	110
om:OMR	new	In the OpenMath2 standard, this element is the main vehicle of the structure sharing representation.	112
omtext	aug	the `type` attribute can now also have the values `axiom`, `definition`, `theorem`, `proposition`, `lemma`, `corollary`, `postulate`, `conjecture`, `false-conjecture`, `obligation`, `assumption`, and `formula`. Furthermore, `omtext` can now have `theory`, `generated-from`, and `generated-via` and `verbalizes` attributes.	124
ordering	aug	Now allows the optional `xml:id` and `terminating` attributes. The latter points to a termination assertion.	141

output	aug	allows an optional `xml:id` attribute	204
pattern	aug	this element is no longer used, the pattern of a recursive equation is determined by the position as the first child.	
path-just	aug	The element can now appear as a top-level element, if it does, the attribute `for` must point to the `axiom-inclusion` element it justifies. It also now allows an optional `xml:id` attribute	184
phrase	new	used to mark up phrases in CMPs and supply them with identifiers and links to context that can be used for presentation and referencing.	126
presentation	cha	The `theory` is not allowed any more, to refer to a symbol outside its theory use its `xml:id` attribute. The element now also allows a mutilingual CMP group, so that it can be used as a notation definition element in mathematical vernacular.	192
private	cha	The `replaces` attribute is now called `reformulates`. The attributes `pto` and `pto-version` have moved to the `data` element. The attribute `type` has been removed and optional attributes `theory`, `generated-from`, and `generated-via` have been added.	202
proof	lib	The `for` attribute is now optional to allow for proofs as objects of mathematical discourse. Furthermore, it can now have `generated-from` and `generated-via` attributes.	161
proofobject	lib	The `for` attribute is now optional to allow for proofs as objects of mathematical discourse. Furthermore, it can now have `generated-from` and `generated-via` attributes.	169
recognizer	cha	The `role` attribute was fixed to `object`. The `commonname` child has been replaced by an initial `metadata` element.	157
ref	aug	`ref` now has an optional `xml:id` attribute that identifies it.	94
selector	cha	The `role` attribute was fixed to `object`. The `commonname` child has been replaced by an initial `metadata` element.	157
solution	cha	the `solution` element now allows arbitrary OMDOC top-level elements as children. Furthermore, it can now have a `theory`, `generated-from`, and `generated-via` attributes.	209

sortdef	cha	The `role` attribute was fixed to `sort`. The `type` from the `adt` element is now on the `sortdef` element. The `commonname` child has been replaced by an initial `metadata` element.	156
dc:subject	aug	The `dc:subject` can now have the optional `dc:id`, and CSS attributes	99
style	aug	The `style` element now allows a `map` element in the body	189
symbol	cha	may no longer contain `selector`, since it only makes sense for constructors in data types. The `kind` attribute has been renamed to `role` for compatibility with OPENMATH2 and can have the additional values `binder`, `attribution`, `semantic-attribution`, and `error` corresponding to the OPENMATH 2 roles. Furthermore, an optional attribute `generated-via` has been added to allow generation via a theory morphism.	136
term	new	the `term` element can appear in mathematical text and contain it. It is used to link technical terms to symbols defined in content dictionaries via its `cd` and `name` attributes.	128
theory	cha	the `theory` element loses the `CMP` and `commonname` children, use the Dublin Core metadata elements `dc:description` and `dc:subject` instead. The `theory` element also gains the optional `cdbase` attribute to specify the disambiguating string prescribed for content dictionaries by the OPENMATH2 standard. The `xml:id` is now optional, it only needs to be specified, if the theory has constitutive elements. Finally, the element has gained the optional attributes `cdurl`, `cdbase`, `cdreviewdate`, `cdversion`, `cdrevision`, and `cdstatus` attributes for encoding the management metadata of OPEN-MATH content dictionaries.	149
dc:title	aug	The `dc:title` can now have the optional `dc:id`, and CSS attributes.	98
tgroup	new	The `tgroup` can be used to structure theories like documents.	149

type	aug	the **type** element now has the optional **just-by** and **theory** attribute. The first one points to an assertion or axiom that justifies the type judgment, the second specifies the home theory. The **system** attribute is now optional. Furthermore, the **type** element can have two math objects as children. If it does, then it is a term declaration, i.e. the first element is interpreted as a mathematical object and the second one is interpreted as its type. Finally, it can now have **generated-from** and **generated-via** attributes.	139
theory-inclusion	aug	the **theory-inclusion** element can now have obligation and decomposition children that justify it. Furthermore, it can now have a **theory**, **generated-from**, and **generated-via** attributes. New optional attributes **conservativity** and **conservativity-just** for stating and justifying conservativity.	178
theory	aug	the **theory** element can now be nested.	149
use	cha	can now contain **element**, **text**, **recurse**, **map**, and **value-of** to specify XML content. We have deprecated the **larg-group** and **rarg-group** attributes, since they were never used.	194
value	aug	this element is no longer used, the value of a recursive equation is determined by the position as the second child.	
with	ren	the role of this element is now taken by the **phrase** element.	126
xslt	cha	the content of this element need not be escaped any more, it is now a valid XSLT fragment.	189

A.2 Changes from 1.0 to 1.1

Version 1.1 was mainly a bug-fix release that has become necessary by the experiments of encoding legacy material in OMDOC. The changes are relatively minor, mostly added optional fields. The only non-conservative changes concern the **private**, **hypothesis**, **sortdef** and **signature** elements. OMDOC files can be upgraded to version 1.1 with the XSLT style sheet http://www.mathweb.org/omdoc/xsl/omdoc1.0adapt1.1.xsl.

element	state	comments	cf.
`attribute`	new	presentation of attributes for XML elements	191
`alternative`	cha	new form of the `alternative-def` element, it can now also used as an alternative to `axiom`. Compared to `alternative-def` it has a new optional attribute `generated-by` to show that an assertion is generated by expanding a some other element like adt.	145
`alternative-def`	dep	new form is `alternative`, since there can be alternative `axiom`s too.	
`argument`	cha	attribute `sort` is now of type IDREF, since it must be local in the definition.	156
`assertion`	aug	more values for the `type` attribute, new optional attribute `generated-by` to show that an assertion is generated by expanding a `definition` or an `adt`. New optional attribute `proofs`.	142
`assertion-just`	dep	this is now `obligation`	
`axiom`	aug	new optional attribute `generated-by` to show that an axiom is generated by expanding a `definition`.	138
`axiom-inclusion`	cha	now allows a CMP group for descriptive text, includes a set of `obligation` elements instead of an `assertion-just`. The `timestamp` attribute is deprecated, use `dc:date` with appropriate `action` instead	180
`CMP`	cha	the attribute `format` is now deprecated, it makes no sense, since we are more strict and consistent about `CMP` content. `CMP` now allows an optional `id` attribute.	122
`code`	cha	Attributes `width` and `height` now in `omlet`, got attributes `classid` and `codebase` from `private`. Attribute `format` moved to `data` children. The multilingual group of `CMP` elements for description is deprecated, use `metadata/dc:description` instead. Child element `data` may appear multiple times (with different values of the `format`).	202
`constructor`	aug	new optional child `recognizer` for a recognizer predicate	156
`Coverage`	dep	this Dublin Core element specifies the place or time which the publication's contents addresses. This does not seem appropriate for the mathematical content of OMDoc.	
`data`	aug	new optional attributes `size` to specify the size of the data file that is referenced by the `href` attribute and `format` for the format the data is in.	203

dc:date	aug	new optional **who** attribute that can be used to specify who did the **action** on this date.	99
Translator	dep	this element is not part of Dublin Core, it got into OMDoc by mistake, we use **dc:contributor** with **role=trl** for this.	98
decomposition	aug	has a new required **id** attribute. It is no longer a child of **theory-inclusion**, but specifies which **theory-inclusion** it justifies by the new required attribute **for**.	184
definition	aug	new optional children **measure** and **ordering** to specify termination of recursive definitions. New optional attribute **generated-by** to show that it is generated by expanding a **definition**.	139
element	new	presentation of XML elements	190
FMP	aug	now allows multiple **conclusion** elements, to represent general Gentzen-type sequents (not only natural deduction.) FMP now allows an optional **id** attribute.	123
hypothesis	cha	new required attribute **discharged-in** to specify the **derive** element that discharges this hypothesis.	163
measure	new	specifies a measure function (as an OMOBJ)	141
metadata	aug	new optional attribute **inherits** that allows to inherit metadata from other declarations	92
method	cha	first child that used to be an om:**OMSTR** or **ref** element is now moved into a required **xref** attribute that holds an URI that points to the element that defines the method. The om:**OMOBJ** content of the other children (they were **parameter** elements) is now directly included in the **method** element.	164
obligation	new	takes over the role of **assertion-just**.	
omgroup	aug	also allows the elements that can only appear in **theory** elements, so that omgroups can also be used for grouping inside **theory** elements. The **type** attribute is now restrained to one of **narrative**, **sequence**, **alternative**, **contrast**.	93
omlet	aug	obtained attributes **width** and **height** from **private**. New optional attributes **action** for the action to be taken when activated, and **data** a URIref to data in a private element. New optional attribute **type** for the type of the applet.	205
omstyle	new	for specifying the style of OMDoc elements	188
omtext	cha	the **from** is deprecated, we only leave the **for** attribute, to specify the referential character of the **type**.	124
ordering	new	specifies a well-founded ordering (as an OMOBJ)	141

parameter	dep	the om:OMOBJ element child is now directly a child of method	
pattern	cha	the child can be an arbitrary OPENMATH element.	
premise	cha	new optional attribute rank for the importance in the inference rule. The old href attribute is renamed to xref to be consistent with other cross-referencing.	
presentation	aug	New attribute xref that allows to inherit the information from another presentation element. New attribute theory to specify the theory the symbol is from; without this, referencing in OMDOC is not unique. The parent attribute has been renamed to role and now takes the values applied, binding, and key, since we want to be less OPENMATH-centric	192
private	cha	new optional attribute for to point to an OMDOC element it provides data for. As a consequence, private elements are no longer allowed in other OMDOC elements, only on top-level. New attribute replaces as a pointer to the OMDOC elements that are replaced by the system-specific information in this element. Old attributes width and height now in omlet. Attribute format moved to data children. The descriptive CMP elements are deprecated, use metadata/dc:description instead. Child element data may appear multiple times (with different values of the format). The attributes classid and codebase are deprecated, since they only make sense on the code element.	q 202
proof	cha	attribute theory is now optional, since the element can appear inside a theory element.	161
proofobject	cha	attribute theory is now optional, since the element can appear inside a theory element.	161
recognizer	new	specifies the recognizer predicate of a sort.	157
recurse	new	recursive calls to presentation in style.	191
ref	cha	attribute kind renamed to type.	94
selector	cha	the old type attribute (had values total and partial) is deprecated, its duty is now carried by an attribute total (values yes and no).	157
signature	dep	for the moment	
sortdef	cha	has a mandatory name attribute, otherwise the defined symbol has no name.	156

`style`	new	allows to specify style information in `presentation` and `omstyle` elements using a simplified OMDOC-internalized version of XSLT.	189
`symbol`	aug	new optional attribute `generated-by` to show that it is generated by expanding a `definition`.	136
`text`	new	presentation of text in `omstyle`.	190
`theory-inclusion`	cha	now allows CMP group for descriptive text, no longer has a `decomposition` child, this is now attached by its `for` attribute. The `timestamp` attribute is deprecated, use `dc:date` with appropriate `action` instead.	178
`type`	aug	can now also appear on top-level. Has an optional `id` attribute for identification, and an optional `for` attribute to point to a `symbol` element it declares type information for.	139
`use`	aug	New attribute `element` allows to specify that the content should be encased in an XML element with the attribute-value pairs specified in the string specified in the attribute `attributes`.	194
`value-of`	new	presentation of values in `style`.	190
`with`	new	used to supply fragments of text in CMPs with `style` and `id` attributes that can be used for presentation and referencing.	126
`xslt`	new	allows to embed XSLT into `presentation` and `omstyle` elements.	189

B

Quick-Reference Table to the OMDOC Elements

Element	p.	Mod.	Required Attribs	Optional Attribs	D C	Content
adt	156	ADT		xml:id, type, style, class, theory, generated-from, generated-via	+	sortdef+
alternative	145	ST	for, entailed-by, entails, entailed-by-thm, entails-thm	xml:id, type, theory, generated-from, generated-via, uniqueness, exhaustivity, consistency, existence, style, class	+	CMP*, (FMP| requation*| (OMOBJ |m:math |legacy)*)
answer	210	QUIZ	verdict	xml:id, style, class	+	CMP*, FMP*
m:apply	115	MML		id, xlink:href	−	bvar?, ⟪*CMel*⟫*
argument	156	ADT	sort		+	selector?
assertion	142	ST		xml:id, type, theory, generated-from, generated-via, style, class	+	CMP*, FMP*
assumption	124	MTXT		xml:id, inductive, style, class	+	CMP*, (OMOBJ |m:math |legacy)?
attribute	191	PRES	name		−	(value-of| text)*
axiom	138	ST	name	xml:id, type, generated-from, generated-via, style, class	+	CMP*, FMP*
axiom-inclusion	180	CTH	from, to	xml:id, style, class, theory, generated-from, generated-via	+	morphism?, (path-just| obligation*)
m:bvar	115	MML		id, xlink:href	−	ci*
m:ci	115	MML		id, xlink:href	−	PCDATA
m:cn	115	MML		id, xlink:href	−	([0-9]|,|.) (*|e([0-9]|,|.)*)?
choice	210	QUIZ		xml:id, style, class	+	CMP*, FMP*

CMP	122	MTXT		xml:lang, xml:id	−	(text| OMOBJ |m:math |legacy | with | term | omlet)*
code	202	EXT		xml:id, for, theory, generated-from, generated-via, requires, style, class	+	input?, output?, effect?, data+
conclusion	124	MTXT		xml:id, style, class	+	CMP*, (OMOBJ |m:math |legacy)?
constructor	156	ADT	name	type, scope, style, class, theory, generated-from, generated-via	+	argument*, recognizer?
dc:contributor	98	DC		xml:id, role, style, class	−	⟪text⟫
dc:creator	98	DC		xml:id, role, style, class	−	⟪text⟫
m:csymbol	115	MML	definitionURL	id, xlink:href	−	EMPTY
data	203	EXT		format, href, size, original	−	<![CDATA[...]]>
dc:date	99	DC		action, who	−	ISO 8601 norm
dd	130	RT		xml:id, style, class, index, verbalizes	+	CMPcontent
di	130	RT		xml:id, style, class, index, verbalizes	+	dt+,dd*
dl	130	RT		xml:id, style, class, index, verbalizes	+	li*
dt	130	RT		xml:id, style, class, index, verbalizes	+	CMPcontent
decomposition	184	DG	links	theory, generated-from, generated-via	−	EMPTY
definition	139	ST	xml:id, for	uniqueness, existence, consistency, exhaustivity, type, generated-from, generated-via, style, class	+	CMP*, (FMP| requation+| OMOBJ |m:math |legacy)?, measure?, ordering?
dc:description	99	DC		xml:lang	−	CMPcontent
derive	162	PF		xml:id, style, class	−	CMP*, FMP?, method?
effect	204	EXT		xml:id, style, class	−	CMP*,FMP*
element	190	PRES	name	xml:id, cr, ns	−	(attribute| element| text| recurse)*
example	146	ST	for	xml:id, type, assertion, proof, style, class, theory, generated-from, generated-via	+	CMP*| (OMOBJ |m:math |legacy)?

exercise	209	QUIZ		xml:id, type, for, from, style, class, theory, generated-from, generated-via	+	CMP*, FMP*, hint?, (solution*|mc*)
FMP	123	MTXT		logic, xml:id	−	(assumption*, conclusion*)|OMOBJ |m:math |legacy
dc:format	100	DC			−	fixed: "application/omdoc+xml"
hint	209	QUIZ		xml:id, style, class, theory, generated-from, generated-via	+	CMP*, FMP*
hypothesis	163	PF		xml:id, style, class, inductive	−	CMP*, FMP*
dc:identifier	100	DC		scheme	−	ANY
ide	130	RT	index	xml:id,sort-by,see, seealso, links, style, class	−	idp*
idp	130	RT		xml:id,sort-by,see, seealso, links, style, class	−	CMPcontent
idt	130	RT		style, class	−	CMPcontent
idx	130	RT		xml:id,sort-by,see, seealso, links, style, class	−	idt?,idp+
ignore	92	DOC		type, comment	−	ANY
imports	150	CTH	from	xml:id, type, style, class	+	morphism?
inclusion	179	CTH	for	xml:id	−	
input	204	EXT		xml:id, style, class	−	CMP*,FMP*
insort	156	ADT	for		−	
dc:language	100	DC			−	ISO 8601 norm
li	130	RT		xml:id, style, class, index, verbalizes	−	Math Vernacular
cc:license	102	CC		jurisdiction	−	permissions, prohibitions, requirements
link	130	RT		xml:id, style, class, index, verbalizes	−	Math Vernacular
m:math	114	MML		id, xlink:href	−	⟪CMel⟫+
mc	210	QUIZ		xml:id, style, class, theory, generated-from, generated-via	−	choice, hint?, answer
measure	141	ST		xml:id	−	OMOBJ |m:math |legacy
metadata	92	DC		inherits	−	(dc-element)*
method	164	PF	xref			(OMOBJ |m:math |legacy| premise | proof | proofobject)*
morphism	175	CTH		xml:id, base, consistency, exhaustivity, type, hiding, style, class	−	requation*, measure?, ordering?
note	130	RT		type,xml:id, style, class, index, verbalizes	−	Math Vernacular

obligation	178	CTH	induced-by, assertion	xml:id	−	EMPTY
om:OMA	108	OM		id, cdbase	−	⟪*OMel*⟫*
om:OMATTR	110	OM		id, cdbase	−	⟪*OMel*⟫
om:OMATP	110	OM		cdbase	−	(OMS, (⟪*OMel*⟫ | om:OMFOREIGN))+
om:OMB	111	OM		id, class, style, class	−	#PCDATA
om:OMBIND	109	OM		id, cdbase	−	⟪*OMel*⟫, om:OMBVAR, ⟪*OMel*⟫?
om:OMBVAR	109	OM			−	(om:OMV | om:OMATTR)+
om:OMFOREIGN	110	OM		id, cdbase	−	ANY
omdoc	90	DOC		xml:id,type, version, style, class, xmlns, theory, generated-from, generated-via	+	(top-level element)*
om:OME	111	OM		xml:id	−	(⟪*OMel*⟫)?
om:OMR	112	OM	href		−	
om:OMF	111	OM		id, dec, hex	−	#PCDATA
omgroup	93	DOC		xml:id, type, style, class, modules, theory, generated-from, generated-via	+	top-level element*
ol	130	RT		xml:id, style, class, index, verbalizes	−	li*
om:OMI	111	OM		id, class, style	−	[0-9]*
omlet	205	EXT		id, argstr, type, function, action, data, style, class	+	ANY
om:OMOBJ	108	OM		id, cdbase, class, style	−	⟪*OMel*⟫?
omstyle	188	PRES	element	for, xml:id, xref, style, class	−	(style|xslt)*
om:OMS	108	OM	cd, name	class, style	−	EMPTY
omtext	124	MTXT		xml:id, type, for, from, style, theory, generated-from, generated-via	+	CMP+, FMP?
om:OMV	108	OM	name	class, style	−	EMPTY
ordering	141	ST		xml:id	−	OMOBJ |m:math |legacy
output	204	EXT		xml:id, style, class	−	CMP*,FMP*
p	129	RT		xml:id, style, class, index, verbalizes	−	Math Vernacular
param	207	EXT	name	value, valuetype	−	EMPTY
path-just	184	DG	local, globals	for, xml:id	−	EMPTY
cc:permissions	103	CC		reproduction, distribution, derivative_works	−	EMPTY
premise	164	PF	xref		−	EMPTY

presentation	192	PRES	for	xml:id, xref, fixity, role, lbrack, rbrack, separator, bracket-style, style, class, precedence, crossref-symbol	−	(use | xslt | style)*
private	202	EXT		xml:id, for, theory, generated-from, generated-via, requires, reformulates, style, class	+	data+
cc:prohibitions	103	CC		commercial_use	−	EMPTY
proof	161	PF		xml:id, for,theory, generated-from, generated-via, style, class	+	(symbol | definition | omtext | derive | hypothesis)*
proofobject	169	PF		xml:id, for, theory, generated-from, generated-via, style, class	+	CMP*, (OMOBJ |m:math |legacy)
dc:publisher	99	DC		xml:id, style, class	−	ANY
ref	94	DOC		xref, type	−	ANY
recognizer	157	ADT	name	type, scope, role, style, class	+	
recurse	191	PRES		select	−	EMPTY
dc:relation	100	DC			−	ANY
requation	140	ST		xml:id, style, class	−	(OMOBJ |m:math |legacy),(OMOBJ |m:math |legacy)
cc:requirements	103	CC		notice, copyleft, attribution	−	EMPTY
dc:rights	100	DC			−	ANY
selector	157	ADT	name	type, scope, role, total, style, class	+	
solution	209	QUIZ		xml:id, for, style, class, theory, generated-from, generated-via	+	(CMP*, FMP*) | proof
sortdef	156	ADT	name	role, scope, style, class	+	(constructor|insert)*
dc:source	100	DC			−	ANY
style	189	PRES	format	xml:lang, requires	−	(element | text | recurse | value-of)*
dc:subject	99	DC		xml:lang	−	CMPcontent
symbol	136	ST	name	role, scope, style, class, generated-from, generated-via	+	type*
table	130	RT		xml:id, style, class, index, verbalizes	−	tr*
term	128	MTXT	cd, name	xml:id, role, style, class	−	CMP content

text	190	PRES			−	#PCDATA
td	130	RT		xml:id, style, class, index, verbalizes	−	Math Vernacular
th	130	RT		xml:id, style, class, index, verbalizes	−	Math Vernacular
theory	149	ST	xml:id	cdbase, style, class	+	(statement|theory)*
theory-inclusion	178	CTH	from, to	xml:id, style, class, theory, generated-from, generated-via	+	(morphism, decomposition?)
tr	130	RT		xml:id, style, class, index, verbalizes	−	(td|th)*
dc:title	98	DC		xml:lang	−	CMPcontent
tgroup	149	DOC		xml:id, type, style, class, modules, generated-from, generated-via	+	top-level or theory-constitutive element*
type	139	ST	system	xml:id, for, style, class	−	CMP*, (OMOBJ |m:math |legacy)
dc:type	100	DC			−	fixed: "Dataset" or "Text" or "Collection"
ul	130	RT		xml:id, style, class, index, verbalizes	−	li*
use	194	PRES	format	xml:lang, requires, fixity, lbrack, rbrack, separator, crossref-symbol, element, attributes	−	(use | xslt | style)*
value-of	190	PRES	select		−	EMPTY
phrase	126	MTXT		xml:id, style, class, index, verbalizes, type	−	CMP content
xslt	189	PRES	format	xml:lang, requires	−	XSLT fragment

C

Quick-Reference Table to the OMDoc Attributes

Attribute	*element*	Values
action	dc:date	unspecified
	specifies the action taken on the document on this date.	
action	omlet	execute, display, other
	specifies the action to be taken when executing the omlet, the value is application-defined.	
actuate	omlet	onPresent, onLoad, onRequest, other
	specifies the timing of the action specified in the action attribute	
assertion	example	
	specifies the assertion that states that the objects given in the example really have the expected properties.	
assertion	obligation	
	specifies the assertion that states that the translation of the statement in the source theory specified by the induced-by attribute is valid in the target theory.	
attributes	use	
	the attribute string for the start tag of the XML element substituted for the brackets (this is specified in the element attribute).	
attribution	cc:requirements	required, not required
	Specifies whether the copyright holder/author must be given credit in derivative works	
base	morphism	
	specifies another morphism that should be used as a base for expansion in the definition of this morphism	
bracket-style	presentation, use	lisp, math
	specifies whether a function application is of the form $f(a,b)$ or (fab)	
cd	om:OMS	
	specifies the content dictionary of an OPENMATH symbol	
cd	term	
	specifies the content dictionary of a technical term	
cdbase	om:*	
	specifies the base URI of the content dictionaries used in an OPENMATH object	
cdreviewdate	theory	
	specifies the date until which the content dictionary will remain unchanged	
cdrevision	theory	
	specifies the minor version number of the content dictionary	
cdstatus	theory	official, experimental, private, obsolete
	specifies the content dictionary status	
cdurl	theory	
	the main URL, where the newest version of the content dictionary can be found	

`cdversion`	`theory`	
	specifies the major version number of the content dictionary	
`comment`	`ignore`	
	specifies a reason why we want to ignore the contents	
`crossref-symbol`	`presentation, use`	`all, brackets, lbrack, no, rbrack, separator, yes`
	specifies whether cross-references to the symbol definition should be generated in the output format.	
`class`	`*`	
	specifies the CSS class	
`commercial_use`	`cc:permissions`	`permitted, prohibited`
	specifies, whether commercial use of the document with this license is permitted	
`consistency`	`morphism, definition`	`OMDoc reference`
	points to an assertion stating that the cases are consistent, i.e. that they give the same values, where they overlap	
`copyleft`	`cc:restrictions`	`required, not_required`
	specifies whether derived works must be licensed with the same license as the current document.	
`cr`	`element`	`yes/no`
	specifies whether an `xlink:href` cross-reference should be set on the result element.	
`cref`	`om:*`	`URI reference`
	extra attribute for cross-references in parallel markup	
`crid`	`element`	`XPATH expression`
	the path to the sub-element that corresponds to the result element.	
`crossref-symbol`	`presentation, use`	`no, yes, brackets, separator, lbrack, rbrack, all`
	specifies which generated presentation elements should carry cross-references to the definition.	
`data`	`omlet`	
	points to a private element that contains the data for this omlet	
`definitionURL`	`m:*`	`URI`
	points to the definition of a mathematical concept	
`derivative_works`	`cc:permissions`	`permitted, not_permitted`
	specifies whether the document may be used for making derivative works.	
`distribution`	`cc:permissions`	`permitted,not_permitted`
	specifies whether distribution of the current document fragment is permitted.	
`element`	`use`	
	the XML element tags to be substituted for the brackets.	
`element`	`omstyle`	
	the XML element, the presentation information contained in the `omstyle` element should be applied to.	
`encoding`	`m:annotation,om:OMFOREIGN`	`MIME type of the content`
	specifies the format of the content	
`entails, entailed-by`	`alternative`	
	specifies the equivalent formulations of a definition or axiom	
`entails-thm, entailed-by-thm`	`alternative`	
	specifies the entailment statements for equivalent formulations of a definition or axiom	
`exhaustivity`	`morphism, definition`	`OMDoc reference`
	points to an assertion that states that the cases are exhaustive.	
`existence`	`definition`	`OMDoc reference`
	points to an assertion that states that the symbol described in an implicit definition exists	
`fixity`	`presentation`	`assoc, infix, postfix, prefix`
	specifies where the function symbol-of a function application should be displayed in the output format	
`function`	`omlet`	

	specifies the function to be called when this omlet is activated.	
format	data	
	specifies the format of the data specified by a data element. The value should e.g. be a MIME type [FB96].	
for	*	
	can be used to reference an element by its unique identifier given in its xml:id attribute.	
formalism	legacy	URI reference
	specifies the formalism in which the content is expressed	
format	legacy	URI reference
	specifies the encoding format of the content	
format	use	cmml, default, html, mathematica, pmml, TeX,...
	specifies the output format for which the notation is specified	
from	imports, theory-inclusion, axiom-inclusion	URI reference
	pointer to source theory of a theory morphism	
from	omtext	URI reference
	points to the source of a relation given by a text type	
generated-from	top-level elements	URI reference
	points to a higher-level syntax element, that generates this statement.	
generated-via	top-level elements,...	URI reference
	points to a theory-morphism, via which it is translated from the element pointed to by the generated-from attribute.	
globals	path-just	
	points to the axiom-inclusions or theory-inclusions that is the rest of the inclusion path.	
hiding	morphism	
	specifies the names of symbols that are in the domain of the morphism	
href	data, link, om:OMR	URI reference
	a URI to an external file containing the data.	
xml:id		
	associates a unique identifier to an element, which can thus be referenced by an for or xref attribute.	
xml:base		
	specifies a base URL for a resource fragment	
index	on RT elements	
	A path identifier to establish multilingual correspondence	
induced-by	obligation	
	points to the statement in the source theory that induces this proof obligation	
inductive	assumption, hypothesis	yes, no
	Marks an assumption or hypothesis inductive.	
inherits	metadata	URI reference
	points to a metadata element from which this one inherits.	
jurisdiction	cc:license	IANA Top level Domain designator
	specifies the country of jurisdiction for a Creative Commons license	
just-by	type	
	points to an assertion that states the type property in question.	
role	symbol, constructor, recognizer, selector, sortdef	object, type, sort, binder, attribution, semantic-attribution, error
	specifies the role (possible syntactic roles) of the symbol in this declaration.	
role	dc:creator,dc:contributor	MARC relators
	specifies the role of a person who has contributed to the document	
role	presentation	applied, binding, key
	specifies which role of the symbol is annotated with notation information	
lbrack	presentation, use	
	the left bracket to use in the notation of a function symbol	
links	decomposition	

	specifies a list of theory- or axiom-inclusions that justify (by decomposition) the theory-inclusion specified in the for attribute.	
local	path-just	
	points to the axiom-inclusion that is the first element in the path.	
logic	FMP	token
	specifies the logical system used to encode the property.	
modules	omdoc, omgroup	module and sub-language shorthands, URI reference
	specifies the modules or OMDoc sub-language used in this document fragment	
name	om:OMS, om:OMV, symbol, term	
	the name of a concept referenced by a symbol, variable, or technical term.	
name	attribute, element	
	the local name of generated element.	
name	param	
	the name of a parameter for an external object.	
notice	cc:requirements	required, not_required
	specifies whether copyright and license notices must be kept intact in distributed copies of this document	
ns	element, attribute	URI
	specifies the namespace URI of the generated element or attribute node	
original	data	local, external
	specifies whether the local copy in the data element is the original or the external resource pointed to by the href attribute.	
parameters	adt	
	The list of formal parameters of a higher-order abstract data type	
precedence	presentation	
	the precedence of a function symbol (for elision of brackets)	
proofs	assertion	
	specifies a list of URIs to proofs of this assertion.	
pto, pto-version	private, code	
	specifies the system and its version this data or code is private to	
rank	premise	
	specifies the rank (importance) of a premise	
rbrack	presentation, use	
	the right bracket to use in the notation of a function symbol	
reformulates	private	
	points to a set of elements whose content is reformulated by the content of the private element for the system.	
reproduction	cc:permissions	permitted,not_permitted
	specifies whether reproduction of the current document fragment is permitted by the licensor	
requires	private, code, use, xslt, style	URI reference
	points to a code element that is needed for the execution of this data by the system.	
role	dc:creator, dc:collaborator	aft, ant, aqt, aui, aut, clb, edt, ths, trc, trl
	the MARC relator code for the contribution of the individual.	
role	phrase, term	
	the role of the phrase annotation	
role	presentation	applied, binding, key
	specifies for which role (as the head of a function application, as a binding symbol, or as a key in a attribution, or as a stand-alone symbol (the default)) of the symbol presentation is intended	
scheme	dc:identifier	scheme name
	specifies the identification scheme (e.g. ISBN) of a resource	
scope	symbol	global, local
	specifies the visibility of the symbol declared. This is a very crude specification, it is better to use theories and importing to specify symbol accessibility.	
select	map, recurse, value-of	XPATH expression
	specifies the path to the sub-expression to act on	

separator	presentation, use	
	the separator for the arguments to use in the notation of a function symbol	
show	omlet	new, replace, embed, other
	specifies the desired presentation of the external object.	
size	data	
	specifies the size the data specified by a data element. The value should be number of kilobytes	
sort	argument	
	specifies the argument sort of the constructor	
style	*	
	specifies a token for a presentation style to be picked up in a presentation element.	
system	type	
	A token that specifies the logical type system that governs the type specified in the type element.	
theory	*	
	specifies the home theory of an OMDoc statement.	
to	theory-inclusion, axiom-inclusion	
	specifies the target theory	
total	selector	no, yes
	specifies whether the symbol declared here is a total or partial function.	
type	adt	free, generated, loose
	defines the semantics of an abstract data type free = no junk, no confusion, generated = no junk, loose is the general case.	
type	assertion	theorem, lemma, corollary, conjecture, false-conjecture, obligation, postulate, formula, assumption, proposition
	tells you more about the intention of the assertion	
type	definition	implicit, inductive, obj, recursive, simple
	specifies the definition principle	
type	derive	conclusion, gap
	singles out special proof steps: conclusions and gaps (unjustified proof steps)	
type	example	against, for
	specifies whether the objects in this example support or falsify some conjecture	
type	ignore	
	specifies the type of error, if ignore is used for in-place error markup	
type	imports	global, local
	local imports only concern the assumptions directly stated in the theory. global imports also concern the ones the source theory inherits.	
type	morphism	
	specifies whether the morphism is recursive or merely pattern-defined	
type	omgroup, omdoc	enumeration, sequence, itemize
	the first three give the text category, the second three are used for generalized tables	
type	omtext	abstract, antithesis, comment, conclusion, elaboration, evidence, introduction, motivation, thesis
	a specification of the intention of the text fragment, in reference to context.	
type	phrase	
	the linguistic or mathematical type of the phrase	
type	ref	include, cite
	specifies whether to replace the ref element by the fragment referenced by href attribute or to merely cite it.	
uniqueness	definition	URI reference
	points to an assertion that states the uniqueness of the concept described in an implicit definition	
value	param	

	specifies the value of the parameter	
valuetype	param	
	specifies the type of the value of the parameter	
verbalizes	on RT elements	URI references
	contains a whitespace-separated list of pointers to OMDOC elements that are verbalized	
verdict	answer	
	specifies the truth or falsity of the answer. This can be used e.g. by a grading application.	
version	omdoc	1.2
	specifies the version of the document, so that the right DTD is used	
version	cc:license	
	specifies the version of the Creative Commons license that applies, if not present, the newest one is assumed	
via	inclusion	
	points to a theory-inclusion that is required for an actualization	
who	dc:date	
	specifies who acted on the document fragment	
xml:lang	CMP, dc:*	ISO 639 code
	the language the text in the element is expressed in.	
xml:lang	use, xslt, style	whitespace-separated list of ISO 639 codes
	specifies for which language the notation is meant	
xlink:*	om:OMR, m:*	URI reference
	specify the link behavior on the elements	
xref	ref, method, premise	URI reference
	Identifies the resource in question	
xref	presentation, omstyle	URI reference
	The element, this URI points to should be in the place of the object containing this attribute.	

D

The RelaxNG Schema for OMDoc

We reprint the modularized RELAXNG schema for OMDOC here. It is available at `http://www.mathweb.org/omdoc/rnc` and consists of separate files for the OMDOC modules, which are loaded by the schema driver `omdoc.rnc` in this directory. We will use the abbreviated syntax for RELAXNG here, since the XML syntax, document type definitions and even XML schemata can be generated from it by standard tools.

The RELAXNG schema consists of the grammar fragments for the modules (see Appendices D.3 to D.14), a definition of the most common attributes that occur in several of the modules (see Appendix D.2), and the sub-language driver files which we will introduce next.

D.1 The Sub-language Drivers

The driver files set up the grammars for the OMDOC sub-languages (see Section 22.3 for a discussion) in layers. The RELAXNG grammar for "Basic OMDOC" sets up the language and loads the relevant modules.

```
   # A RelaxNG schema for Open Mathematical documents (OMDoc 1.2: OMDoc Basic)
   #   SYSTEM http://www.mathweb.org/omdoc/rnc/omdoc−basic.rnc
   #   PUBLIC −//OMDoc//RNC OMDoc Basic V1.2//EN
   # See the documentation and examples at http://www.mathweb.org/omdoc
 5 # (c) 2004 Michael Kohlhase, released under the GNU Public License (GPL)

   start = omdoc

   include "omdocattribs.rnc"
10 include "omdocmobj.rnc"
   include "omdocdoc.rnc"
   include "omdocdc.rnc"
   include "omdoccc.rnc"
   include "omdocmtxt.rnc"
15 include "omdocrt.rnc"
```

The RELAXNG grammar for "Content Dictionary OMDOC" adds modules PRES and ST.

```
# A RelaxNG for Open Mathematical documents (OMDoc 1.2: OMDoc Content Dictionaries)
#    SYSTEM http://www.mathweb.org/omdoc/rnc/omdoc−cd.rnc
#    PUBLIC −//OMDoc//RNC OMDoc CD 1.2//EN
# See the documentation and examples at http://www.mathweb.org/omdoc
# (c) 2004 Michael Kohlhase, released under the GNU Public License (GPL)

include "omdoc−basic.rnc"
include "omdocpres.rnc"
include "omdocst.rnc"
```

The RELAXNG grammar for "Educational OMDOC" adds modules PF and QUIZ to that:

```
# A RelaxNG for Open Mathematical documents (OMDoc 1.2: OMDoc Education)
#    SYSTEM http://www.mathweb.org/omdoc/rnc/omdoc−education.rnc
#    PUBLIC −//OMDoc//RNC OMDoc Education V1.2//EN
# See the documentation and examples at http://www.mathweb.org/omdoc
# (c) 2004 Michael Kohlhase, released under the GNU Public License (GPL)

include "omdoc−mathweb.rnc"
include "omdocquiz.rnc"
```

The RELAXNG grammar for "Educational OMDOC" starts with "Content Dictionary OMDOC" adds modules PF and EXT:

```
# A RelaxNG for Open Mathematical documents (OMDoc 1.2: OMDoc MathWeb)
# SYSTEM http://www.mathweb.org/omdoc/rnc/omdoc−mathweb.rnc
# PUBLIC −//OMDoc//RNC OMDoc MathWeb V1.2//EN
# See the documentation and examples at http://www.mathweb.org/omdoc
# (c) 2004 Michael Kohlhase, released under the GNU Public License (GPL)

include "omdoc−cd.rnc"
include "omdocext.rnc"
include "omdocpf.rnc"
```

The RELAXNG grammar for "Educational OMDOC" starts with "Content Dictionary OMDOC" adds modules PF and EXT:

```
# A RelaxNG schema for Open Mathematical documents (OMDoc 1.2: OMDoc Specification)
#    SYSTEM http://www.mathweb.org/omdoc/rnc/omdoc−spec.rnc
#    PUBLIC −//OMDoc//RNC OMDoc Spec V1.2//EN
# See the documentation and examples at http://www.mathweb.org/omdoc
# (c) 2004 Michael Kohlhase, released under the GNU Public License (GPL)

# default namespace omdoc = "http://www.mathweb.org/omdoc"

include "omdoc−cd.rnc"
include "omdoccth.rnc"
include "omdocdg.rnc"
include "omdocpf.rnc"
include "omdocadt.rnc"
```

Finally, the The RELAXNG grammar for full OMDOC only needs to add modules EXT and QUIZ:

```
# A RelaxNG schema for Open Mathematical documents (OMDoc 1.2)
#    SYSTEM http://www.mathweb.org/omdoc/rnc/omdoc.rnc
#    PUBLIC −//OMDoc//RNC OMDoc V1.2//EN
# See the documentation and examples at http://www.mathweb.org/omdoc
# (c) 2004 Michael Kohlhase, released under the GNU Public License (GPL)

include "omdoc−spec.rnc"
include "omdocext.rnc"
include "omdocquiz.rnc"
```

D.2   Common Attributes

The RELAXNG grammar for OMDOC separates out declarations for commonly used attributes.

```
   # A RelaxNG schema for Open Mathematical documents (OMDoc 1.2) Common attributes
   #   SYSTEM http://www.mathweb.org/omdoc/rnc/omdocmobj.rnc
   #   PUBLIC −//OMDoc//RNC OMDoc ATTRIBS V1.2//EN
   # See the documentation and examples at http://www.mathweb.org/omdoc
5  # (c) 2004 Michael Kohlhase, released under the GNU Public License (GPL)

   default namespace omdoc = "http://www.mathweb.org/omdoc"
   namespace local = ""

10 # all the explicitly namespaced attributes, except xml:lang, which
   # is handled explicitly
   nonlocal−attribs = attribute * − ( local: * | xml:*) {xsd:string}

   # the attributes for CSS and PRES styling
15 css. attribs = attribute style {xsd:string}?, attribute class {xsd:string}?

   omdocref = xsd:anyURI       # an URI reference pointing to an OMDoc fragment
   omdocrefs = list {xsd:anyURI*} # a whitespace−separated list of omdocref

20 xref. attrib = attribute xref {omdocref}
   idrest. attribs = css. attribs , nonlocal−attribs*, attribute xml:base {xsd:anyURI}?
   id. attrib = attribute xml:id {xsd:ID}?, idrest. attribs

   omdoc.toplevel. attribs = id. attrib , attribute generated−from {omdocref}?
25
   # The current XML−recommendation doesn't yet support the
   # three−letter short names for languages (ISO 693−2). So
   # the following section will be using the two−letter
   # (ISO 693−1) encoding for the languages.
30 #
   #       en : English,   de : German,   fr : French,
   #       la : Latin,     it : Italian ,  nl : Dutch,
   #       ru : Russian,   pl : Polish ,   es : Spanish,
   #       tr : Turkish,   zh : Chinese,   ja : Japanese,
35 #       ko : Korean    ...
   iso639 = "aa" | "ab" | "af" | "am" | "ar" | "as" |
   "ay" | "az" | "ba" | "be" | "bg" | "bh" | "bi" | "bn" | "bo" | "br" | "ca" | "co"
   | "cs" | "cy" | "da" | "de" | "dz" | "el" | "en" | "eo" | "es" | "et" | "eu" |
40 "fa" | "fi" | "fj" | "fo" | "fr" | "fy" | "ga" | "gd" | "gl" | "gn" | "gu" | "ha"
   | "he" | "hi" | "hr" | "hu" | "hy" | "ia" | "ie" | "ik" | "id" | "is" | "it" |
   "iu" | "ja" | "jv" | "ka" | "kk" | "kl" | "km" | "kn" | "ko" | "ks" | "ku" | "ky"
   | "la" | "ln" | "lo" | "lt" | "lv" | "mg" | "mi" | "mk" | "ml" | "mn" | "mo" |
   "mr" | "ms" | "mt" | "my" | "na" | "ne" | "nl" | "no" | "oc" | "om" | "or" | "pa"
45 | "pl" | "ps" | "pt" | "qu" | "rm" | "rn" | "ro" | "ru" | "rw" | "sa" | "sd" |
   "sg" | "sh" | "si" | "sk" | "sl" | "sm" | "sn" | "so" | "sq" | "sr" | "ss" | "st"
   | "su" | "sv" | "sw" | "ta" | "te" | "tg" | "th" | "ti" | "tk" | "tl" | "tn" |
   "to" | "tr" | "ts" | "tt" | "tw" | "ug" | "uk" | "ur" | "uz" | "vi" | "vo" | "wo"
   | "xh" | "yi" | "yo" | "za" | "zh" | "zu"
50
   xml.lang. attrib   = attribute xml:lang {iso639}?
```

D.3   Module MOBJ: Mathematical Objects and Text

The RNC module MOBJ includes the representations for mathematical objects and defines the `legacy` element (see Chapter 13 for a discussion). It includes the standard RELAXNG schema for OPENMATH (we have reprinted it in Appendix E.1) adding the OMDOC identifier and CSS attributes to all elements. If also includes a schema for MATHML (see Appendix E.2).

```
     # A RelaxNG schema for Open Mathematical documents (OMDoc 1.2) Module MOBJ
     #   SYSTEM http://www.mathweb.org/omdoc/rnc/omdocmobj.rnc
     #   PUBLIC −//OMDoc//RNC OMDoc MOBJ V1.2//EN
     # See the documentation and examples at http://www.mathweb.org/omdoc
 5   # (c) 2004 Michael Kohlhase, released under the GNU Public License (GPL)

     default namespace omdoc = "http://www.mathweb.org/omdoc"

     namespace om = "http://www.openmath.org/OpenMath"
10
     # we iclude the OpenMath 2 schema, but we also allow CSS attributes, etc.
     include "openmath2.rnc" {common.attributes = attribute id {xsd:ID}?,idrest.attribs}

     # we include the MathML2 schema
15   include "mathml2/mathml2.rnc"

     # the legacy element, it can encapsulate the non−migrated formats
     legacy = (ss | element legacy {id. attrib ,
                               attribute formalism {xsd:anyURI}?,
20                             attribute format {xsd:anyURI},
                               Anything}) # to allow everything

     omdocmobj.class = legacy | OMOBJ | math
```

D.4   Module MTXT: Mathematical Text

The RNC module MTXT provides infrastructure for mathematical vernacular (see Chapter 14 for a discussion).

```
     # A RelaxNG schema for Open Mathematical documents (OMDoc 1.2) Module MTXT
     #   SYSTEM http://www.mathweb.org/omdoc/dtd/omdocmtext.dtd
     #   PUBLIC −//OMDoc//RNC OMDoc MTXT V1.2//EN
     # See the documentation and examples at http://www.mathweb.org/omdoc
 5   # (c) 2004 Michael Kohlhase, released under the GNU Public License (GPL)

     default namespace omdoc = "http://www.mathweb.org/omdoc"

     omdoc.class |= omtext
10
     #attribute for is a whitespace−separated list of  URIrefs
     for. attrib = attribute for {omdocrefs}
     fori . attrib = attribute for {omdocrefs}?
     from.attrib = attribute from {omdocref}
15   verbalizes . attrib = attribute verbalizes {omdocrefs}
     parallel . attribs = verbalizes. attrib ?, attribute index {xsd:NMTOKEN}?
     omdocmtxt.MC.content = metadata?,CMP∗
     omdocmtxt.MCF.content = omdocmtxt.MC.content,FMP∗

20   # what can go into a mathematical text (to be extended in other modules)
     omdoc.mtext.class = text | phrase | term | omdocmobj.class

     rsttype = "abstract" | "introduction" | "annote" |
            "conclusion" | " thesis " | " comment" | "antithesis" |
25          "elaboration" | " motivation" | "evidence" | "note" | " notation"

     statementtype = "axiom" | "definition" | "example" | "proof" |
```

```
            "derive" | " hypothesis"

30  assertiontype = "theorem" | "lemma" | "corollary" | "proposition" |
                    "conjecture" | " false −conjecture" | "obligation " |
                    "postulate" | " formula" | "assumption" | "rule"

    omtext = element omtext {omdoc.toplevel.attribs,
35                          attribute type {(rsttype | statementtype | assertiontype)}?,
                            attribute for {omdocref}?,
                            attribute from {omdocref}?,
                            verbalizes. attrib ?,
                            metadata?,CMP+,FMP*}
40  # attribute 'for' is a URIref, to omdocdoc.class's it is needed by the 'type' attribute

    CMP = (ss| element CMP {xml.lang.attrib, id.attrib, (omdoc.mtext.class)*})

    phrase = (ss| element phrase {id.attrib , parallel . attribs ,
45                              attribute type {xsd:string}?,
                               (omdoc.mtext.class)*})
    # identifies a text passage and allows to attatch style and role information to it

    term = (ss| element term {id.attrib ,
50                          attribute role {text}?,
                            attribute cd {xsd:NCName},
                            attribute name {xsd:NCName},
                            (omdoc.mtext.class)*})

55  FMP = (ss| element FMP {id.attrib, attribute logic {xsd:NMTOKEN}?,
                           ((assumption*,conclusion*)|omdocmobj.class)})

    # If FMP contains a omdocmobj.class then this is the assertion,
    # if it contains (assumption*,conclusion*), then it is a
60  # logical sequent (A1 ,..., An |− C1,...,Cm):
    # all the Ai entail one of the Ci

    assumption = (ss| element assumption {id.attrib,
                                attribute inductive {"yes" | " no"}?,
65                              (omdocmobj.class)})
    conclusion = (ss| element conclusion {id. attrib , ( omdocmobj.class?)})
```

D.5 Module DOC: Document Infrastructure

The RNC module DOC specifies the document infrastructure of OMDoc
documents (see Chapter 11 for a discussion).

```
    # A RelaxNG for Open Mathematical documents (OMDoc 1.2) Module DOC
    #   SYSTEM http://www.mathweb.org/omdoc/rnc/omdocdoc.rnc
    #   PUBLIC −//OMDoc//RNC OMDoc DOC V1.2//EN
    # See the documentation and examples at http://www.mathweb.org/omdoc
5   # (c) 2004 Michael Kohlhase, released under the GNU Public License (GPL)

    default namespace omdoc = "http://www.mathweb.org/omdoc"
    # extend the stuff that can go into a mathematical text
    omdoc.mtext.class |= ignore | ref
10
    ss = ignore | ref
    omdoc.class |= ss | omgroup
    omdoc.meta.class |= notAllowed

15  metadata = element metadata {id.attrib,
                                attribute inherits {omdocref}?,
```

```
                           (omdoc.meta.class)*}

        Anything = (AnyElement|text)*
20      AnyElement = element * {AnyAttribute,(text|AnyElement)*}
        AnyAttribute = attribute * { text }*

        # this element can be used in lieu of a comment, it is read
        # by the style sheet, (comments are not) and can therefore
25      # be transformed by them

        ignore = element ignore {attribute type {xsd:string}?,
                                 attribute comment {xsd:string}?,
                                 Anything}
30
        ref = element ref {id.attrib,
                           xref.attrib,
                           attribute type {xsd:string}?}

35      # the types supported (there may be more over time) are
        # - 'include' (the default) for in−text replacement
        # - 'cite' for a reference with a generated label

        group.attribs = attribute type {xsd:anyURI}?, attribute modules {xsd:anyURI}?
40
        group.elts = metadata?,(omdoc.class)*

        # grouping defines the structure of a document
        omgroup = element omgroup {group.attribs,omdoc.toplevel.attribs,group.elts}
45
        # finally the definition of the OMDoc root element
        omdoc = element omdoc {omdoc.toplevel.attribs,group.attribs,
                               attribute version {xsd:string {pattern = "1.2"}}?,
                               group.elts}
```

D.6   Module DC: Dublin Core Metadata

The RNC module DC includes an extension of the Dublin Core vocabulary
for bibliographic metadata, see Sections 12.1 and 12.2 for a discussion.

```
        # A RelaxNG schema for Open Mathematical documents (OMDoc 1.2) Module DC
        #   SYSTEM http://www.mathweb.org/omdoc/rnc/omdocdc.rnc
        #   PUBLIC −//OMDoc//RNC OMDoc DC V1.2//EN
        # See the documentation and examples at http://www.mathweb.org/omdoc
5       # (c) 2004 Michael Kohlhase, released under the GNU Public License (GPL)

        default namespace dc = "http://purl.org/dc/elements/1.1/"

        # Persons in Dublin Core Metadata
10      omdocdc.person.content = text
        # the rest of Dublin Core content
        omdocdc.rest.content = (text | AnyElement)*

        omdoc.meta.class |= ss | dc.contributor | dc.creator | dc.rights
15                          | dc.subject | dc.title | dc.description | dc.publisher
                            | dc.date | dc.type | dc.format | dc.identifier
                            | dc.source | dc.language | dc.relation

        # the MARC relator set; see http://www.loc.gov/marc/relators
20      dcrole = attribute role {"act" | "adp" | "aft" | "ann" | "ant" | "app" | "aqt" |
                    "arc" | "arr" | "art" | "asg" | "asn" | "att" | "auc" | "aud" | "aui" |
                    "aus" | "aut" | "bdd" | "bjd" | "bkd" | "bkp" | "bnd" | "bpd" | "bsl" |
```

```
         "ccp" | "chr" | "clb" | "cli " | "cll " | "clt " | "cmm" | "cmp" | "cmt" |
         "cnd" | "cns" | "coe" | "col" | "com" | "cos" | "cot" | "cov" | "cpc" |
25       "cpe" | "cph" | "cpl" | "cpt" | "cre" | "crp" | "crr" | "csl" | "csp" |
         "cst" | "ctb" | "cte" | "ctg" | "ctr" | "cts" | "ctt" | "cur" | "cwt" |
         "dfd" | "dfe" | "dft" | "dgg" | "dis" | "dln" | "dnc" | "dnr" | "dpc" |
         "dpt" | "drm" | "drt" | "dsr" | "dst" | "dte" | "dto" | "dub" | "edt" |
         "egr" | "elt" | "eng" | "etr" | "exp" | "fac" | "flm" | "fmo" | "fnd" |
30       "fpy" | "frg" | "hnr" | "hst" | "ill " | "ilu " | "ins" | "inv" | "itr" |
         "ive " | "ivr " | "lbt" | "lee " | "lel " | "len" | "let" | "lie " | "lil " |
         "lit " | "lsa " | "lse " | "lso " | "ltg " | "lyr" | "mdc" | "mod" | "mon" |
         "mrk" | "mte" | "mus" | "nrt" | "opn" | "org" | "orm" | "oth" | "own" |
         "pat" | "pbd" | "pbl" | "pfr" | "pht" | "plt " | "pop" | "ppm" | "prc" |
35       "prd" | "prf" | "prg" | "prm" | "pro" | "prt" | "pta" | "pte" | "ptf" |
         "pth" | "ptt" | "rbr" | "rce " | "rcp" | "red" | "ren" | "res " | "rev" |
         "rpt" | "rpy" | "rse " | "rsp" | "rst " | "rth" | "rtm" | "sad" | "sce " |
         "scl " | "scr " | "sec" | "sgn" | "sng" | "spk" | "spn" | "spy" | "srv" |
         "stl " | "stn" | "str " | "ths" | "trc " | "trl " | "tyd" | "tyg" | "voc" |
40       "wam" | "wdc" | "wde" | "wit"}?
      dclang = id.attrib , xml.lang.attrib

      # first the Dublin Core Metadata model of the
      # Dublin Metadata initiative (http://purl.org/dc)
45
      dc.contributor = element contributor {dclang,dcrole,omdocdc.person.content}
      dc.creator = element creator {dclang,dcrole,omdocdc.person.content}
      dc.title = element title {dclang,(omdoc.mtext.class)*}
      dc.subject = element subject {dclang,(omdoc.mtext.class)*}
50    dc.description = element description {dclang,(omdoc.mtext.class)*}
      dc.publisher = element publisher {id.attrib,omdocdc.rest.content}
      dc.type = element type {("Dataset" | "Text" | "Collection")}
      dc.format = element format {("application/omdoc+xml")}
      dc.source = element source {omdocdc.rest.content}
55    dc.language = element language {omdocdc.rest.content}
      dc.relation = element relation {omdocdc.rest.content}
      dc.rights = element rights {omdocdc.rest.content}
      dc.date = element date {attribute action {xsd:NMTOKEN}?,
                              attribute who {omdocref}?,
60                            xsd:dateTime}
      dc.identifier = element identifier {attribute scheme {xsd:NMTOKEN},text}
```

D.7 Module ST: Mathematical Statements

The RNC module ST deals with mathematical statements like assertions and examples in OMDOC and provides an infrastructure for mathematical theories as contexts, for the OMDOC elements that fix the meaning for symbols, see Chapter 15 for a discussion.

```
      # A RelaxNG schema for Open Mathematical documents (OMDoc 1.2) Module ST
      #   SYSTEM http://www.mathweb.org/omdoc/rnc/omdocst.rnc
      #   PUBLIC -//OMDoc//RNC OMDoc ST V1.2//EN
      # See the documentation and examples at http://www.mathweb.org/omdoc
5     # (c) 2004 Michael Kohlhase, released under the GNU Public License (GPL)

      default namespace omdoc = "http://www.mathweb.org/omdoc"

      omdocst.scope.attrib = attribute scope {"global" | "local"}?
10
      omdocst.constitutive.class    = symbol | axiom | definition | imports
      omdocst.nonconstitutive.class = assertion | type | alternative | example | theory
```

```
     theory−unique = xsd:NCName
15   just−by.attrib = attribute just−by {omdocref}

     omdoc.class |= omdocst.nonconstitutive.class

     omdocst.constitutive. attribs = id. attrib , attribute generated−from {omdocref}?
20
     sym.role. attrib = attribute role {"type" | "sort" | "object" |
                                        "binder" | "attribution" |
                                        "semantic−attribution" | "error"}

25   symbol = element symbol {omdocst.scope.attrib,
                             attribute name {theory−unique}?,
                             omdocst.constitutive. attribs ,
                             sym.role. attrib ?,
                             metadata?,type∗}
30
     axiom = element axiom {omdocst.constitutive.attribs, fori. attrib ,
                           attribute type {xsd:string}?,omdocmtxt.MCF.content}


     #informal definitions
35   def.informal = attribute type {"informal"}?


     #simple definitions
     exists . attrib = attribute existence {omdocref}
     def.simple = (attribute type {"simple"},exists. attrib ?,(omdocmobj.class))
40
     #implicit definitions
     unique.attrib = attribute uniqueness {omdocref}
     def. implicit = (attribute type {"implicit"}, exists . attrib ?, unique.attrib ?, FMP∗)

45   #definitions by (recursive) equations
     exhaust. attrib = attribute exhaustivity {omdocref}
     consist . attrib = attribute consistency {omdocref}
     def.eq = attribute type {"pattern"|"inductive"}?,exhaust.attrib?, consist . attrib ?,
              requation+,measure?,ordering?
50
     #all definition forms, add more by extending this.
     defs. all = def.informal|def.simple|def. implicit |def.eq


     # Definitions contain CMPs, FMPs and concept specifications.
55   # The latter define the set of concepts defined in this element.
     # They can be reached under this name in the content dictionary
     # of the name specified in the theory attribute of the definition .
     definition = element definition {omdocst.constitutive. attribs , for. attrib ,
                                     omdocmtxt.MC.content,(defs.all)}
60
     requation = (ss| element requation {id. attrib ,omdocmobj.class,omdocmobj.class})
     measure = (ss| element measure {id.attrib,omdocmobj.class})
     ordering = (ss| element ordering {id. attrib , attribute terminating {omdocref}?,omdocmobj.class})

65   # the non−constitutive statements, they need a theory attribute
     omdoc.toplevel.attribs &= attribute theory {omdocref}?

     ded.status. class = " satisfiable " | " counter−satisfiable " | " no−consequence" |
                         "theorem" | "conter−theorem" | "contradictory−axioms" |
70                       "tautologous−conclusion" | " tautology" | "equivalent" |
                         "conunter−equivalent" | "unsatisfiable−conclusion" | " unsatisfiable"


     assertion = element assertion {omdoc.toplevel.attribs ,
                                    attribute type {assertiontype}?,
75                                  attribute status {ded.status. class }?,
                                    attribute just−by {omdocrefs}?,
                                    omdocmtxt.MCF.content}
     # the assertiontype has no formal meaning yet, it is solely for human consumption.
     # 'just−by' is a list of URIRefs that point to proof objects, etc that justifies the status.
```

```
80    type = element type {omdoc.toplevel.attribs, just−by.attrib?,
                            attribute system {omdocref}?,
                            omdocmtxt.MC.content,
                            (omdocmobj.class),
85                          (omdocmobj.class)?}

      alternative = element alternative {omdoc.toplevel.attribs , for. attrib ,
                                         omdocmtxt.MC.content,(defs.all),
                                         ((attribute equivalence {omdocref},
90                                          attribute equivalence−thm {omdocref}) |
                                          (attribute entailed−by {omdocref},
                                           attribute entails {omdocref},
                                           attribute entailed−by−thm {omdocref},
                                           attribute entails −thm {omdocref}))}
95    # just−by, points to the theorem justifying well−definedness
      # entailed−by, entails, point to other (equivalent definitions
      # entailed−by−thm, entails−thm point to the theorems justifying
      # the entailment relation)

100   example = element example {omdoc.toplevel.attribs, for.attrib,
                                 attribute type {"for" | "against" }?,
                                 attribute assertion {omdocref}?,
                                 omdocmtxt.MC.content,
                                 (omdocmobj.class)∗}
105   theory = element theory {id.attrib,
                              attribute cdurl {xsd:anyURI}?,
                              attribute cdbase {xsd:anyURI}?,
                              attribute cdreviewdate {xsd:date}?,
                              attribute cdversion {xsd:nonNegativeInteger}?,
110                           attribute cdrevision {xsd:nonNegativeInteger}?,
                              attribute cdstatus {" official " | " experimental"
                                               |"private" |   "obsolete"}?,
                              metadata?,
                              (omdoc.class | omdocst.constitutive. class | tgroup)∗}
115
      omdocsth.imports.model = id.attrib,from.attrib,metadata?
      imports = (ss| element imports {omdocsth.imports.model})

      tgroup = element tgroup {omdocst.constitutive.attribs,group.attribs,
120                                       metadata?,
                                         (omdoc.class | omdocst.constitutive. class | tgroup)∗}
```

D.8 Module ADT: Abstract Data Types

The RNC module ADT specifies the grammar for abstract data types in
OMDOC, see Chapter 16 for a discussion.

```
    # A RelaxNG schema for Open Mathematical documents (OMDoc 1.2) Module ADT
    #   SYSTEM http://www.mathweb.org/omdoc/rnc/omdocadt.rnc
    #   PUBLIC −//OMDoc//RNC OMDoc ADT V1.2//EN
    # See the documentation and examples at http://www.mathweb.org/omdoc
5   # (c) 2004 Michael Kohlhase, released under the GNU Public License (GPL)

    default namespace omdoc = "http://www.mathweb.org/omdoc"
    omdoc.class |= adt

10  omdocadt.sym.attrib = id.attrib,omdocst.scope.attrib,attribute name {xsd:NCName}

    # adts are abstract data types, they are short forms for groups of symbols
    # and their definitions , therefore , they have much the same attributes.
```

```
15  adt = element adt {omdoc.toplevel.attribs,
                        attribute parameters {list {xsd:NCName*}}?, metadata?, sortdef+}

    adttype = "loose" | "generated" | " free"
    sortdef = (ss| element sortdef {omdocadt.sym.attrib,
20                                  attribute role {"sort"}?,
                                    attribute type {adttype}?,
                                    metadata?,(constructor | insort )*, recognizer?})
    insort = (ss| element insort { attribute for {omdocref}})
    # for is a reference to a sort symbol element

25
    constructor = (ss| element constructor {omdocadt.sym.attrib,
                                            sym.role. attrib ?,
                                            metadata?,argument*})
    recognizer = (ss| element recognizer {omdocadt.sym.attrib,
30                                         sym.role. attrib ?,
                                          metadata?})


    argument = (ss| element argument {type,selector?})
35  # sort is a reference to a sort symbol element p

    selector = (ss| element selector {omdocadt.sym.attrib,
                                      sym.role. attrib ?,
                                      attribute total {"yes" | " no"}?,
40                                    metadata?})
```

D.9   Module PF: Proofs and Proof objects

The RNC module PF deals with mathematical argumentations and proofs in
OMDoc, see Chapter 17 for a discussion.

```
    # A RelaxNG schema for Open Mathematical documents (OMDoc 1.2) Module PF
    #   SYSTEM http://www.mathweb.org/omdoc/rnc/omdocpf.rnc
    #   PUBLIC −//OMDoc//RNC OMDoc PF V1.2//EN
    # See the documentation and examples at http://www.mathweb.org/omdoc
5   # (c) 2004 Michael Kohlhase, released under the GNU Public License (GPL)

    default namespace omdoc = "http://www.mathweb.org/omdoc"

    omdocpf.opt.content |= proof | proofobject
10  omdoc.class         |= proof | proofobject

    proof = element proof {omdoc.toplevel.attribs, fori . attrib ,
                            metadata?,(omtext|symbol|definition|derive|hypothesis)*}
    proofobject = element proofobject {omdoc.toplevel.attribs, fori . attrib ,
15                                      metadata?,(omdocmobj.class)}
    omdocpf.just.content = method?, premise*, (proof | proofobject)*

    derive.type. attr = attribute type {("conclusion" | "gap")}

20  derive     = (ss| element derive {id. attrib ,derive.type. attr ?,
                                      omdocmtxt.MCF.content,method?})
    hypothesis = (ss| element hypothesis {id. attrib ,
                                          attribute inductive {"yes" | " no"}?,
                                          omdocmtxt.MCF.content})
25
    method = (ss| element method {xref.attrib?, (omdocmobj.class|premise|proof|proofobject)*})
    # 'xref' is a pointer to the element defining the method
```

```
     premise = (ss| element premise {xref. attrib ,
30                     attribute rank {xsd:string {pattern = "0|[1−9][0−9*]"}}?})

     # The rank of a premise specifies its importance in the inference rule.
     # Rank 0 (the default) is a real premise, whereas positive rank signifies
     # sideconditions of varying degree.
```

D.10   Module CTH: Complex Theories

The RNC presented in this section deals with the module CTH of complex
theories (see Chapter 18 for a discussion).

```
    # A RelaxNG schema for Open Mathematical documents (OMDoc 1.2) Module CTH
    #   SYSTEM http://www.mathweb.org/omdoc/rnc/omdoccth.rnc
    #   PUBLIC −//OMDoc//RNC OMDoc CTH V1.2//EN
    # See the documentation and examples at http://www.mathweb.org/omdoc
5   # (c) 2004 Michael Kohlhase, released under the GNU Public License (GPL)

    default namespace omdoc = "http://www.mathweb.org/omdoc"

    omdocst.constitutive. class |= inclusion
10  omdocsth.imports.model &= morphism?,
                        attribute type { "local" | "global" }?,
                        attribute conservativity {"conservative" | "monomorphism" | "definitional"}?,
                        attribute conservativity −just {omdocref}?

15  omdoc.toplevel. attribs      &= attribute generated−via {omdocref}?
    omdocst.constitutive. attribs &= attribute generated−via {omdocref}?

    omdoc.class |= theory−inclusion | axiom−inclusion
    omdoccth.theory − inclusion.justification = obligation∗
20  omdoccth.axiom − inclusion.justification = obligation∗

    fromto. attrib = from.attrib , attribute to {omdocref}
    # attributes 'to' and 'from' are URIref

25  morphism − (ss| element morphism {id.attrib,attribute hiding {omdocrefs}?, attribute base {omdocrefs}?,def.eq?})
    # base points to some other morphism it extends

    inclusion = element inclusion {id. attrib , attribute via {omdocref}}
    # via points to a theory−inclusion
30
    theory−inclusion = element theory−inclusion {omdoc.toplevel.attribs,fromto.attrib,
                                    metadata?,morphism?,
                                    (omdoccth.theory−inclusion.justification)}

35  axiom−inclusion = element axiom−inclusion {omdoc.toplevel.attribs,fromto.attrib,
                                    metadata?,morphism?,
                                    (omdoccth.axiom−inclusion.justification)}

    obligation = (ss| element obligation {id. attrib ,
40                      attribute induced−by {omdocref},
                        attribute assertion {omdocref}})

    # attribute 'assertion' is a URIref, points to an assertion
    # that is the proof obligation induced by the axiom or definition
45  # specified by 'induced−by'.
```

D.11   Module RT: Rich Text Structure

The RNC module RT provides text structuring elements for mathematical text below the level of mathematical statements (see Section 14.6 for a discussion).

```
    # A RelaxNG schema for Open Mathematical documents (OMDoc 1.2) Module DOC
    #   SYSTEM http://www.mathweb.org/omdoc/rnc/omdocdoc.rnc
    #   PUBLIC −//OMDoc//RNC OMDoc RT V1.2//EN
    # See the documentation and examples at http://www.mathweb.org/omdoc
5   # (c) 2004 Michael Kohlhase, released under the GNU Public License (GPL)

    default namespace omdoc = "http://www.mathweb.org/omdoc"

    omdoc.mtext.class |= ss|ul|ol|dl|p|note|link|table|idx
10
    omdocrt.common.attrib = id.attrib, fori. attrib , parallel . attribs

    ul = element ul {omdocrt.common.attrib, metadata?,li+}
    ol = element ol {omdocrt.common.attrib, metadata?,li+}
15  dl = element dl {omdocrt.common.attrib, metadata?,di+}
    li = element li {omdocrt.common.attrib, metadata?,(omdoc.mtext.class)*}
    di = element di {omdocrt.common.attrib, metadata?,dt+,dd*}
    dt = element dt {omdocrt.common.attrib, metadata?,(omdoc.mtext.class)*}
    dd = element dd {omdocrt.common.attrib, metadata?,(omdoc.mtext.class)*}
20

    p = element p {omdocrt.common.attrib, (omdoc.mtext.class)*}
    note = element note {omdocrt.common.attrib,
                         attribute type {xsd:NMTOKEN}?,
25                       (omdoc.mtext.class)*}


    # a simplified html table
    table = element table {omdocrt.common.attrib, tr+}
30  tr = (ss| element tr {omdocrt.common.attrib, (td|th)+})
    td = (ss| element td {omdocrt.common.attrib, (omdoc.mtext.class)*})
    th = (ss| element th {omdocrt.common.attrib, (omdoc.mtext.class)*})

    link = element link {omdocrt.common.attrib,
35                        attribute href {xsd:anyURI},
                          (omdoc.mtext.class)*}

    # index
    index.att = attribute sort−by {text}?,
40              attribute see {omdocrefs}?,
                attribute seealso {omdocrefs}?,
                attribute links { list {xsd:anyURI*}}?
    idx = element idx {(id. attrib | xref. attrib ), idt ?, ide+}
    ide = element ide {attribute index {xsd:NCName},index.att,idp*}
45  idt = element idt { idrest . attribs ,omdoc.mtext.class*}
    idp = element idp {index.att,omdoc.mtext.class*}
```

D.12   Module EXT: Applets and Non-XML Data

The RNC module EXT provides an infrastructure for applets, program code, and non-XML data like images or measurements (see Chapter 20 for a discussion).

```
   # A RelaxNG schema for Open Mathematical documents (OMDoc 1.2) Module EXT
   #    SYSTEM http://www.mathweb.org/omdoc/rnc/omdocext.rnc
   #    PUBLIC −//OMDoc//RNC OMDoc EXT V1.2//EN
   # See the documentation and examples at http://www.mathweb.org/omdoc
5  # (c) 2004 Michael Kohlhase, released under the GNU Public License (GPL)

   default namespace omdoc = "http://www.mathweb.org/omdoc"

   omdoc.mtext.class |= omlet
10 omdocext.class  = private | code | omlet
   omdoc.class |= omdocext.class

   omdocext.private.attrib = fori . attrib , attribute requires {omdocref}?

15 private = element private {omdoc.toplevel.attribs ,omdocext.private.attrib,
                             attribute reformulates {omdocref}?,
                    metadata?,data+}
   # reformulates is a URIref to the omdoc elements that are reformulated by the
   # system−specific information in this element
20 code = element code {omdoc.toplevel.attribs,omdocext.private.attrib,
                        (metadata?,data+,input?,output?,effect?)}
   input  = (ss| element input {id. attrib , ( omdocmtxt.MCF.content)})
   output = (ss| element output {id.attrib , ( omdocmtxt.MCF.content)})
   effect = (ss| element effect {id. attrib , ( omdocmtxt.MCF.content)})
25
   data = (ss| element data {id.attrib ,
                            attribute format {xsd:string}?,
                            attribute href {xsd:anyURI}?,
                            attribute size { xsd:string }?,
30              attribute pto {xsd:string }?,
                            attribute pto−version {xsd:string}?,
                            attribute original {"external" | " local"}?,
                Anything})

35 omlet = (ss| element omlet {id.attrib ,
                            attribute action    {"display" | " execute" | " other"}?,
                            attribute show      {"new" | "replace" | "embed" | "other"}?,
                            attribute actuate  {"onPresent" | "onLoad" | "onRequest" | "other"}?,
                            (omdoc.mtext.class | param)∗,
40              (attribute data {xsd:anyURI}|(private|code))})

   param = (ss| element param {id.attrib ,
                attribute name     {xsd:string },
                attribute value    {xsd:string }?,
45 attribute valuetype {"data" | "ref" | " object"}?,
                (omdocmobj.class)?})
```

D.13   Module PRES: Adding Presentation Information

The RNC module PRES provides a sub-language for defining notations for mathematical symbols and for styling OMDoc elements (see Chapter 19 for a discussion).

```
   # A RelaxNG for Open Mathematical documents (OMDoc 1.2) Module PRES
   #    SYSTEM http://www.mathweb.org/omdoc/rnc/omdocpres.rnc
   #    PUBLIC −//OMDoc//RNC OMDoc PRES 1.2//EN
   # See the documentation and examples at http://www.mathweb.org/omdoc
5  # (c) 2004 Michael Kohlhase, released under the GNU Public License (GPL)

   default namespace omdoc = "http://www.mathweb.org/omdoc"
```

```
       # we include the OpenMath 2 schema, but we also allow CSS attributes, etc.
10     include "xslt10.rnc"

       #xslt10 = external "xslt10.rnc"

       omdoc.class |= presentation |omstyle
15
       crossref.attrib = attribute crossref−symbol
                             {"no" | "yes" | "brackets" | "separator" | "lbrack" | "rbrack" | "all"}
       fixity.attrib = attribute fixity {"prefix" | "infix" | "postfix" | "assoc" | "infixl" | "infixr"}
       format.attrib = attribute format {xsd:string}, attribute requires {omdocref}?, xml.lang.attrib
20     bracket−style.attrib = attribute bracket−style {"lisp" | "math"}
       lbrack.attrib = attribute lbrack {xsd:string}
       rbrack.attrib = attribute rbrack {xsd:string}
       separator.attrib = attribute separator {xsd:string}
       precedence.attrib = attribute precedence {xsd:string {pattern = "0|[1−9][0−9]*"}}
25     role.attrib = attribute role {"applied" | "binding" | "key"}

       presentation = element presentation {omdoc.toplevel.attribs,
                                            attribute for {omdocref},
                                            role.attrib?,
30                                           (xref.attrib?|
                                            (fixity.attrib?,
                                            lbrack.attrib?,
                                            rbrack.attrib?,
                                            separator.attrib?,
35                                           bracket−style.attrib?,
                                            precedence.attrib?,
                                            crossref.attrib?,
                                            CMP*,
                                            (use | xslt | style)*))}
40
       omdocpres.use.mix = elt | txt | recurse | value−of | map

       use = (ss| element use {format.attrib,
              bracket−style.attrib?,
45            fixity.attrib?,
              lbrack.attrib?,rbrack.attrib?,separator.attrib?,
              attribute element {xsd:string}?,
              attribute attributes {xsd:string}?,
              crossref.attrib?,
50            (text | omdocpres.use.mix)*})
       # the attributes in the <use> element overwrite those in the
       # <presentation> element, therefore, they do not have defaults

       omstyle = element omstyle {omdoc.toplevel.attribs,
55            attribute for {omdocref}?,
              attribute element {xsd:string}?,
              (xslt|style)*}

       xslt = (ss| element xslt {format.attrib,xref.attrib?,template.model})
60
       style = (ss| element style {format.attrib, (omdocpres.use.mix)*})
       # this element contains mock xslt expressed in the elements below

       elt = (ss| element element {attribute name {xsd:NMTOKEN},
65                                  attribute crid {xsd:string}?,
                                    attribute cr {"yes" |"no"}?,
                                    attribute ns {xsd:anyURI}?,
                                    (attrb | omdocpres.use.mix)*})

70     map = (ss| element map {attribute select {xsd:string}?,
                               attribute lbrack {xsd:string}?,
                               attribute rbrack {xsd:string}?,
                               attribute precedence {xsd:string}?,
```

```
75              separator?,
                (omdocpres.use.mix)*})

    separator = (ss| element separator {(omdocpres.use.mix)*})

    attrb = (ss| element attribute {attribute name {xsd:NMTOKEN},
80                                   attribute ns {xsd:anyURI}?,
                                     (attribute select {xsd:string} |
                                     (txt | value−of)*)})

    txt = (ss| element text {text})
85
    value−of = (ss| element value−of {attribute select {xsd:string}})
    recurse = (ss| element recurse {attribute select {xsd:string}?})
```

D.14   Module QUIZ: Infrastructure for Assessments

The RNC module QUIZ provides a basic infrastructure for various kinds of exercises (see Chapter 21 for a discussion).

```
    # A RelaxNG schema for Open Mathematical documents (OMDoc 1.2) Module QUIZ
    #    SYSTEM http://www.mathweb.org/omdoc/rnc/omdocquiz.rnc
    #    PUBLIC −//OMDoc//RNC OMDoc QUIZ V1.2//EN
    # See the documentation and examples at http://www.mathweb.org/omdoc
5   # (c) 2004 Michael Kohlhase, released under the GNU Public License (GPL)

    default namespace omdoc = "http://www.mathweb.org/omdoc"
    omdoc.class |= exercise | hint | mc | solution

10  exercise = element exercise {id. attrib , fori . attrib ,
                                 omdocmtxt.MCF.content,
                                 hint*,
                                 (solution*|mc*)}

15  omdocpf.opt.content = notAllowed

    hint = element hint {omdoc.toplevel.attribs, fori . attrib , omdocmtxt.MCF.content}
    solution = element solution {omdoc.toplevel. attribs , fori . attrib ,metadata?,(omdoc.class)*}
    mc = element mc {omdoc.toplevel.attribs,fori.attrib ,choice,hint?,answer}
20
    choice = (ss| element choice {id. attrib ,omdocmtxt.MCF.content})

    answer = (ss| element answer {id.attrib,
                                  attribute verdict {"true" | " false"}?,
25                                omdocmtxt.MCF.content})
```

E

The RelaxNG Schemata for Mathematical Objects

For completeness we reprint the RELAXNG schemata for the external formats OMDOC makes use of.

E.1 The RelaxNG Schema for OpenMath

For completeness we reprint the RELAXNG schema for OPENMATH, the original can be found in the OPENMATH2 standard [BCC+04].

```
# RELAX NG Schema for OpenMath 2

default namespace om = "http://www.openmath.org/OpenMath"

#start = OMOBJ

# OpenMath object constructor
OMOBJ = element OMOBJ { compound.attributes,
                        attribute version { xsd:string }?,
                        omel }

# Elements which can appear inside an OpenMath object
omel =
    OMS | OMV | OMI | OMB | OMSTR | OMF | OMA | OMBIND | OME | OMATTR |OMR

# things which can be variables
omvar = OMV | attvar

attvar = element OMATTR { common.attributes,(OMATP , (OMV | attvar))}


cdbase = attribute cdbase { xsd:anyURI}?

# attributes common to all elements
common.attributes = (attribute id { xsd:ID })?

# attributes common to all elements that construct compount OM objects.
compound.attributes = common.attributes,cdbase

# symbol
OMS = element OMS { common.attributes,
                    attribute name {xsd:NCName},
                    attribute cd {xsd:NCName},
```

```
                                    cdbase }
35
     # variable
     OMV = element OMV { common.attributes,
                              attribute name { xsd:NCName} }

40   # integer
     OMI = element OMI { common.attributes,
                          xsd:string {pattern = "\s*(−\s?)?[0−9]+(\s[0−9]+)*\s*"}}
     # byte array
     OMB = element OMB { common.attributes, xsd:base64Binary }
45
     # string
     OMSTR = element OMSTR { common.attributes, text }

     # IEEE floating point number
50   OMF = element OMF { common.attributes,
                          ( attribute dec { xsd:double } |
                            attribute hex { xsd:string {pattern = "[0−9A−F]+"}}) }

     # apply constructor
55   OMA = element OMA { compound.attributes, omel+ }

     # binding constructor
     OMBIND = element OMBIND { compound.attributes, omel, OMBVAR, omel }

60   # variables used in binding constructor
     OMBVAR = element OMBVAR { common.attributes, omvar+ }

     # error constructor
     OME = element OME { common.attributes, OMS, (omel|OMFOREIGN)* }
65
     # attribution constructor and attribute pair constructor
     OMATTR = element OMATTR { compound.attributes, OMATP, omel }

     OMATP = element OMATP { compound.attributes, (OMS, (omel | OMFOREIGN) )+ }
70
     # foreign constructor
     OMFOREIGN = element OMFOREIGN {
        compound.attributes, attribute encoding {xsd:string}?,
        (omel|notom)* }
75
     # Any elements not in the om namespace
     # (valid om is allowed as a descendant)
     notom =
       (element * − om:* {attribute * { text }*,(omel|notom)*}
80     | text)

     # reference constructor
     OMR = element OMR { common.attributes,
                          attribute href { xsd:anyURI }
85                       }
```

E.2 The RelaxNG Schema for MathML

For completeness, we reprint the RELAXNG schema for MATHML. It comes in three parts, the schema driver, and the parts for content- and presentation MATHML which we will present in the next two subsections.

```
     # A RelaxNG schema for MathML2
     #   SYSTEM http://www.mathweb.org/omdoc/rnc/mathml2.rnc
     # (c) 2005 Michael Kohlhase, released under the GNU Public License (GPL)

 5   default namespace m = "http://www.w3.org/1998/Math/MathML"
     namespace a = "http://relaxng.org/ns/compatibility/annotations/1.0"
     namespace xlink = "http://www.w3.org/1999/xlink"
     namespace local = ""

10   non−mathml−attribs = attribute * − (local:*|xlink:href) {xsd:string}

     MathML.Common.attrib = attribute class {xsd:NMTOKENS}?,
                 attribute style {xsd:string}?,
                 attribute id {xsd:ID}?,
15               attribute xlink:href {xsd:anyURI}?,
                 non−mathml−attribs*


     include "mathml2−presentation.rnc"
20   include "mathml2−content.rnc"

     Presentation−expr.class = PresExpr.class | ContExpr.class

     Content−expr.class = ContExpr.class | PresExpr.class
25
     PresExpr.class = Presentation−token.class |
                 Presentation−layout.class |
                 Presentation−script.class |
                 Presentation−table.class |
30               mspace | maction | merror | mstyle

     ContExpr.class = Content−tokens.class |
                 Content−arith.class |
                 Content−functions.class |
35               Content−logic.class |
                 Content−constants.class |
                 Content−sets.class |
                 Content−relations.class |
                 Content−elementary−functions.class |
40               Content−calculus.class |
                 Content−linear−algebra.class |
                 Content−vector−calculus.class |
                 Content−statistics.class |
                 Content−constructs.class |
45               semantics

     Browser−interface.attrib = attribute baseline {xsd:string}?,
                 [a:default = "scroll"]
                 attribute overflow {"scroll" | "elide" | "truncate" | "scale"}?,
50               attribute altimg {xsd:anyURI}?,
                 attribute alttext {xsd:string}?,
                 attribute type {xsd:string}?,
                 attribute name {xsd:string}?,
                 attribute height {xsd:string}?,
55               attribute width {xsd:string}?

     math.attlist = Browser−interface.attrib,
                 attribute macros {xsd:string}?,
                 [a:default = "inline"]
```

```
60              attribute display {"block" | " inline"}?,
                MathML.Common.attrib

     math.content = PresExpr.class | ContExpr.class

65   math = element math {math.attlist,math.content*}
```

E.2.1   Presentation MATHML

```
     # A RelaxNG schema for MathML2 Presentation Elements
     #   SYSTEM http://www.mathweb.org/omdoc/rnc/mathml2−content.rnc
     # (c) 2005 Michael Kohlhase, released under the GNU Public License (GPL)

5    default namespace m = "http://www.w3.org/1998/Math/MathML"
     namespace a = "http://relaxng.org/ns/compatibility/annotations/1.0"

     # Simple sizes
     simple−size= "small" | "normal" | "big"
10
     # Centering values
     centering.values = "left " | " center " | " right"

     # The named spaces
15   # this is also used in the value of the "width" attribute on the "mpadded" element
     named−space = "veryverythinmathspace" |
                   "verythinmathspace" |
               "thinmathspace" |
                   "mediummathspace" |
20                 "thickmathspace" |
                   "verythickmathspace" |
                   "veryverythickmathspace"

     # Thickness
25   thickness = "thin" | "medium" | "thick"

     # number with units used to specified lengths
     length−with−unit =
         xsd:string #{pattern="(−?([0−9]+|[0−9]*\.[0−9]+)*(em|ex|px|in|cm|mm|pt|pc|%))|0"}
30   length−with−optional−unit =
         xsd:string #{pattern="−?([0−9]+|[0−9]*\.[0−9]+)*(em|ex|px|in|cm|mm|pt|pc|%)?"}

     # This is just " infinity " that can be used as a length
     infinity = " infinity "
35
     # colors defined as RGB
     RGB−color = xsd:string {pattern="#(([0−9]|[a−f]){3}|([0−9]|[a−f]){6})"}

     # This schema module defines sets of attributes common to several elements
40   # of presentation MathML.

     # The mathematics style attributes. These attributes are valid on all
     #     presentation token elements except "mspace" and "mglyph", and on no
     #     other elements except "mstyle".
45
     Token−style.attrib = attribute mathvariant
                     {"normal" | "bold" | " italic " | " bold−italic" | " double−struck" |
                         "bold−fraktur" | "script" | " bold−script" | "fraktur" |
                     "sans−serif" | " bold−sans−serif" | "sans−serif−italic" |
50               "sans−serif−bold−italic" | "monospace"}?,
                         attribute mathsize {simple−size | length−with−unit}?,
         # For both of the following attributes the types should be more restricted
                         attribute mathcolor {xsd:string}?,
                 attribute mathbackground {xsd:string}?
55
```

These operators are all related to operators. They are valid on "mo" and "mstyle".

Operator.attrib =
this attribute value is normally inferred from the position of
the operator in its "<mrow>"
    attribute form {"prefix" | " infix " | " postfix"}?,
    # set by dictionnary, else it is "thickmathspace"
    attribute lspace {length−with−unit | named−space}?,
    # set by dictionary, else it is "thickmathspace"
    attribute rspace {length−with−unit | named−space}?,
    # set by dictionnary, else it is " false "
    attribute fence {xsd:boolean}?,
    # set by dictionnary, else it is " false "
    attribute separator {xsd:boolean}?,
    # set by dictionnary, else it is " false "
    attribute stretchy {xsd:boolean}?,
    # set by dictionnary, else it is "true"
    attribute symmetric {xsd:boolean}?,
    # set by dictionnary, else it is " false "
    attribute movablelimits {xsd:boolean}?,
    # set by dictionnary, else it is " false "
    attribute accent {xsd:boolean}?,
    # set by dictionnary, else it is " false "
    attribute largeop {xsd:boolean}?,
    attribute minsize {length−with−unit | named−space}?,
    attribute maxsize {length−with−unit | named−space | infinity | xsd:float}?

mglyph = element mglyph {attribute alt {xsd:string}?,
                         attribute fontfamily {xsd:string}?,
                    attribute index { xsd:positiveInteger }?}

This is the XML schema module for the token elements of the
presentation part of MathML.

Glyph−alignmark.class = malignmark|mglyph

# "mi" is supposed to have a default value of its "mathvariant" attribute set to " italic "
mi = element mi {Token−style.attrib, MathML.Common.attrib,(Glyph−alignmark.class|text)*}

# "mo"
mo = element mo {Operator.attrib,Token−style.attrib,MathML.Common.attrib,
                (text|Glyph−alignmark.class)*}

# "mn"
mn = element mn {Token−style.attrib, MathML.Common.attrib,(text|Glyph−alignmark.class)*}

# "mtext"
mtext = element mn {Token−style.attrib, MathML.Common.attrib,(text|Glyph−alignmark.class)*}

# ms (the values of "lquote" or "rquote" are not restricted to be one character strings ...)
ms = element ms {[a:default="""] attribute lquote {xsd:string}?,
                 [a:default="""] attribute rquote {xsd:string}?,
                 Token−style.attrib,MathML.Common.attrib,
                 (text|Glyph−alignmark.class)*}

And the group of any token
Presentation−token.class = mi | mo | mn | mtext | ms

This is an XML Schema module for the presentation elements of MathML
dealing with subscripts and superscripts.

# "msub"
msub = element msub {attribute subscriptshift {length−with−unit}?, MathML.Common.attrib,
                     Presentation−expr.class,(Presentation−expr.class)}

# "msup"
msup = element msup {attribute supscriptshift {length−with−unit}?, MathML.Common.attrib,

```
                              Presentation−expr.class,Presentation−expr.class}

125   # "msubsup"
      msubsup = element msubsup {MathML.Common.attrib,
                              attribute subscriptshift {length−with−unit}?,
                              attribute supscriptshift {length−with−unit}?,
                              Presentation−expr.class,Presentation−expr.class}

130
      # "munder"
      munder = element munder {MathML.Common.attrib,
                              attribute accentunder {xsd:boolean}?,
                              Presentation−expr.class,Presentation−expr.class}

135
      # "mover"
      mover = element mover {MathML.Common.attrib,
                              attribute accent {xsd:boolean}?,
                              Presentation−expr.class,Presentation−expr.class}

140
      # "munderover"
      munderover = element munderover {MathML.Common.attrib,
                              attribute accentunder {xsd:boolean}?,
                              attribute accent {xsd:boolean}?,
                              Presentation−expr.class,
145             Presentation−expr.class,
                Presentation−expr.class}

      # "mmultiscripts", "mprescripts" and "none"
150
      Presentation−expr−or−none.class = Presentation−expr.class | none

      mmultiscripts = element mmultiscripts{MathML.Common.attrib,
                              Presentation−expr.class,
155             (Presentation−expr−or−none.class,
                Presentation−expr−or−none.class)*,
                (mprescripts,
                (Presentation−expr−or−none.class,
                Presentation−expr−or−none.class)*)?}
160   none = element none {empty}
      mprescripts = element mprescripts {empty}

      Presentation−script.class = msub|msup|msubsup|munder|mover|munderover|mmultiscripts

165   mspace = element mspace {[a:defaultValue = "0em"]
                        attribute width {length−with−unit | named−space}?,
                [a:defaultValue = "0ex"]
                        attribute height {length−with−unit}?,
                [a:defaultValue = "0ex"]
170             attribute depth {length−with−unit}?,
                [a:defaultValue="auto"]
                        attribute linebreak {"auto" | "newline" | "indentingnewline" |
                                "nobreak" | "goodbreak" | "badbreak"}?,
                MathML.Common.attrib}

175
      # This is the XML schema module for the layout elements of the
      # presentation part of MathML.

      # "mrow"
180   mrow = element mrow {MathML.Common.attrib,(Presentation−expr.class)*}

      # "mfrac"

      mfrac = element mfrac {attribute bevelled {xsd:boolean}?,
185             [a:defaultValue = "center"]
                        attribute denomalign {centering.values}?,
                [a:defaultValue = "center"]
                attribute numalign {centering.values}?,
                [a:defaultValue="1"]
```

190

```
                    attribute linethickness {length−with−optional−unit|thickness}?,
                         MathML.Common.attrib,
                      Presentation−expr.class,Presentation−expr.class}
        # "msqrt"
        msqrt = element msqrt {MathML.Common.attrib,(Presentation−expr.class)*}
195
        # "mroot"
        mroot = element mroot {MathML.Common.attrib,Presentation−expr.class,Presentation−expr.class}

        # "mpadded"
200     mpadded−space = xsd:string {pattern="(\+|−)?([0−9]+|[0−9]*\.[0−9]+)(((%?)*(width|lspace|
                              height|depth))|(em|ex|px|in|cm|mm|pt|pc))"}

        # MaxF: definition from spec seems wrong,
        #      fixing to ([+|−] unsigned−number (%[pseudo−unit]|pseudo−unit|h−unit)) | namedspace | 0
205
        mpadded−width−space = xsd:string {pattern="((\+|−)?([0−9]+|[0−9]*\.[0−9]+)(((%?) *(width|
              lspace|height|depth)?)|(width|lspace|height|depth)|(em|ex|px|in|cm|mm|pt|pc)))|
              ((veryverythin|verythin|thin|medium|thick|verythick|veryverythick)mathspace)|0"}
        mpadded = element mpadded {attribute width {mpadded−width−space},
210                    # should have default=0 below but '0' is not in value space
                       # see bug #425
                       attribute lspace {mpadded−space}?,
                       attribute height {mpadded−space}?,
                       attribute depth {mpadded−space}?,
215                    MathML.Common.attrib,
                       (Presentation−expr.class)*}

        # "mphantom"
        mphantom = element mphantom.attlist {MathML.Common.attrib,Presentation−expr.class*}
220
        # "mfenced"
        mfenced = element mfenced {[a:defaultValue= "("] attribute open {xsd:string}?,
                         [a:defaultValue = ")"] attribute close {xsd:string }?,
                         [a:defaultValue = ","] attribute separators {xsd:string }?,
225                      MathML.Common.attrib,
                         (Presentation−expr.class)*}

        # "menclose"
        menclose = element menclose {[a:defaultValue="longdiv"]
230                       attribute notation {"actuarial"|"longdiv"|"radical"|
                                    "box"|"roundedbox"|"circle"|
                            "left "|"right"|"top"|"bottom"|
                            "updiagonalstrike"|"downdiagonalstrike"|
                            " verticalstrike "|" horizontalstrike "}?,
235                      MathML.Common.attrib,
                         (Presentation−expr.class)*}

        # And the group of everything
        Presentation−layout.class = mrow|mfrac|msqrt|mroot|mpadded|mphantom|mfenced|menclose
240
        # This is an XML Schema module for tables in MathML presentation.

        Table−alignment.attrib = [a:defaultValue = "baseline"]
                    attribute rowalign
245              {xsd:string {pattern="(top|bottom|center|baseline|axis)(top|bottom|center|baseline|
                                                                        axis)*"}}?,
                   [a:defaultValue = "center"]
                   attribute columnalign
                      {xsd:string {pattern="(left|center| right )( left |center| right )*"}}?,
250              attribute groupalign {xsd:string}?

        mtr = element mtr {Table−alignment.attrib, MathML.Common.attrib,mtd+}

        mlabeledtr = element mlabeledtr {Table−alignment.attrib,MathML.Common.attrib,mtd*}
255
        # "mtd"
```

```
      mtd = element mtd {Table−alignment.attrib,
                          [a:defaultValue="1"] attribute columnspan {xsd:positiveInteger}?,
                  [a:defaultValue="1"] attribute rowspan {xsd:positiveInteger}?,
260               MathML.Common.attrib,
                  Presentation−expr.class*}

      # "mtable"
      mtable = element mtable {Table−alignment.attrib,
265                            [a:defaultValue="axis"] attribute  align {xsd:string }?,
                              [a:defaultValue="true"]
                  attribute  alignmentscope {xsd:string {pattern="(true|false)( true| false )*"}}?,
                  [a:defaultValue="auto"] attribute columnwidth {xsd:string}?,
                  [a:defaultValue="auto"] attribute width {xsd:string}?,
270               [a:defaultValue="1.0ex"] attribute rowspacing {xsd:string}?,
                  [a:defaultValue="0.8em"] attribute columnspacing {xsd:string}?,
                  [a:defaultValue="none"] attribute rowlines {xsd:string }?,
                  [a:defaultValue="none"] attribute columnlines {xsd:string}?,
                  [a:defaultValue="none"] attribute frame {"none" | "solid" | "dashed"}?,
275               [a:defaultValue="0.4em 0.5ex"] attribute framespacing {xsd:string}?,
                  [a:defaultValue="false"] attribute equalrows {xsd:boolean}?,
                  [a:defaultValue="false"] attribute equalcolumns {xsd:boolean}?,
                  [a:defaultValue="false"] attribute displaystyle {xsd:boolean}?,
                  [a:defaultValue="right"]
280            attribute side {" left"|"right"|" leftoverlap "|" rightoverlap"}?,
                  [a:defaultValue="0.8em"] attribute minlabelspacing {length−with−unit}?,
                  MathML.Common.attrib,
                  (mtr|mlabeledtr)*}

285   # "maligngroup"
      maligngroup = element maligngroup {
          attribute groupalign {" left " | " center " | " right " | " decimalpoint"}?,
          MathML.Common.attrib}

290   # "malignmark"

      malignmark = element malignmark {[a:defaultValue="left"] attribute edge {"left" | "right"}?,
                              MathML.Common.attrib}
      Presentation−table.class = mtable|maligngroup|malignmark
295
      # "mstyle"
      mstyle = element mstyle {attribute scriptlevel {xsd:integer }?,
                              attribute  displaystyle {xsd:boolean}?,
                  [a:defaultValue="0.71"] attribute  scriptsizemultiplier  {xsd:decimal}?,
300               [a:defaultValue="8pt"] attribute scriptminsize {length−with−unit}?,
                  attribute color {xsd:string }?,
                  [a:defaultValue="transparent"] attribute background {xsd:string}?,
                  [a:defaultValue="0.0555556em"] attribute veryverythinmathspace {length−with−unit}?,
                  [a:defaultValue="0.111111em"] attribute verythinmathspace {length−with−unit}?,
305               [a:defaultValue="0.166667em"] attribute thinmathspace {length−with−unit}?,
                     [a:defaultValue="0.222222em"] attribute mediummathspace {length−with−unit}?,
                     [a:defaultValue="0.277778em"] attribute thickmathspace {length−with−unit}?,
                     [a:defaultValue="0.333333em"] attribute verythickmathspace {length−with−unit}?,
                     [a:defaultValue="0.388889em"] attribute veryverythickmathspace {length−with−unit}?,
310               [a:defaultValue="1"] attribute linethickness {length−with−optional−unit|thickness}?,
                  Operator.attrib,Token−style.attrib,MathML.Common.attrib,
                  Presentation−expr.class*}

      # This is the XML Schema module for the MathML "merror" element.
315
      merror = element merror {MathML.Common.attrib,Presentation−expr.class*}

      # This is the XML Schema module for the MathML "maction" element.

320   maction = element maction {attribute actiontype {xsd:string}?,
                              [a:defaultValue="1"] attribute  selection  { xsd:positiveInteger }?,
                  MathML.Common.attrib,
                  Presentation−expr.class*}
```

E.2.2   Content MathML

```
     # A RelaxNG schema for MathML2 Content Elements
     #   SYSTEM http://www.mathweb.org/omdoc/rnc/mathml2−content.rnc
     # (c) 2005 Michael Kohlhase, released under the GNU Public License (GPL)

5    default namespace m = "http://www.w3.org/1998/Math/MathML"
     namespace a = "http://relaxng.org/ns/compatibility/annotations/1.0"

     Definition.attrib = attribute encoding {xsd:string}?,
                         attribute definitionURL {xsd:anyURI}?
10
     # This is the XML schema module for the token elements of the content part of MathML.

     Content−token.content = text|Presentation−expr.class
     # the content of "cn" may have <sep> elements in it
15
     sep = element sep {empty}

     cn = element cn {######attribute base {xsd:positiveInteger [1,...,36]},
                      attribute type {"e−notation"|"integer"|"rational"|"real" |
20                                      "complex−cartesian"|"complex−polar"|"constant" }?,
                 Definition.attrib,
                 MathML.Common.attrib,
                 (text|sep|Presentation−expr.class)*}
     # "ci"
25   ci = element ci {attribute type {xsd:string}?,
                 Definition.attrib,
                 MathML.Common.attrib,
                 Content−token.content}
     # "csymbol"
30
     csymbol = element csymbol {Definition.attrib, MathML.Common.attrib, Content−token.content}

     # And the group of everything
     Content−tokens.class = cn|ci|csymbol
35
     # This is an XML Schema module for the "arithmetic" operators of content MathML.

     Arith.type = Definition.attrib, MathML.Common.attrib

40   abs = element abs {Arith.type}
     conjugate = element conjugate {Arith.type}
     arg = element arg {Arith.type}
     real = element real {Arith.type}
     imaginary = element imaginary{Arith.type}
45
     floor = element floor {Arith.type}
     ceiling = element floor {Arith.type}

     power = element power {Arith.type}
50   root = element root {Arith.type}

     minus = element minus {Arith.type}
     plus = element plus {Arith.type}
     sum = element sum {Arith.type}
55   times = element times {Arith.type}
     product = element product {Arith.type}

     max = element max {Arith.type}
     min = element min {Arith.type}
60
     factorial = element factorial {Arith.type}
     quotient = element quotient {Arith.type}
     divide = element divide {Arith.type}
     rem = element rem {Arith.type}
65   gcd = element gcd {Arith.type}
     lcm = element lcm {Arith.type}
```

Content−arith.class = abs|conjugate|factorial|arg|real|imaginary|
floor | ceiling |quotient|divide|rem|minus|
70   plus|times|power|root|max|min|gcd|lcm|
sum|product
This is an XML Schema module for operators dealing with functions in content MathML.

75 Functions.type = Definition.attrib , MathML.Common.attrib

# "compose"
compose = element compose {Functions.type}

Domain, codomain and image
80 domain = element domain {Functions.type}
codomain = element codomain {Functions.type}
image = element image {Functions.type}

"domainofapplication"
85 domainofapplication = element domainofapplication{Definition.attrib,
MathML.Common.attrib,
Content−expr.class}
# identity
ident = element ident {Functions.type}
90

Content−functions.class = compose|domain|codomain|image|domainofapplication|ident

This is an XML Schema module for the logic operators of content MathML.

95 Logic.type = Definition.attrib ,MathML.Common.attrib

and = element and {Logic.type}
or = element or {Logic.type}
xor = element xor {Logic.type}
100 not = element not {Logic.type}
exists = element exists {Logic.type}
forall = element forall {Logic.type}
implies = element implies {Logic.type}

105 Content−logic.class = and|or|xor|not|exists | forall |implies

This is an XML Schema module for the basic constructs of content MathML.

apply = element apply {MathML.Common.attrib,Content−expr.class*}
110
interval = element interval {MathML.Common.attrib,
[a:defaultvalue = "closed"]
attribute closure {"closed"|"open"|"open−closed"|"closed−open"}?,
(Content−expr.class),(Content−expr.class)}
115 inverse = element inverse {Definition . attrib , MathML.Common.attrib}
condition = element condition {Definition. attrib , Content−expr.class+}

declare = element declare {attribute type {xsd:string}?,
attribute scope {xsd:string}?,
120   attribute nargs {xsd:nonNegativeInteger}?,
attribute occurrence {"prefix"|" infix"|"function−model"}?,
Definition . attrib ,
Content−expr.class+}

125 lambda = element lambda {MathML.Common.attrib,Content−expr.class+}

#"piecewise" and its inner elements
otherwise = element otherwise {Content−expr.class,MathML.Common.attrib}
piece = element piece {MathML.Common.attrib,Content−expr.class+}
130 piecewise = element piecewise {MathML.Common.attrib,piece*,(otherwise,piece)*}
bvar = element bvar {MathML.Common.attrib,Content−expr.class+}
degree = element degree {MathML.Common.attrib,Content−expr.class+}

135 Content−constructs.class = apply|interval| inverse |condition|declare |lambda|piecewise|bvar|degree

```
      # This is the XML Schema module for the basic constants of MathML content.

      Constant.type = Definition.attrib , MathML.Common.attrib
140
      # Basic sets
      naturalnumbers= element naturalnumbers {Constant.type}
      primes= element primes{Constant.type}
      integers = element integers {Constant.type}
145   rationals = element rationals{Constant.type}
      reals = element reals {Constant.type}
      complexes = element complexes {Constant.type}

      #Empty set
150   emptyset = element emptyset {Constant.type}

      # Basic constants
      exponentiale = element exponentiale {Constant.type}
      imaginaryi = element imaginaryi {Constant.type}
155   pi = element pi {Constant.type}
      eulergamma = element eulergamma {Constant.type}

      # Boolean constants
      true = element true {Constant.type}
160   false = element false {Constant.type}

      # Infinty
      infinit  = element infinity {Constant.type}

165   # NotANumber
      notanumber = element notanumber {Constant.type}

      Content−constants.class = naturalnumbers|primes|integers|rationals|reals|
                   complexes|emptyset|exponentiale|imaginaryi|pi|
170                eulergamma|true|false| infinit |notanumber

      # This is an XML Schema module for the elementary functions in content MathML.

      Elementary−functions.type = Definition.attrib,MathML.Common.attrib
175
      # Exp and logs

      exp= element exp {Elementary−functions.type}
      ln = element ln {Elementary−functions.type}
180   log = element log {Elementary−functions.type}

      # special element of the base of logarithms

      logbase = element logbase {MathML.Common.attrib,Content−expr.class}
185
      # Trigonometric functions

      sin = element sin {Elementary−functions.type}
      cos = element cos {Elementary−functions.type}
190   tan = element tan {Elementary−functions.type}
      sec = element sec {Elementary−functions.type}
      csc = element csc {Elementary−functions.type}
      cot = element cot {Elementary−functions.type}

195   arcsin = element arcsin {Elementary−functions.type}
      arccos = element arccos {Elementary−functions.type}
      arctan = element arctan {Elementary−functions.type}
      arccot = element arccot {Elementary−functions.type}
      arccsc = element arccsc {Elementary−functions.type}
200   arcsec = element arcsec {Elementary−functions.type}

      # Hyperbolic trigonometric functions

      sinh = element sinh {Elementary−functions.type}
```

```
205   cosh = element cosh {Elementary−functions.type}
      tanh = element tanh {Elementary−functions.type}
      sech = element sech {Elementary−functions.type}
      csch = element csch {Elementary−functions.type}
      coth = element coth {Elementary−functions.type}
210   arccosh = element arccosh {Elementary−functions.type}
      arccoth = element arccoth {Elementary−functions.type}
      arccsch = element arccsch {Elementary−functions.type}
      arcsech = element arcsech {Elementary−functions.type}
      arcsinh = element arcsinh {Elementary−functions.type}
215   arctanh = element arctanh {Elementary−functions.type}

      # And the group of everything

      Content−elementary−functions.class =
220        exp|ln|log|logbase|sin|cos|tan|sec|csc|cot|
           arcsin|arccos|arctan|arcsec|arccsc|arccot|
           sinh|cosh|tanh|sech|csch|coth|
           arccosh|arccoth|arccsch|arcsech|arcsinh|arctanh

225   # This is an XML Schema module for the relational operators of content MathML.

      # a common type for all this
      Relations.type = Definition.attrib , MathML.Common.attrib

230   eq = element eq {Relations.type}
      neq = element neq {Relations.type}
      leq = element leq {Relations.type}
      lt = element lt {Relations.type}
      geq = element geq {Relations.type}
235   gt = element gt {Relations.type}
      equivalent = element equivalent {Relations.type}
      approx = element approx {Relations.type}
      factorof = element factorof {Relations.type}

240   # And the group of everything
      Content−relations.class = eq|neq|leq| lt |geq|gt|equivalent |approx|factorof

      # "annotation"
      annotation = element annotation {attribute encoding {xsd:string}?,MathML.Common.attrib,text}
245
      # "annotation−xml"
      anyElement = element * {(attribute * {text}|text| anyElement)*}
      annotation−xml = element annotation−xml {Definition.attrib,MathML.Common.attrib, anyElement}

250   # "semantics"
      semantics = element semantics {attribute encoding {xsd:string}?,
                             attribute definitionURL {xsd:anyURI}?,
                    MathML.Common.attrib,
                    Content−expr.class,
255                 (annotation|annotation−xml)*}

      # This is an XML Schema module for the part of content MathML dealing with sets and lists.

      # "set" ("type" could be "multiset" or "normal" or anything else)
260   set = element set {attribute type {xsd:string}?,
                    MathML.Common.attrib, Content−expr.class*}

      # "list"
      lst = element list {attribute order {"lexicographic"|"numeric"}?,
265                 MathML.Common.attrib,
                    Content−expr.class*}
      # "union"
      union = element union {Definition.attrib, MathML.Common.attrib,Content−expr.class*}
      intersect = element intersect {Definition . attrib , MathML.Common.attrib,Content−expr.class*}
270   in = element in {Definition. attrib , MathML.Common.attrib}
      notin = element notin {Definition. attrib , MathML.Common.attrib}
      subset = element subset {Definition.attrib , MathML.Common.attrib}
      prsubset = element prsubset {Definition.attrib , MathML.Common.attrib}
```

```
       notsubset = element notsubset {Definition.attrib, MathML.Common.attrib}
275    notprsubset = element notprsubset {Definition.attrib, MathML.Common.attrib}
       setdiff = element setdiff {Definition.attrib, MathML.Common.attrib}
       card = element card {Definition.attrib, MathML.Common.attrib}
       cartesianproduct = element cartesianproduct {Definition.attrib, MathML.Common.attrib}

280    # And the group of everything

       Content−sets.class = set| lst |union| intersect | in |notin|subset|
                    prsubset|notsubset|notprsubset| setdiff |card|cartesianproduct

285    # This is an XML Schema module for the linear algebra part of content MathML.

       # "vector"

       vector = element vector {MathML.Common.attrib,Content−expr.class∗}
290    matrix = element matrix {MathML.Common.attrib,matrixrow+}
       matrixrow = element matrixrow {MathML.Common.attrib,Content−expr.class+}
       determinant = element determinant {Definition.attrib,MathML.Common.attrib}
       transpose = element transpose {Definition.attrib,MathML.Common.attrib}
       mselector = element selector {Definition.attrib,MathML.Common.attrib}
295    vectorproduct = element vectorproduct {Definition.attrib,MathML.Common.attrib}
       scalarproduct = element scalarproduct {Definition.attrib,MathML.Common.attrib}
       outerproduct = element outerproduct {Definition.attrib,MathML.Common.attrib}

       Content−linear−algebra.class = vector|matrix|determinant|transpose|mselector|
300                    vectorproduct|scalarproduct|outerproduct

       # This' is an XML Schema module for the calculus operators of content MathML.

       calculus .type = Definition.attrib , MathML.Common.attrib
305
       int = element int {calculus .type}
       diff = element diff {calculus .type}
       partialdiff = element partialdiff {calculus .type}
       limit = element limit {calculus .type}
310    lowlimit = element lowlimit {calculus.type,Content−expr.class+}
       uplimit = element uplimit {calculus.type,Content−expr.class+}
       tendsto = element tendsto {calculus.type,attribute type {xsd:string}?}

       Content−calculus.class = int| diff | partialdiff | limit |lowlimit |uplimit |tendsto
315
       # This is an XML Schema module for the vector calculus operators of content MathML.

       divergence = element divergence {Definition.attrib ,MathML.Common.attrib}
       grad = element grad {Definition.attrib,MathML.Common.attrib}
320    curl = element curl {Definition. attrib ,MathML.Common.attrib}
       laplacian = element laplacian {Definition. attrib ,MathML.Common.attrib}

       # And the group of everything

325    Content−vector−calculus.class = divergence|grad|curl|laplacian

       # This is an XML Schema module for the statistical operators of content MathML.

       mean = element mean {Definition.attrib,MathML.Common.attrib}
330    sdev = element sdev {Definition.attrib, MathML.Common.attrib}
       variance = element variance {Definition. attrib ,MathML.Common.attrib}
       median = element median {Definition.attrib,MathML.Common.attrib}
       mode = element mode {Definition.attrib,MathML.Common.attrib}
       moment = element moment {Definition.attrib,MathML.Common.attrib}
335    momentabout = element momentabout {Definition.attrib,MathML.Common.attrib,
                                   Content−expr.class+}

       Content−statistics. class = mean|sdev|variance|median|mode|moment|momentabout
```

Bibliography

[ABC+03] Ron Ausbrooks, Stephen Buswell, David Carlisle, Stéphane Dalmas, Stan Devitt, Angel Diaz, Max Froumentin, Roger Hunter, Patrick Ion, Michael Kohlhase, Robert Miner, Nico Poppelier, Bruce Smith, Neil Soiffer, Robert Sutor, and Stephen Watt. Mathematical Markup Language (MathML) version 2.0 (second edition). W3c recommendation, World Wide Web Consortium, 2003. Available at http://www.w3.org/TR/MathML2.

[ABD+01] Murray Altheim, Frank Boumphrey, Sam Dooley, Shane McCarron, Sebastian Schnitzenbaumer, and Ted Wugofski (eds.). Modularization of xhtml. W3c recommendation, The World Wide Web Consortium, 2001. available at http://www.w3.org/TR/xhtml-modularization.

[ABF+03] Serge Autexier, Christoph Benzmüller, Armin Fiedler, Helmut Horacek, and Quoc Bao Vo. Assertion level proof representation with underspecification. In Fairouz Kamareddine, editor, *Proceedings of MKM Symposium*, Heriot-Watt, Edinburgh, November 2003.

[ABFL06] Serge Autexier, Christoph Benzmüller, Armin Fiedler, and Henri Lesourd. Integrating proof assistants as reasoning and verification tools into a scientific wysiwig editor. *Proceedings of UITP'05*, 2006.

[ABI+96] Peter B. Andrews, Matthew Bishop, Sunil Issar, Dan Nesmith, Frank Pfenning, and Hongwei Xi. TPS: A theorem-proving system for classical type theory. *Journal of Automated Reasoning*, 16:321–353, 1996.

[Abn96] S. Abney. Partial parsing via finite-state cascades, 1996. http://cite seer.ifi.unizh.ch/abney96partial.html.

[ABT04] Andrea Asperti, Grzegorz Bancerek, and Andrej Trybulec, editors. *Mathematical Knowledge Management, MKM'04*, number 3119 in LNCS. Springer Verlag, 2004.

[AF05] Serge Autexier and Armin Fiedler. Textbook proofs meet formal logic - the problem of underspecification and granularity. In Michael Kohlhase, editor, *Proceedings of MKM'05*, volume 3863 of *LNAI*, IUB Bremen, Germany, june 2005. Springer.

[AH05] Serge Autexier and Dieter Hutter. Formal software development in maya. In Dieter Hutter and Werner Stephan, editors, *Festschrift in Honor of J. Siekmann*, volume 2605 of *LNAI*. Springer, february 2005.

[AHL+00] S. Autexier, D. Hutter, B. Langenstein, H. Mantel, G. Rock, A. Schairer, W. Stephan, R. Vogt, and A. Wolpers. Vse: Formal methods meet industrial needs. *International Journal on Software Tools for Technology Transfer, Special issue on Mechanized Theorem Proving for Technology*, 3(1), september 2000.

[AHMS99] S. Autexier, D. Hutter, H. Mantel, and A. Schairer. System description: INKA 5.0 - a logical voyager. In H. Ganzinger, editor, *16th International Conference on Automated Deduction, CADE-16*, volume 1732 of *Lecture Notes in Artificial Intelligence*, Trento, 1999. Springer.

[AHMS00] Serge Autexier, Dieter Hutter, Heiko Mantel, and Axel Schairer. Towards an evolutionary formal software-development using CASL. In C. Choppy and D. Bert, editors, *Proceedings Workshop on Algebraic Development Techniques, WADT-99*, number 1827 in LNCS. Springer, 2000.

[AK02] Andrea Asperti and Michael Kohlhase. Mathml in the MOWGLI project. In *Second International Conference on MathML and Technologies for Math on the Web*, Chicago, USA, 2002.

[AKC03] Andrea Asperti, Michael Kohlhase, and Claudio Sacerdoti Coen. Prototype n. d2.b document type descriptors: Omdoc proofs. Mowgli deliverable, The MoWGLI Project, 2003.

[AM02] Serge Autexier and Till Mossakowski. Integrating holcasl into the development graph manager maya. In Alessandro Armando, editor, *Frontiers of Combinning Systems (FROCOS'02)*, number 2309 in LNAI, pages 2–17. Springer, 2002.

[And02] Peter B. Andrews. *An Introduction to Mathematical Logic and Type Theory: To Truth Through Proof*. Kluwer Academic Publishers, second edition, 2002.

[And06] Andrea Kohlhase. What if PowerPoint became emPowerPoint (through CPoint)? In Caroline M. Crawford, editor, *SITE 2006*, pages 2934–2939. SITE, AACE, 2006. Orlando (USA), 2006-03-20/24.

[APCS01] Andrea Asperti, Luca Padovani, Claudio Sacerdoti Coen, and Irene Schena. HELM and the semantic math-web. In Paul B. Jackson Richard. J. Boulton, editor, *Theorem Proving in Higher Order Logics: TPHOLs'01*, volume 2152 of *LNCS*, pages 59–74. Springer, 2001.

[Asp03] Andrea Asperti, editor. *Mathematical Knowledge Management, MKM'03*, number 2594 in LNAI. Springer Verlag, 2003.

[Aut03] Serge Autexier. *Hierarchical Contextual Reasoning*. PhD thesis, Saarland University, 2003.

[Aut05] Serge Autexier. The core calculus. In Robert Nieuwenhuis, editor, *Proceedings of the 20th International Conference on Automated Deduction (CADE-20)*, volume 3632 of *LNAI*, Tallinn, Estonia, july 2005. Springer.

[Bar80] Hendrik P. Barendregt. *The Lambda-Calculus: Its Syntax and Semantics*. North-Holland, 1980.

[Bau99] Judith Baur. Syntax und Semantik mathematischer Texte — ein Prototyp. Master's thesis, Saarland University, 1999.

[BB01] P. Baumgartner and A. Blohm. Automated deduction techniques for the management of personalized documents. In Buchberger and Caprotti [BC01b]. http://www.risc.uni-linz.ac.at/institute/conferences/MKM2001/Procee dings/.

[BC01a] Henk Barendregt and Arjeh M. Cohen. Electronic communication of mathematics and the interaction of computer algebra systems and proof assistants. *Journal of Symbolic Computation*, 32:3–22, 2001.

[BC01b] Bruno Buchberger and Olga Caprotti, editors. *Electronic Proceedings of the First International Workshop on Mathematical Knowledge Management: MKM'2001*, 2001. http://www.risc.uni-linz.ac.at/institute/conferences/MKM2001/Proceedings/.

[BCC⁺04] Stephen Buswell, Olga Caprotti, David P. Carlisle, Michael C. Dewar, Marc Gaetano, and Michael Kohlhase. The Open Math standard, version 2.0. Technical report, The Open Math Society, 2004. http://www.openmath.org/standard/om20.

[BCD⁺02] R. Bradford, R. M. Corless, J. H. Davenport, D. J. Jeffrey, and S. M. Watt. Reasoning about the elementary functions of complex analysis. *Annals of Mathematics and Artificial Intelligence*, 36:303 – 318, 2002.

[BCF⁺97] C. Benzmüller, L. Cheikhrouhou, D. Fehrer, A. Fiedler, X. Huang, M. Kerber, M. Kohlhase, K. Konrad, E. Melis, A. Meier, W. Schaarschmidt, J. Siekmann, and V. Sorge. ΩMEGA: Towards a mathematical assistant. In McCune [McC97], pages 252–255.

[BDD⁺99] Stephen Buswell, Stan Devitt, Angel Diaz, Patrick Ion, Robert Miner, Nico Poppelier, Bruce Smith, Neil Soiffer, Robert Sutor, and Stephen Watt. Mathematical Markup Language (MathML) 1.01 specification. W3c recommendation, World Wide Web Consortium (W3C), 1999. Available at http://www.w3.org/TR/REC-MathML.

[Ber91] Paul Bernays. *Axiomatic Set Theory*. Dover Publications, 1991.

[BF06] Jon Borwein and William M. Farmer, editors. *Mathematical Knowledge Management, MKM'06*, number 4108 in LNAI. Springer Verlag, 2006. in press.

[BLFM98] Tim Berners-Lee, R. Fielding, and L. Masinter. Uniform resource identifiers (uri), generic syntax. RFC 2717, Internt Engineering Task Force, 1998. available at http://www.ietf.org/rfc/rfc2717.txt.

[Blo56] B.S. Bloom, editor. *Taxonomy of educational objectives: The classification of educational goals: Handbook I, cognitive domain*. Longmans, Green, New York, Toronto, 1956.

[BM79] R. S. Boyer and J S. Moore. *A Computational Logic*. ACM monograph series. Academic Press, New York, 1979.

[Bos98] Cascading style sheets, level 2; CSS2 specification, 1998. available as http://www.w3.org/TR/1998/REC-CSS2-19980512.

[Bou74] Nicolas Bourbaki. *Algebra I. Elements of Mahtcmatics*. Springer Verlag, 1974.

[BPSM97] Tim Bray, Jean Paoli, and C. M. Sperberg-McQueen. Extensible Markup Language (XML). W3C Recommendation TR-XML, World Wide Web Consortium, December 1997. Available at http://www.w3.org/TR/REC-xml/.

[BPSM⁺04] Tim Bray, Jean Paoli, C. M. Sperberg-McQueen, Eve Maler, François Yergeau, and John Cowan. Extensible Markup Language (XML) 1.1. W3C Recommendation REC-xml11-20040204, World Wide Web Consortium, 2004. available at http://www.w3.org/TR/2004/REC-xml11-20040204/.

[Bra99] Namespaces in xml, 1999. available at http://www.w3.org/TR/REC-xml-names.

[Bug05] Bugzilla. web page at http://www.bugzilla.org, seen 2005.

[BvHHS90] A. Bundy, F. van Harmelen, C. Horn, and A. Smaill. The Oyster-Clam system. In M. E. Stickel, editor, *10th International Conference on Automated Deduction*, pages 647–648. Springer-Verlag, 1990. Lecture Notes in Artificial Intelligence No. 449. Also available from Edinburgh as DAI Research Paper 507.

[CAB⁺86] Robert L. Constable, S. Allen, H. Bromly, W. Cleaveland, J. Cremer, R. Harper, D. Howe, T. Knoblock, N. Mendler, P. Panangaden, J. Sasaki, and S. Smith. *Implementing Mathematics with the Nuprl Proof Development System*. Prentice-Hall, Englewood Cliffs, New Jersey, 1986.

[CCC⁺00] Olga Caprotti, Arjeh M. Cohen, Hans Cuypers, Manfred N. Riem, and Hans Sterk. Using OPENMATH servers for distributing mathematical computations. In Wei Chi Yang, Sung-Chi Chu, and Jen-Chung Chuan, editors, *ATCM 2000: Proceedings of the Fifth Asian Technology Conference in Mathematics,*, pages 325–336, Chiang-Mai, Thailand, 2000. ATCM, Inc.

[CCR00] Olga Caprotti, Arjeh M. Cohen, and Manfred Riem. Java Phrasebooks for Computer Algebra and Automated Deduction. *Bulletin of the ACM Special Interest Group on Symbolic and Automated Mathematics (SIGSAM)*, 34(2):43–48, 2000.

[CCS99] Arjeh Cohen, Hans Cuypers, and Hans Sterk. *Algebra Interactive!* Springer Verlag, 1999. Interactive Book on CD.

[CGG⁺92] Bruce W. Char, Keith O. Geddes, Gaston H. Gonnet, Benton L. Leong, Michael B. Monagan, and Stephen M. Watt. *First leaves: a tutorial introduction to Maple V*. Springer Verlag, Berlin, 1992.

[CGM⁺04] R. Conejo, E. Guzman, E. Millan, M. Trella, J. L. Perez de-la Cruz, and A. Rios. SIETTE: A Web-Based Tool for Adaptive Teaching. *International Journal of Artificial Intelligence in Education (IJAIED 2004)*, 14:29–61, 2004.

[CIMP01] David Carlisle, Patrick Ion, Robert Miner, and Nico Poppelier. Mathematical Markup Language (MathML) version 2.0. W3c recommendation, World Wide Web Consortium, 2001. Available at http://www.w3.org/TR/MathML2.

[CKOS03] Edmund Clarke, Michael Kohlhase, Joël Ouaknine, and Klaus Sutner. System description: Analytica 2. In Thérèse Hardin and Renaud Rioboo, editors, *Proceedings of the 11th Symposium on the Integration of Symbolic Computation and Mechanized Reasoning (Calculemus-2000)*, pages 69–73, 2003. avaialble at http://ftp.lip6.fr/lip6/reports/2003/lip6.2003.010.pdf.

[Cla97] James Clark. Comparison of SGML and XML. World Wide Web Consortium Note, 1997. http://www.w3.org/TR/NOTE-sgml-xml.html.

[Cla99a] Associating style sheets with XML documents version 1.0, 1999. available as http://www.w3.org/TR/xml-stylesheet.

[Cla99b] XML path language (XPath) version 1.0, 1999. available at http://www.w3.org/TR/xpath.

[Cla05] James Clark. nXML mode. web page at http://www.thaiopensource.com/nxml-mode/, seen 2005.

[CM98] A.M. Cohen and L. Meertens. The ACELA project: Aims and plans. In N. Kajler, editor, *Computer-Human interaction in Symbolic Computation*, Texts and Monographs in Symbolic Computation, pages 7–23. Springer Verlag, 1998.

[CML05] Chemical Markup Language CML. web page at http://www.ch.cam.ac.uk/CUCL/staff/pm.html, seen July2005.

[Coe05] Claudio Sacerdoti Coen. Explanation in natural language of $\bar{\lambda}\mu\bar{\mu}$-terms. In Kohlhase [Koh05b].

[CoF04] CoFI (The Common Framework Initiative). CASL *Reference Manual*. LNCS 2960 (IFIP Series). Springer, 2004.

[Coma] Userland Com. XML Remote Procedure Call Specification. http://www.xmlrpc.com/.

[Comb] The Mizar Library Committee. Mizar mathematical library. web page at http://www.mizar.org/library.

[Con01] IMS Global Learning Consortium. Learnig resource metadata specification, 2001. http://www.imsglobal.org/metadata/.

[Cor] Microsoft Corp. Microsoft internet explorer. web page at http://www.microsoft.com/windows/ie.

[Cow04] XML information set (second edition), February 2004. Available at http://www.w3.org/TR/xml-infoset.

[Crea] Creative Commons. web page at http://creativecommons.org.

[Creb] Metadata Commons Worldwide. web page at http://creativecommons.org/learn/technology/metadata.

[Crec] Creative Commons Worldwide. web page at http://creativecommons.org/worldwide.

[Dah01] Ingo Dahn. Slicing book technology - providing online support for textbooks. In *The 20th ICDE World Conference on Open Learning and Distance Education*, 2001.

[dB80] Nicolaas Govert de Bruijn. A survey of the project AUTOMATH. In R. Hindley and J. Seldin, editors, *To H.B. Curry: Essays in Combinator Logic, Lambda Calculus and Formalisms*, pages 579–606. Academic Press, 1980.

[dB94] N. G. de Bruijn. The mathematical vernacular, a language for mathematics with typed sets. In R. P Nederpelt, J. H. Geuvers, and R. C. de Vrijer, editors, *Selected Papers on Automath*, volume 133 of *Studies in Logic and the Foundations of Mathematics*, pages 865 – 935. Elsevier, 1994.

[Dea99] Stephen Deach. Extensible stylesheet language (xsl) specification. W3c working draft, W3C, 1999. Available at http://www.w3.org/TR/WD-xsl.

[Des05] Deskzilla. web page at http://www.deskzilla.com, seen 2005.

[dH01] Joris Van der Hoeven. Gnu $\text{T}_{\text{E}}\text{X}_{\text{MACS}}$: A free, structured, wysiwyg and technical text editor. *Cahiers GUTenberg*, pages 39–40, May 2001.

[DJ02] Jun Fujisawa Dean Jackson, Jon Ferraiolo. Scalable vector graphics (svg) 1.1 specification. W3c candidate recommendation, World Wide Web Consortium (W3C), 2002. http://www.w3.org/TR/2002/CR-SVG11-20020430.

[DMD02] Steven J. DeRose, Eve Maler, and Ron Jr. Daniel. Xpointer xpointer() scheme. W3c working draft, World Wide Web Constortium W3C, 19. December 2002.

[DMOT01] Steve DeRose, Eve Maler, David Orchard, and Ben Trafford. XML linking language (XLink). W3c recommendation, W3C, 2001. Available at http://www.w3.org/TR/xlink.

[DOM] Document object model DOM. web page at http://www.w3.org/DOM/.

[DUB03a] The DCMI Usage Board. DCMI metadata terms. DCMI recommendation, Dublin Core Metadata Initiative, 2003.
http://dublincore.org/documents/dcmi-terms/.

[DUB03b] The DCMI Usage Board. DCMI type vocabulary. DCMI recommendation, Dublin Core Metadata Initiative, 2003.
http://dublincore.org/documents/dcmi-type-vocabulary/.

[DuC97] Bob DuCharme. Formatting documents with dsssl specifications and jade. *The SGML Newsletter*, 10(5):6–10, 1997.

[Duc98] Denys Duchier. The NEGRA tree bank. Private communication, 1998.

[DW05] Mark Davis and Ken Whistler. Unicode collation algorithm. available at http://www.unicode.org/reports/tr10/, 2005. Unicode Technical Standard #10.

[Far93] William M. Farmer. Theory interpretation in simple type theory. In *HOA'93, an International Workshop on Higher-order Algebra, Logic and Term Rewriting*, volume 816 of *LNCS*, Amsterdam, The Netherlands, 1993. Springer Verlag.

[FB96] N. Freed and N. Borenstein. Multipurpose internet mail extensions (mime) part two: Media types. RFC 2046:
http://www.faqs.org/rfcs/rfc2046.html, 1996.

[FGT93] William M. Farmer, Joshua D. Guttman, and F. Javier Thayer. IMPS: An Interactive Mathematical Proof System. *Journal of Automated Reasoning*, 11(2):213–248, October 1993.

[FH97] Amy P. Felty and Douglas J. Howe. Hybrid interactive theorem proving using NuPRL and HOL. In McCune [McC97], pages 351–365.

[FH01]   Armin Fiedler and Helmut Horacek. Argumentation in explanations to logical problems. In Vassil N. Alexandrov, Jack J. Dongarra, Benjoe A. Juliano, Renè S. Renner, and C. J. Kenneth Tan, editors, *Computational Science — ICCS 2001*, number 2074 in LNCS, pages 969–978, San Francisco, CA, 2001. Springer Verlag.

[FHJ+99a]   A. Franke, S. Hess, C. Jung, M. Kohlhase, and V. Sorge. Agent-Oriented Integration of Distributed Mathematical Services. *Journal of Universal Computer Science*, 5(3):156–187, March 1999. Special issue on Integration of Deduction System.

[FHJ+99b]   Andreas Franke, Stephan M. Hess, Christoph G. Jung, Michael Kohlhase, and Volker Sorge. Agent-oriented integration of distributed mathematical services. *Journal of Universal Computer Science*, 5:156–187, 1999.

[Fie97]   Armin Fiedler. Towards a proof explainer. In Siekmann et al. [SPH97], pages 53–54.

[Fie99]   Armin Fiedler. Using a cognitive architecture to plan dialogs for the adaptive explanation of proofs. In Thomas Dean, editor, *Proceedings of the 16th International Joint Conference on Artificial Intelligence (IJCAI)*, pages 358–363, Stockholm, Sweden, 1999. Morgan Kaufmann.

[Fie01a]   Armin Fiedler. Dialog-driven adaptation of explanations of proofs. In Bernhard Nebel, editor, *Proceedings of the 17th International Joint Conference on Artificial Intelligence (IJCAI)*, pages 1295–1300, Seattle, WA, 2001. Morgan Kaufmann.

[Fie01b]   Armin Fiedler. *User-adaptive Proof Explanation*. Phd thesis, Naturwissenschaftlich-Technische Fakultät I, Universität des Saarlandes, Saarbrücken, Germany, 2001.

[FK99]   Andreas Franke and Michael Kohlhase. System description: MATHWEB, an agent-based communication layer for distributed automated theorem proving. In Harald Ganzinger, editor, *Automated Deduction — CADE-16*, number 1632 in LNAI, pages 217–221. Springer Verlag, 1999.

[Fou]   The Apache Software Foundation. Xalan-java. web page at http://xml.apache.org/xalan-j.

[FSF91]   Free Software Foundation FSF. Gnu general public license. Software License available at http://www.gnu.org/copyleft/gpl.html, 1991.

[FSF99]   Free Software Foundation FSF. GNU lesser general public license. Software License available at http://www.gnu.org/copyleft/lesser.html, 1999.

[GB92]   J. A. Goguen and R. M. Burstall. Institutions: Abstract model theory for specification and programming. *Journal of the Association for Computing Machinery*, 39:95–146, 1992. Predecessor in: LNCS 164, 221–256, 1984.

[Gen35]   Gerhard Gentzen. Untersuchungen über das logische Schließen I & II. *Mathematische Zeitschrift*, 39:176–210, 572–595, 1935.

[GHMN03]   Martin Gudgin, Marc Hadley, Jean-Jacques Moreau, and Henrik Frystyk Nielsen. Soap 1.2 part 1: Adjuncts, 2003. available at http://www.w3.org/TR/2003/REC-soap12-part2-20030624.

[GM93]   M. J. C. Gordon and T. F. Melham. *Introduction to HOL – A theorem proving environment for higher order logic*. Cambridge University Press, 1993.

[GMMW03a]   Paul Grosso, Eve Maler, Jonathan Marsh, and Norman Walsh. Xpointer element() scheme. W3c recommendation, World Wide Web Constortium W3C, 25 March 2003.

[GMMW03b]   Paul Grosso, Eve Maler, Jonathan Marsh, and Norman Walsh. Xpointer framework. W3c recommendation, World Wide Web Constortium W3C, 25 March 2003.

[GMUC03] Georgi Goguadze, Erica Melis, Carsten Ullrich, and Paul Cairns. Problems and solutions for markup for mathematical examples and exercises*. In Asperti [Asp03], pages 80–92.

[Gol90] C. F. Goldfarb. *The SGML Handbook*. Oxford University Press, 1990.

[GP03] Georgi Goguadze and Alberto González Palomo. Adapting mainstream editors for semantic authoring of mathematics. Presented at the Mathematical Knowledge Management Symposium, Heriot-Watt University, Edinbourgh, Scotland, November 2003.

[GPE05] G.Goguadze, A.González Palomo, and E.Melis. Interactivity of exercises in activemath. In *Accepted to the International Conference on Computers in Education (ICCE 2005)*, Singapore, 2005.

[GR02] J. Goguen and G. Rosu. Institution morphisms. *Formal aspects of computing*, 13:274–307, 2002.

[Gra96] Peter Graf. *Term Indexing*. Number 1053 in LNCS. Springer Verlag, 1996.

[Gro99] The Open eBook Group. Open ebook[tm] publication structure 1.0. Draft recommendation, The OpenEBook Initiative, 1999. Available at http://www.openEbook.org.

[Gro00] The W3C HTML Working Group. XHTML 1.0 The Extensible Hyper-Text Markup Language (Second Edition) A Reformulation of HTML 4 in XML 1.0. W3c recommendation, World Wide Web Consortium (W3C), 2000. http://www.w3.org/TR/xhtml1.

[GUM+04] G. Goguadze, C. Ullrich, E. Melis, J. Siekmann, Ch. Gross, and R. Morales. LeActiveMath Structure and Metadata Model. Deliverable D6, LeActiveMath Consortium, 2004. accessible from http://www.leactivemath.org/.

[Har01] Eliotte Rusty Harold. *XML Bible*. Hungry Minds, gold edition edition, 2001.

[Har03] Eliotte Rusty Harold. *Effective XML*, chapter 15. Addison Wesley, 2003.

[HF96] Xiaorong Huang and Armin Fiedler. Presenting machine-found proofs. In McRobbie and Slaney [MS96], pages 221–225.

[HF97] Xiaorong Huang and Armin Fiedler. Proof verbalization in *PROVERB*. In Siekmann et al. [SPH97], pages 35–36.

[HKW96] Reiner Hähnle, Manfred Kerber, and Christoph Weidenbach. Common syntax of dfg-schwerpunktprogramm "deduktion". Interner Bericht 10/96, Universität Karlsruhe, Fakultät für Informatik, 1996.

[HS96] Dieter Hutter and Claus Sengler. INKA - The Next Generation. In McRobbie and Slaney [MS96], pages 288–292.

[Hua96] Xiaorong Huang. *Human Oriented Proof Presentation: A Reconstructive Approach*. Number 112 in DISKI. Infix, Sankt Augustin, Germany, 1996.

[Hut00] Dieter Hutter. Management of change in verification systems. In *Proceedings Automated Software Engineering (ASE-2000)*. IEEE Press, 2000.

[IAN] Root-zone whois information. http://www.iana.org/cctld/cctld-whois.htm.

[IEE02] IEEE Learning Technology Standards Committee. 1484.12.1-2002 IEEE standard for Learning Object Metadata, 2002.

[Inc03] Unicode Inc., editor. *The Unicode Standard, Version 4.0*. Addison-Wesley, 2003.

[JEN] Jena — a semantic web framework for java. web page at http://jena.sf.net.

[JSP] JavaServer Pages. web page at http://java.sun.com/products/jsp.

[KA03] Michael Kohlhase and Romeo Anghelache. Towards collaborative content management and version control for structured mathematical knowledge. In Asperti [Asp03], pages 147–161.

[KAB⁺04] E. Klieme, H. Avenarius, W. Blum, P. Döbrich, H. Gruber, M. Prenzel, K. Reiss, K. Riquarts, J. Rost, H. Tenorth, and H. J. Vollmer. The development of national educational standards - an expertise. Technical report, Bundesministerium für Bildung und Forschung / German Federal Ministry of Education and Research, 2004.

[Kay] Michael Kay. Saxon, the xslt and xquery processor. web page at http://saxon.sf.net.

[KBHL⁺03] B. Krieg-Brückner, D. Hutter, A. Lindow, C. Lüth, A. Mahnke, E. Melis, P. Meier, A. Poetzsch-Heffter, M. Roggenbach, G. Russell, J.-G. Smaus, and M. Wirsing. Multimedia Instruction in Safe and Secure Systems. In M. Wirsing and R. Hennicker D. Pattinson, editors, *Recent Trends in Algebraic Development Techniques, 16th International Workshop (WADT 2002)*, volume 2755 of *Lecture Notes in Computer Science*, pages 82–117. Springer-Verlag Heidelberg, 2003.

[KBKB⁺04] B. Krieg-Brückner, B. Krämer, D. Basin, J. Siekmann, and M. Wirsing. Multimedia Instruction in Safe and Secure Systems. Final report, University Bremen, 2004. BMBF project 01NM070, 2001-2004.

[KBLL⁺04] B. Krieg-Brückner, A. Lindow, C. Lüth, A. Mahnke, and G. Russell. Semantic Interrelation of Documents via an Ontology. In G. Engels and S. Seehusen, editors, *DeLFI 2004, Tagungsband der 2. e-Learning Fachtagung Informatik, 6.-8. September 2004, Paderborn, Germany*, volume P-52 of *Lecture Notes in Informatics*, pages 271–282. Springer-Verlag Heidelberg, 2004.

[KD03a] Michael Kohlhase and Stan Devitt. Bound variables in mathml. W3C Note, 2003. available at http://www.w3.org/TR/mathml-bvar/.

[KD03b] Michael Kohlhase and Stan Devitt. Structured types in mathml 2.0. W3C Note, 2003. available at http://www.w3.org/TR/mathml-types/.

[KF00] M. Kohlhase and A. Franke. MBASE: Representing knowledge and context for the integration of mathematical software systems. *Journal of Symbolic Computation*, 2000.

[KF01] Michael Kohlhase and Andreas Franke. MBase: Representing knowledge and context for the integration of mathematical software systems. *Journal of Symbolic Computation; Special Issue on the Integration of Computer algebra and Deduction Systems*, 32(4):365–402, 2001.

[KK04] Andrea Kohlhase and Michael Kohlhase. CPoint: Dissolving the author's dilemma. In Asperti et al. [ABT04], pages 175–189.

[KK05] Andrea Kohlhase and Michael Kohlhase. An exploration in the space of mathematical knowledge. In Kohlhase [Koh05b], pages 17–32.

[KK06] Andrea Kohlhase and Michael Kohlhase. Communities of practice in mkm: An extensional model. In Borwein and Farmer [BF06]. in press.

[KM96] M. Kaufmann and J S. Moore. ACL2: An industrial strength version of Nqthm. In *Compass'96: Eleventh Annual Conference on Computer Assurance*, page 23, Gaithersburg, Maryland, 1996. National Institute of Standards and Technology.

[Knu84] Donald E. Knuth. *The TₑXbook*. Addison Wesley, 1984.

[Koha] Michael Kohlhase. CodeML: An open markup format the content and presentatation of program code. Internet Draft at https://svn.mathweb.org/repos/mathweb.org/trunk/omdoc/projects/codeml/doc/spec/codeml.pdf.

[Kohb] Michael Kohlhase. OMDOC: An open markup format for mathematical documents (latest released version). Specification, http://www.mathweb.org/omdoc/spec.pdf.

[Kohc] Michael Kohlhase. The OMDoc Document Type Definition. http://www.mathweb.org/omdoc/dtd/omdoc.dtd.

[Kohd] Michael Kohlhase. OMDoc mailing lists.
http://www.mathweb.org/omdoc/resources/mailing-lists.html.

[Kohe] Michael Kohlhase. The OMDoc RelaxNG schema.
http://www.mathweb.org/omdoc/rnc/omdoc.rnc.

[Kohf] Michael Kohlhase. The OMDoc XML schema.
http://www.mathweb.org/omdoc/rnc/omdoc.xsd.

[Kohg] Michael Kohlhase. XSL style sheets for OMDoc.
http://www.mathweb.org/omdoc/xsl/.

[Koh05a] Andrea Kohlhase. Overcoming Proprietary Hurdles: CPoint as Invasive Editor. In Fred de Vries and Graham Attwell and Raymond Elferink and Alexandra Tödt, editor, *Open Source for Education in Europe: Research and Practise*, pages 51–56. Open Universiteit of the Netherlands, Heerlen, 2005.

[Koh05b] Michael Kohlhase, editor. *Mathematical Knowledge Management, MKM'05*, number 3863 in LNAI/LNCS forthcoming. Springer Verlag, 2005.

[Koh05c] Michael Kohlhase. Semantic markup for TEX/LATEX. Manuscript, available at http://kwarc.eecs.iu-bremen.de/software/stex, 2005.

[Koh05d] Michael Kohlhase. Inference rules. OMDoc Content Dictionary at https://svn.mathweb.org/repos/mathweb.org/trunk/omdoc/examples/logics/inference-rules.omdoc, seen Jan 2005.

[Koh06] Michael Kohlhase. OMDOC *An open markup format for mathematical documents (Version 1.2)*. LNAI. Springer Verlag, 2006. to appear, manuscript at http://www.mathweb.org/omdoc/pubs/omdoc1.2.pdf.

[KR93] Hans Kamp and Uwe Reyle. *From Discourse to Logic*. Kluwer, Dordrecht, 1993.

[KZ95] D. Kapur and H. Zhang. An overview of rewrite rule laboratory (RRL). *J. Computer and Mathematics with Applications*, 29(2):91–114, 1995.

[Lam94] Leslie Lamport. *LATEX: A Document Preparation System, 2/e*. Addison Wesley, 1994.

[Lee98] Tim Berner's Lee. The semantic web. W3C Architecture Note, 1998. http://www.w3.org/DesignIssues/Semantic.html.

[Len04] Richard Lennox. Development of an RDF/XML based data model for bibliographic data. Dissertation for Bachelor of Science in Computer Science, 2004. http://richardlennox.net/dissertation.pdf.

[LC01] Bo Leuf and Ward Cunningham. *The Wiki Way: Collaboration and Sharing on the Internet*. Addison-Wesley Professional, 2001.

[Lib04] P. Libbrecht. Authoring web content in activemath: From developer tools and further. In Alexandra Christea and Franca Garzotto, editors, *Proceedings of the Second International Workshop on Authoring Adaptive and Adaptable Educational Hypermedia, AH-2004: Workshop Proceedings, Part II, CS-Report 04-19*, pages 455–460. Technische Universiteit Eindhoven, 2004.

[Lom05] Cyprien Lomas. 7 things you should know about social bookmarking. http://www.educause.edu/ir/library/pdf/ELI7001.pdf, 2005.

[LS99] Ora Lassila and Ralph R. Swick. Resource description framework (rdf) model and syntax specification. W3c recommendation, World Wide Web Consortium (W3C), 1999. http://www.w3.org/TR/1999/REC-rdf-syntax.

[MAF+01] E. Melis, J. Buedenbender E. Andres, Adrian Frischauf, G. Goguadze, P. Libbrecht, M. Pollet, and C. Ullrich. The ACTIVEMATH learning environment. *Artificial Intelligence and Education*, 12(4), winter 2001 2001.

[MAH06] Till Mossakowski, Serge Autexier, and Dieter Hutter. Extending development graphs with hiding. *Journal of Logic and Algebraic Programming, special issue on Algebraic Specification and Development Techniques*, 2006.

[MAR03] MARC code list for relators, sources, description conventions, 2003. Web Version at http://www.loc.gov/marc/relators.

[Mat] Using the mathweb.org subversion repository. Web page at
http://www.mathweb.org/svn.html.

[Max] Maxima - a gpl cas based on doe-macsyma.
web page at http://maxima.sourceforge.net.

[MBa] Mbase. http://mbase.mathweb.org:8000.

[MBG+03] Erica Melis, Jochen Büdenbender, George Goguadze, Paul Libbrecht,
and Carsten Ullrich. Knowledge representation and management in activemath. *Annals of Mathematics and Artificial Intelligence*, 38:47–64, 2003. see
http://www.activemath.org.

[McC97] William McCune, editor. *Proceedings of the 14th Conference on Automated Deduction*, number 1249 in LNAI, Townsville, Australia, 1997. Springer
Verlag.

[Mei00] Andreas Meier. System description: TRAMP: Transformation of machine-
found proofs into ND-proofs at the assertion level. In David McAllester, editor, *Automated Deduction – CADE-17*, number 1831 in LNAI, pages 460–464.
Springer Verlag, 2000.

[Mes89] J. Meseguer. General logics. In *Logic Colloquium 87*, pages 275–329. North
Holland, 1989.

[MG04] E. Melis and G. Goguadze. Towards adaptive generation of faded examples.
In *International Conference on Intelligent Tutoring Systems*, number 3220 in
LNCS, pages 762–771. Springer-Verlag, 2004.

[MGDT05] Till Mossakowski, Joseph Goguen, Razvan Diaconescu, and Andrzej
Tarlecki. What is a logic? In Jean-Yves Beziau, editor, *Logica Universalis*,
pages 113–133. Birkhäuser, 2005.

[MGH+05] Erica Melis, Giorgi Goguadze, Martin Homik, Paul Libbrecht, Carsten
Ullrich, and Stefan Winterstein. Semantic-aware components and services of
activemath. *British Journal of Educational Technology*, 2005.

[Mil] Bruce Miller. LaTeXML: A L^AT_EX to XML converter. Web Manual at
http://dlmf.nist.gov/LaTeXML/.

[Mit03] Nilo Mitra. Soap 1.2 part 0: Primer, 2003. available at
http://www.w3.org/TR/2003/REC-soap12-part0-20030624.

[Miz06] Mizar language. web page at http://mizar.org/language, seen III 2006.

[MKB04] A. Mahnke and B. Krieg-Brückner. Literate Ontology Development.
In Robert Meersman, Zahir Tari, and Angelo Corsaro et al., editors, *On the
Move to Meaningful Internet Systems 2004: OTM 2004 Workshops*, volume
3292 of *Lecture Notes in Computer Science*, pages 753–757. Springer; Berlin;
http://www.springer.de, 2004.

[MKH05] E. Melis, P. Kärger, and M. Homik. Interactive Concept Mapping in
ActiveMath (iCMap). In Djamshid Tavangarian Jörg M. Haake, Ulrich Lucke,
editor, *Delfi 2005: 3. Deutsche eLearning Fachtagung Informatik*, volume 66 of
LNI, pages 247–258. Gesellschaft für Informatik e.V. (GI), 2005.

[MLUM05] S. Manzoor, P. Libbrecht, C. Ullrich, and E. Melis. Authoring presentation for openmath. In Michael Kohlhase et al., editor, *Proceedings of the Fourth
International Conference on Mathematical Knowledge Management (MKM-05)*,
pages –, Bremen, Germany, 2005. International University of Bremen.

[MMLW] T. Mossakowski, Christian Maeder, Klaus Lüttich, and Stefan Wölfl. The
heterogeneous tool set. Submitted for publication.

[Mon] MONET, an EU funded project. web page at http://www.monet.nag.co.uk.

[Mos02] Till Mossakowski. Heterogeneous development graphs and heterogeneous
borrowing. In Mogens Nielsen and Uffe Engberg, editors, *Foundations of
Software Science and Computation Structures (FOSSACS02)*, number 2303 in
LNCS, pages 310–325. Springer, 2002.

[Mos04] T. Mossakowski. Hetcasl - heterogeneous specification. language summary, 2004.

[Mos05] T. Mossakowski. Heterogeneous specification and the heterogeneous tool set. Habilitation thesis, University of Bremen, 2005.

[Moz] The mozart programming system.

[MP04] M. Mavrikis and A. González Palomo. Mathematical, interactive exercise generation from static documents. *Electronic Notes in Computer Science*, 93:183–201, 2004.

[MS96] M.A. McRobbie and J.K. Slaney, editors. *Proceedings of the 13th Conference on Automated Deduction*, number 1104 in LNAI, New Brunswick, NJ, USA, 1996. Springer Verlag.

[MSLK01] M. Murata, S. St. Laurent, and D. Kohn. XML media types. RFC 3023, January 2001. http://www.faqs.org/rfcs/rfc3023.html.

[Mül05] Normen Müller. OMDoc-repräsentation von programmen und beweisen in VeriFun. Master's thesis, Programmiermethodik, Technische Universität Darmstadt, 2005.

[MVW05] Jonathan Marsh, Daniel Veillard, and Norman Walsh. xml:id version 1.0. Recommendation, W3C, 2005. http://www.w3.org/TR/xml-id/.

[NS81] Alan Newell and Herbert A. Simon. Computer science as empirical inquiry: Symbols and search. *Communications of the Association for Computing Machinery*, 19:113–126, 1981.

[OAI02] The open archives initiative protocol for metadata harvesting, June 2002. Available at http://www.openarchives.org/OAI/openarchives protocol.html.

[Odl95] A.M. Odlyzko. Tragic loss or good riddance? the impending demise of traditional scholarly journals. *International Journal of Human-Computer Studies*, 42:71–122, 1995.

[OM] OPENMATH. web page at http://www.openmath.org.

[OMC] OPENMATH content dictionaries. web page at http://www.openmath.org/cd/.

[OMDa] The omdoc subversion repository. Repository at https://svn.mathweb.org/repos/mathweb.org/trunk/omdoc.

[OMDb] The OMDoc wiki. http://www.mathweb.org/omdoc/wiki/.

[Org] The Mozilla Organization. Mozilla. web page at http://www.mozilla.org.

[ORS92] S. Owre, J. M. Rushby, and N. Shankar. PVS: a prototype verification system. In D. Kapur, editor, *Proceedings of the 11th Conference on Automated Deduction*, volume 607 of *LNCS*, pages 748–752, Saratoga Spings, NY, USA, 1992. Springer Verlag.

[Pala] Alberto González Palomo. Algebra.

[Palb] Alberto González Palomo. Qmath history. http://www.matracas.org/qmath/history.html.

[Pau94] Lawrence C. Paulson. *Isabelle: A Generic Theorem Prover*. LNCS. Springer Verlag, 1994.

[PB04] F. Piroi and B. Buchberger. An environment for building mathematical knowledge libraries, 2004. http://citeseer.ifi.unizh.ch/piroi04environment.html.

[PDF06] Adobe Systems Inc. *PDF Reference fifth edition, Adobe Portable Document Format Version 1.6*, seen Jan 2006. available at http://partners.adobe.com/public/developer/pdf/index_reference.html.

[Pfe91] Frank Pfenning. Logic programming in the LF logical framework. In Gérard P. Huet and Gordon D. Plotkin, editors, *Logical Frameworks*. Cambridge University Press, 1991.

[Pfe01] Frank Pfenning. Logical frameworks. In Alan Robinson and Andrei Voronkov, editors, *Handbook of Automated Reasoning*, volume I and II. Elsevier Science and MIT Press, 2001.

[Pie80] John R. Pierce. *An Introduction to Information Theory. Symbols, Signals and Noise*. Dover Publications Inc., 1980.

[PN90] Lawrence C. Paulson and Tobias Nipkow. Isabelle tutorial and user's manual. Technical Report 189, Computer Laboratory, University of Cambridge, January 1990.

[PS99] F. Pfenning and C. Schürmann. System description: Twelf — A metalogical framework for deductive systems. In H. Ganzinger, editor, *Proceedings of the 16th International Conference on Automated Deduction (CADE-16)*, pages 202–206, Trento, Italy, 1999. Springer-Verlag LNAI 1632.

[PSBKK04] Manfred Pinkal, Jörg Siekmann, Christoph Benzmüller, and Ivana Kruijff-Korbayova. Dialog: Natural language-based interaction with a mathematics assistance system. Project proposal in the Collaborative Research Centre SFB 378 on Resource-adaptive Cognitive Processes, 2004.

[Rei87] Glenn C. Reid. *PostScript, Language, Program Design*. Addison Wesley, 1987.

[RHJ98] Dave Raggett, Arnaud Le Hors, and Ian Jacobs. HTML 4.0 Specification. W3C Recommendation REC-html40, World Wide Web Consortium, April 1998. Available at http://www.w3.org/TR/PR-xml.html.

[ROM] ROML, The RIACA OPENMATH Library. web page at http://crystal.win.tue.nl/download/.

[RSG98a] J.D.C. Richardson, A. Smaill, and I. Green. System description: Proof planning in higher-order logic with lambda-clam. In C. Kirchner and H. Kirchner, editors, *Conference on Automated Deduction (CADE'98)*, volume 1421 of *Lecture Notes in Computer Science*, pages 129–133. Springer-Verlag, 1998.

[RSG98b] Julian D.C. Richardson, Alan Smaill, and Ian M. Green. System description: Proof planning in higher-order logic with λclam. In Claude Kirchner and Hélène Kirchner, editors, *Proceedings of the 15th Conference on Automated Deduction*, number 1421 in LNAI. Springer Verlag, 1998.

[Rud92] Piotr Rudnicki. An overview of the mizar project. In *Proceedingsof the 1992 Workshop on Types and Proofs as Programs*, pages 311–332, 1992.

[SBA05] Jörg Siekmann, Christoph Benzmüller, and Serge Autexier. Computer supported mathematics with omega. *Journal of Applied Logic, special issue on Mathematics Assistance Systems*, december 2005.

[SBB+02] Jörg Siekmann, Christoph Benzmüller, Vladimir Brezhnev, Lassaad Cheikhrouhou, Armin Fiedler, Andreas Franke, Helmut Horacek, Michael Kohlhase, Andreas Meier, Erica Melis, Markus Moschner, Immanuel Normann, Martin Pollet, Volker Sorge, Carsten Ullrich, Claus-Peter Wirth, and Jürgen Zimmer. Proof development with OMEGA. In Andrei Voronkov, editor, *Proceedings of the 18th International Conference on Automated Deduction (CADE-18)*, number 2392 in LNAI, pages 144–149, Copenhagen, Denmark, 2002. Springer.

[SBB+06] Sebastian Schaffert, Diana Bischof, Tobias Bürger, Andreas Gruber, Wolf Hilzensauer, and Sandra Schaffert. Learning with semantic wikis. http://www.wastl.net/download/paper/Schaffert06_SemWikiLearning.pdf, 2006.

[SBC+00] Jörg Siekmann, Christoph Benzmüller, Lassaad Cheikhrouhou, Armin Fiedler, Andreas Franke, Helmut Horacek, Michael Kohlhase, Andreas Meier, Erica Melis, Martin Pollet, Volker Sorge, Carsten Ullrich, and Jürgen Zimmer.

Adaptive course generation and presentation. In P. Brusilovski, editor, *Proceedings of ITS-2000 workshop on Adaptive and Intelligent Web-Based Education Systems*, Montreal, 2000.

[Sch04] Klaus Schneider. *Verification of Reactive Systems*. Springer Verlag, 2004.

[Sch06] Sebastian Schaffert. IkeWiki: A semantic wiki for collaborative knowledge management. Technical report, Salzburg Research Forschungsgesellschaft, 2006.

[Sci] Design Science. Mathplayer ¡display math in your browser¿. web page at http://www.dessci.com/en/products/mathplayer.

[SHB⁺99] Jörg Siekmann, Stephan M. Hess, Christoph Benzmüller, Lassaad Cheikhrouhou, Armin Fiedler, Helmut Horacek, Michael Kohlhase, Karsten Konrad, Andreas Meier, Erica Melis, Martin Pollet, and Volker Sorge. $\mathcal{LOUI}$: *Lovely* ΩMEGA *User Interface*. *Formal Aspects of Computing*, 11:326–342, 1999.

[Smo95] G. Smolka. The Oz programming model. In Jan van Leeuwen, editor, *Computer Science Today*, volume 1000 of *LNCS*, pages 324–343. Springer-Verlag, Berlin, 1995.

[SPH97] J. Siekmann, F. Pfenning, and X. Huang, editors. *Proceedings of the First International Workshop on Proof Transformation and Presentation*, Schloss Dagstuhl, Germany, 1997.

[SS98] G. Sutcliffe and C. Suttner. The TPTP problem library: CNF release v1.2.1. *Journal of Automated Reasoning*, 21(2):177–203, 1998.

[SSY94] Geoff Stucliffe, Christian Suttner, and Theodor Yemenis. The TPTP problem library. In Alan Bundy, editor, *Proceedings of the 12th Conference on Automated Deduction*, number 814 in LNAI, Nancy, France, 1994. Springer Verlag.

[Sta02] Richard M. Stallman. *GNU Emacs Manual*. GNU Press, 15 edition, 2002. online at http://www.gnu.org/manual/emacs-21.2.

[Sut01] G. Sutcliffe. The CADE-17 ATP system competition. *Journal of Automated Reasoning*, 27(3):227–250, 2001.

[Sut06] Klaus Sutner. Converting MATHEMATICA notebooks to OMDoc. In OMDoc *An open markup format for mathematical documents (Version 1.2)* [Koh06]. to appear, manuscript at http://www.mathweb.org/omdoc/pubs/omdoc1.2.pdf.

[SZS04] G. Sutcliffe, J. Zimmer, and S. Schulz. TSTP Data-Exchange Formats for Automated Theorem Proving Tools. In W. Zhang and V. Sorge, editors, *Distributed Constraint Problem Solving and Reasoning in Multi-Agent Systems*, number 112 in Frontiers in Artificial Intelligence and Applications, pages 201–215. IOS Press, 2004.

[TB06] Robert Tolksdorf and Elena Paslaru Bontas. Towards wikis as semantic hypermedia. http://stoc.ag-nbi.de/papers/semwikidesign.pdf, 2006.

[Tea] Coq Development Team. *The Coq Proof Assistant Reference Manual*. INRIA. see http://coq.inria.fr/doc/main.html.

[Tho91] Simon Thompson. *Type Theory and Functional Programming*. International Computer Science Series. Addison-Wesley, 1991.

[Tob] Richard Tobin Rxp an XML parser available under the GPL. System Home page at http://www.cogsci.ed.ac.uk/~richard/rxp.html.

[Ull04] C. Ullrich. Description of an instructional ontology and its application in web services for education. In *Poster Proceedings of the 3rd International Semantic Web Conference, ISWC2004*, pages 93–94, Hiroshima, Japan, November 2004.

[Ull05] C. Ullrich. Tutorial planning: Adapting course generation to today's needs. In M. Grandbastian, editor, *Young Researcher Track Proceedings of 12th International Conference on Artificial Intelligence in Education*, pages 155–160, Amsterdam, 2005.

[ULWM04] C. Ullrich, P. Libbrecht, S. Winterstein, and M. Mühlenbrock. A flexible and efficient presentation-architecture for adaptive hypermedia: Description and technical evaluation. In Kinshuk, C. Looi, E. Sutinen, D. Sampson, I. Aedo, L. Uden, and E. Kähkönen, editors, *Proceedings of the 4th IEEE International Conference on Advanced Learning Technologies (ICALT 2004)*, pages 21–25, 2004.

[Vat] Irène Vatton. Welcome to amaya. web page at http://www.w3.org/Amaya.

[Veia] Daniel Veillard. The XML c parser and toolkit of gnome; libxml. System Home page at http://xmlsoft.org.

[Veib] Daniel Veillard. The xslt c library for gnome: libxslt. web page at http://xmlsoft.org/XSLT/.

[VKS+06] Max Völkel, Malte Kiesel, Sebastian Schaffert, Björn Decker, and Eyal Oren. Semantic wiki state of the art paper. http://wiki.ontoworld.org/index.php/Semantic_Wiki_State_of_The_Art_Paper, 2006.

[VKVH06] Max Völkel, Markus Krötzsch, Denny Vrandečić, and Heiko Haller. Semantic Wikipedia. In *Proceedings of the 15th international conference on World Wide Web, WWW 2006, Edinburgh, Scotland, May 23-26, 2006*, May 2006.

[Vli03] Eric van der Vlist. *Relax NG*. O'Reilly, 2003.

[W3C] World Wide Web Consortium. web page at http://w3.org.

[Wei97] Christoph Weidenbach. SPASS: Version 0.49. *Journal of Automated Reasoning*, 18(2):247–252, 1997. Special Issue on the CADE-13 Automated Theorem Proving System Competition.

[Wir] Wiris cas. web page at http://www.wiris.com/overview/products/wiris-cas.html.

[WM99] Norman Walsh and Leonard Muellner. *DocBook: The Definitive Guide*. O'Reilly, 1999.

[Wol00] Stephen Wolfram. Mathematical notation, past and future. In *International MathML Conference*, 2000. http://www.stephenwolfram.com/publications/talks/mathml.

[Wol02] Stephen Wolfram. *The Mathematica Book*. Cambridge University Press, 2002.

[WR10] Alfred North Whitehead and Bertrand Russell. *Principia Mathematica*, volume I. Cambridge University Press, Cambridge, Great Britain; second edition, 1910.

[WS02] Christoph Walther and Stephan Schweitzer. The VeriFun Tutorial. Technical Report VFR 02/04, Programmiermethodik, Technische Universität Darmstadt, 2002.

[XML] XML schema. Web page at http://www.w3.org/XML/Schema.

[XML04] XML schema part 2: Datatypes second edition, October 2004. http://www.w3.org/TR/xmlschema-2/.

[XSL99] Xslt transformations (xslt) version 1.0. W3c recommendation, W3C, 1999. Available at http://www.w3.org/TR/xslt.

[Yac] The yacas computer algebra system. web page at http://www.xs4all.nl/~apinkus/yacas.html.

[Zim04] Jürgen Zimmer. A Framework for Agent-based Brokering of Reasoning Services. In Raul Monroy, Gustavo Arroyo-Figueroa, Luis Enrique Sucar, and Juan Humberto Sossa Azuela, editors, *MICAI*, volume 2972 of *Lecture Notes in Computer Science*. Springer, 2004.

[ZK02] Jürgen Zimmer and Michael Kohlhase. System Description: The Mathweb Software Bus for Distributed Mathematical Reasoning. In Andrei Voronkov, editor, *Proceedings of the 18th International Conference on Automated Deduction*, number 2392 in LNAI, pages 247–252. Springer Verlag, 2002.

Index

Lecture Notes in Artificial Intelligence (LNAI)

Vol. 3913: O. Boissier, J. Padget, V. Dignum, G. Lindemann, E. Matson, S. Ossowski, J.S. Sichman, J. Vázquez-Salceda (Eds.), Coordination, Organizations, Institutions, and Norms in Multi-Agent Systems. XII, 259 pages. 2006.

Vol. 3910: S.A. Brueckner, G.D.M. Serugendo, D. Hales, F. Zambonelli (Eds.), Engineering Self-Organising Systems. XII, 245 pages. 2006.

Vol. 3904: M. Baldoni, U. Endriss, A. Omicini, P. Torroni (Eds.), Declarative Agent Languages and Technologies III. XII, 245 pages. 2006.

Vol. 3900: F. Toni, P. Torroni (Eds.), Computational Logic in Multi-Agent Systems. XVII, 427 pages. 2006.

Vol. 3899: S. Frintrop, VOCUS: A Visual Attention System for Object Detection and Goal-Directed Search. XIV, 216 pages. 2006.

Vol. 3898: K. Tuyls, P.J. 't Hoen, K. Verbeeck, S. Sen (Eds.), Learning and Adaption in Multi-Agent Systems. X, 217 pages. 2006.

Vol. 3891: J.S. Sichman, L. Antunes (Eds.), Multi-Agent-Based Simulation VI. X, 191 pages. 2006.

Vol. 3890: S.G. Thompson, R. Ghanea-Hercock (Eds.), Defence Applications of Multi-Agent Systems. XII, 141 pages. 2006.

Vol. 3885: V. Torra, Y. Narukawa, A. Valls, J. Domingo-Ferrer (Eds.), Modeling Decisions for Artificial Intelligence. XII, 374 pages. 2006.

Vol. 3881: S. Gibet, N. Courty, J.-F. Kamp (Eds.), Gesture in Human-Computer Interaction and Simulation. XIII, 344 pages. 2006.

Vol. 3874: R. Missaoui, J. Schmidt (Eds.), Formal Concept Analysis. X, 309 pages. 2006.

Vol. 3873: L. Maicher, J. Park (Eds.), Charting the Topic Maps Research and Applications Landscape. VIII, 281 pages. 2006.

Vol. 3864: Y. Cai, J. Abascal (Eds.), Ambient Intelligence in Everyday Life. XII, 323 pages. 2006.

Vol. 3863: M. Kohlhase (Ed.), Mathematical Knowledge Management. XI, 405 pages. 2006.

Vol. 3862: R.H. Bordini, M. Dastani, J. Dix, A.E.F. Seghrouchni (Eds.), Programming Multi-Agent Systems. XIV, 267 pages. 2006.

Vol. 3849: I. Bloch, A. Petrosino, A.G.B. Tettamanzi (Eds.), Fuzzy Logic and Applications. XIV, 438 pages. 2006.

Vol. 3848: J.-F. Boulicaut, L. De Raedt, H. Mannila (Eds.), Constraint-Based Mining and Inductive Databases. X, 401 pages. 2006.

Vol. 3847: K.P. Jantke, A. Lunzer, N. Spyratos, Y. Tanaka (Eds.), Federation over the Web. X, 215 pages. 2006.

Vol. 3835: G. Sutcliffe, A. Voronkov (Eds.), Logic for Programming, Artificial Intelligence, and Reasoning. XIV, 744 pages. 2005.

Vol. 3830: D. Weyns, H. V.D. Parunak, F. Michel (Eds.), Environments for Multi-Agent Systems II. VIII, 291 pages. 2006.

Vol. 3817: M. Faundez-Zanuy, L. Janer, A. Esposito, A. Satue-Villar, J. Roure, V. Espinosa-Duro (Eds.), Nonlinear Analyses and Algorithms for Speech Processing. XII, 380 pages. 2006.

Vol. 3814: M. Maybury, O. Stock, W. Wahlster (Eds.), Intelligent Technologies for Interactive Entertainment. XV, 342 pages. 2005.

Vol. 3809: S. Zhang, R. Jarvis (Eds.), AI 2005: Advances in Artificial Intelligence. XXVII, 1344 pages. 2005.

Vol. 3808: C. Bento, A. Cardoso, G. Dias (Eds.), Progress in Artificial Intelligence. XVIII, 704 pages. 2005.

Vol. 3802: Y. Hao, J. Liu, Y.-P. Wang, Y.-m. Cheung, H. Yin, L. Jiao, J. Ma, Y.-C. Jiao (Eds.), Computational Intelligence and Security, Part II. XLII, 1166 pages. 2005.

Vol. 3801: Y. Hao, J. Liu, Y.-P. Wang, Y.-m. Cheung, H. Yin, L. Jiao, J. Ma, Y.-C. Jiao (Eds.), Computational Intelligence and Security, Part I. XLI, 1122 pages. 2005.

Vol. 3789: A. Gelbukh, Á. de Albornoz, H. Terashima-Marín (Eds.), MICAI 2005: Advances in Artificial Intelligence. XXVI, 1198 pages. 2005.

Vol. 3782: K.-D. Althoff, A. Dengel, R. Bergmann, M. Nick, T.R. Roth-Berghofer (Eds.), Professional Knowledge Management. XXIII, 739 pages. 2005.

Vol. 3763: H. Hong, D. Wang (Eds.), Automated Deduction in Geometry. X, 213 pages. 2006.

Vol. 3755: G.J. Williams, S.J. Simoff (Eds.), Data Mining. XI, 331 pages. 2006.

Vol. 3735: A. Hoffmann, H. Motoda, T. Scheffer (Eds.), Discovery Science. XVI, 400 pages. 2005.

Vol. 3734: S. Jain, H.U. Simon, E. Tomita (Eds.), Algorithmic Learning Theory. XII, 490 pages. 2005.

Vol. 3721: A.M. Jorge, L. Torgo, P.B. Brazdil, R. Camacho, J. Gama (Eds.), Knowledge Discovery in Databases: PKDD 2005. XXIII, 719 pages. 2005.

Vol. 3720: J. Gama, R. Camacho, P.B. Brazdil, A.M. Jorge, L. Torgo (Eds.), Machine Learning: ECML 2005. XXIII, 769 pages. 2005.

Vol. 3717: B. Gramlich (Ed.), Frontiers of Combining Systems. X, 321 pages. 2005.

Vol. 3702: B. Beckert (Ed.), Automated Reasoning with Analytic Tableaux and Related Methods. XIII, 343 pages. 2005.

Vol. 3698: U. Furbach (Ed.), KI 2005: Advances in Artificial Intelligence. XIII, 409 pages. 2005.

Vol. 3690: M. Pĕchouček, P. Petta, L.Z. Varga (Eds.), Multi-Agent Systems and Applications IV. XVII, 667 pages. 2005.

Vol. 3684: R. Khosla, R.J. Howlett, L.C. Jain (Eds.), Knowledge-Based Intelligent Information and Engineering Systems, Part IV. LXXIX, 933 pages. 2005.

Vol. 3683: R. Khosla, R.J. Howlett, L.C. Jain (Eds.), Knowledge-Based Intelligent Information and Engineering Systems, Part III. LXXX, 1397 pages. 2005.

Vol. 3682: R. Khosla, R.J. Howlett, L.C. Jain (Eds.), Knowledge-Based Intelligent Information and Engineering Systems, Part II. LXXIX, 1371 pages. 2005.